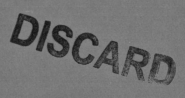

fresh

SUSANNE FREIDBERG

fresh

a perishable history

The Belknap Press of
Harvard University Press
Cambridge, Massachusetts
London, England
2009

Library of Congress Cataloging-in-Publication Data

Freidberg, Susanne.
  Fresh : a perishable history / Susanne Freidberg.
    p. cm.
  Includes bibliographical references and index.
  ISBN 978-0-674-03291-0 (alk. paper)
    1. Food—Quality. 2. Food handling.
3. Food—Labeling. 4. Perishable goods. I. Title.
TP372.5.F74 2009
664—dc22   2008052221

Contents

fresh

Introduction

Is it fresh? As consumers we ask this question about all kinds of food—the lettuce in the supermarket, the fish on the menu, the milk in the fridge. It's a sensible question, asked to avoid disappointment and maybe even danger. But what exactly *is* fresh? This question is not so simple. In the United States, not even the authorities responsible for food labeling can provide clear answers. And it's not as though they haven't tried. In July 2000, for example, the Food and Drug Administration (FDA) invited the public to a meeting in Chicago to discuss "the use of the term 'fresh.'" The FDA needed to decide whether perishables could be labeled "fresh" if they were treated with nonthermal technologies such as ultraviolet light.

An FDA official began the meeting with a review of the existing legislation, which forbade use of the word "fresh" on labels for frozen, heated, chemically treated, or "otherwise preserved" foods —with a few important exceptions. Pasteurized milk could be labeled "fresh," for example, because consumers expected milk to be pasteurized. But juice subject to the same process could not make the same claim. "Fresh frozen" was OK under some circumstances; "fresh" irradiated or waxed fruit was OK; even "fresh" cooked crabmeat was OK because, as the official explained, "consumers cannot obtain raw crabmeat." And refrigerated foods could of course be called fresh, even if they were weeks old.

The official then concluded with a set of questions that fueled debate for the next several hours. Could freshness be measured? Should it be defined by taste or by internal qualities? Did the agency need another term to describe foods that were processed but "fresh-like"? How much information about a product's freshness did consumers need or want on a label? And finally, if the FDA redefined the legal meaning of "fresh," who would profit or lose out? Not surprisingly, the last question directly concerned most of the meeting's speakers. Small fruit-juice makers warned that a looser use of "fresh" would not only mislead consumers but also destroy their niche as producers. A consultant to the processing industry argued that any food that tasted fresh deserved the label. A maker of bagged salads described fresh as anything "alive and respiring" (a definition better suited to salads than, say, to poultry). And a lobbyist for the American Fresh Juice Council suggested that the entire debate was futile. "Fresh is not a measurement," he said. "Fresh is a state of being."[1]

The lobbyist hadn't come all the way from Florida to wax philosophical. He had come to fight any regulatory changes that might hurt the fruit growers and juice makers who hired him (ultimately he succeeded). The meaning of fresh mattered very much to them; as the lobbyist said, it was "the cornerstone of our industry." Indeed, theirs was an industrial freshness: mass-produced, nationally distributed, and constantly refrigerated. Far from a natural state, it depended on a host of carefully coordinated technologies, from antifungal sprays to bottle caps to climate-controlled semi trucks. It was a "state of being" that Americans of the not-so-distant past—and people in many other countries still today—would have considered nowhere near fresh.[2]

On the surface, few food qualities seem as unquestionably good as freshness. Dig a little deeper, and few qualities appear more complex and contested. At bottom, the history of freshness reveals much about our uneasy appetites for modern living, especially in the United States. This book traces that history by way of a tour through an ordinary refrigerator. What is it about the word "fresh" that makes marketers so keen to put it on every possible label? Its appeal, I think, lies in the anxieties and dilemmas borne of industrial capitalism and the culture of mass consumption. This culture promotes novelty and nostalgia, obsolescence and shelf life, indulgence and discipline. It surrounds us with great abundance, but not with much that feels authentic or healthful. It leaves many people yearning to connect to nature and community but too busy to spend much time in either. Above all, it's a culture that encourages us to consume both as often as possible and in ever better, more enlightened ways.

Of all the qualities we seek in food, freshness best satisfies all these modern appetites. It offers both proof of our progress and an antidote to the ills that progress brings—at least for a little while. An ingenious range of technologies now protects our fresh food supply against spoilage. But the shelf life of satisfaction remains short. This is the larger perishable history waiting inside the fridge. It's the story of all the forces that create both demand for freshness and doubts about what it means.

My local bookstore is full of histories of fresh food, from apples to sushi. Most are stories of decline, with mass production and marketing destroying taste, diversity, and sometimes species. Many end with anecdotes about the rare places where such foods can still be found in a genuinely fresh state. This book starts by asking, what's behind the common storyline? To answer this question I explore the history of several common perishables. They don't all rank among the most common refrigerator staples, but they do help to illustrate two of the book's basic premises. The first is simply that freshness means different things in different foods. At one level, this is obvious. Everybody knows that some foods last longer than others, and that a few actually improve with age. Less familiar is the biology beneath this variety, and how it has historically influenced the ways humans preserve and move their most fragile foods. So each of the chapters about individual foods includes a short section on the basic science of spoilage. This will explain why, for example, fresh beef became the first global perishable, and why eggs under certain conditions can last for months.

The book's second premise is that biology alone can't explain what "fresh" means to people. As part of my research I pored through mountains of old refrigeration and food industry trade journals. I read home economics texts, women's magazines, technical manuals, cookbooks, and the memoirs of farmers, merchants, social reformers, and inventors. I spent a lot of time looking at food art, advertising, and packaging. I also traveled to several countries and talked to hundreds of people, including both professionals (among them fishermen, dairy farmers, chefs, and assorted salespeople) and many strangers I met while writing in cafés and bars (where I heard every possible bad joke about freshness—repeatedly). They didn't all agree about what particular fresh foods were supposed to look like or where they should come from. But all these conversations and written sources made clear that people value freshness in ways that can't be boiled down to nutrition or taste. Each of the following chapters, then, uses the history of different foods and technologies to explore what else "fresh" means.

Consider the basic rationale behind the refrigerator: keeping food fresh is good because otherwise it goes bad. Over millennia, people developed all kinds of ways to keep perishables good. But most methods were home-based or artisanal, and most transformed foods into states clearly different from fresh.[3] In the late nineteenth century, the business of stopping spoilage became less visible and more controversial. It moved into chilled warehouses, packing plants, railcars, and steamships. It left consumers to wonder about the real age of foods that looked good despite coming from distant places and seasons. Often they turned out not so good at all. Sometimes machinery was to blame, sometimes the merchants who used it. Regardless, consumers wanted labels to distinguish fresh from cold-stored foods. Whether or not they worried about the latter's safety, many saw cold storage itself as a tool of cheats, speculators, and would-be food monopolies. They considered it immoral to prolong freshness for commercial gain.

These days, labels advise us to refrigerate all foods we want to keep fresh—including those preserved in cans and bottles until we open them. This linguistic shift reflects more than refrigeration's evolution from novelty to household necessity. It reflects also the idea that freshness depends less on time or distance than on the technology that protects it. In the history of food handling, this marks a radical change, and one that doesn't apply only to refrigeration. Now many technologies—from shrink-wrap to irradiation—keep our perishables looking as good as new for longer periods and over longer distances.

How did this change in the meaning of freshness occur? Not smoothly or evenly, as the next few chapters show. For now what matters is that refrigeration alone can't take all the credit. In other words, this isn't just a story about how people came to accept and then depend on what we now call the cold chain. This technological shift has to be understood in light of larger changes in where and how people lived, and in how they understood the value of perishable foods.

Ideas about health and nutrition count among the most important changes. These days we are constantly reminded that freshness is good for us. Health experts advise consumers to shop the supermarket's perimeter, where most perishables are kept.[4] They tell us to seek out fresh foods both for what they contain (vitamins, minerals, and fiber) and for what they don't ("empty" calories, as well as excessive sodium and other additives). News that such foods might also contain salmonella, mercury, or *E. coli* has left many consumers confused—for if spinach and tuna aren't healthful, what's left? But past generations knew all too well that perishables could kill as well as cure. Milk and meat carried tuberculosis, dirty vegetables dysentery. Fruit was often blamed for cholera and sometimes banned from urban markets.[5]

With the discovery of first bacteria and later vitamins at the turn of the twentieth century, scientific opinion about fresh foods changed dramatically. By the 1910s and 1920s, consumers were inundated with advice about the benefits of fresh foods for children, "brain workers," and aspiring flappers.[6] While some of this advice came from traditional sources—doctors and social reformers, columns in women's magazines—more and more showed up in advertisements for the fresh foods themselves. Largely unregulated, ads for these foods often made bold (if unsupported) references to the latest scientific findings. Consumers learned that Sunkist oranges would cure bad moods and indigestion, while the "mystery vitamin" in iceberg lettuce would melt fat, boost energy, and preserve youth.

Alongside the nutritional claims, advertising stressed that fresh foods were pure and natural. For consumers in the early twentieth century, this mattered. Thanks partly to the passage of national food safety laws in the United States and Europe, groceries were generally cleaner than they'd been even fifty years before, when staples ranging from tea to flour to milk were either filthy, purposely adulterated, or both.[7] But a new generation of industrially processed foods left consumers wondering about what had been lost or added inside the factory. The very fact that fresh produce, meats, and milk needed refrigeration to protect against quick spoilage—unlike, say, the canned corned beef immortalized by Upton Sinclair's novel *The Jungle*—added to their credibility as natural foods. So did their appearance. They looked like they had come straight from a farm or perhaps a butcher's shop, not from a factory.

Note that fewer and fewer consumers came from farms themselves. Increasingly removed from the day-to-day work of growing and marketing food, they forgot that it took, in fact, a lot of work. This shaped views not only of farm life but also of fresh foods. It became easier for consumers to overlook the sharp contrast between the virtuous image of the foods themselves and the harsh conditions under which they were often produced. It became easier for them to imagine that fruits and vegetables naturally looked as uniformly good as the ones displayed in the supermarket. And eventually it became profitable for processors to trim, slice, and package all kinds of refrigerated foods, and to market them as convenient *yet still fresh*. "Fresh cut" fruits and bagged salads are among the more familiar examples from the produce shelves. They've had value added and mess and drudgery stripped away. They're nature made simple—to eat, anyway. But no-fuss fresh still needs human labor somewhere. Later chapters consider how the ideal of fresh as natural has helped to obscure that basic fact.

The qualities that make freshness seem natural have also traditionally made it a marker of wealth and social status. In all but the most egalitarian of pre-industrial societies, the rich generally ate not just better but also fresher food than the poor. They had the land and labor to raise it, and the money to buy whatever they could not produce. Rome's senators stocked artificial ponds with exotic fish and had barges full of live seafood brought from distant waters. England's landed gentry dined on fresh game from their hunting grounds; French aristocrats adorned their tables with peaches and pears from private orchards. Some also built greenhouses and chilled storerooms, so as to have fruit even when it was costly and scarce. The conspicuous consumption of freshness didn't just reflect social hierarchies; it could also help keep them intact. In many societies, community feasts of first-run salmon, slaughtered bulls, and game meat reinforced the authority of chiefs and feudal lords. Leaders couldn't stop these valued foods from spoiling, but they could control their distribution while fresh.[8]

Feasts also reinforced ideas about the strength and vitality inherent in the fresh flesh of certain mammals and fish. This meaning of freshness proved among the most important forces behind the development of new technologies and trades in the late nineteenth and early twentieth centuries. As the buying power of workers in the United States and Western Europe increased, so did their expectations. They wanted the fresh butcher's meat that the better off enjoyed, not cured pork and beef. They wanted oranges and eggs in winter, and milk delivered to their doorsteps. Freshness wasn't the only food quality associated with the good life; workers also wanted whiter bread and better coffee. But market demand for perishables—combined with the belief that national well being depended on the availability of certain perishables, such as red meat and clean milk—fueled both government and industry investment. Britain's fleet of refrigerated steamships, Argentina's meatpacking plants, the United States' municipal cold storage houses: all were built in an era when the provision of affordable freshness testified to a nation's progress. Later in the twentieth century, new wealth in parts of Asia and Latin America—Hong Kong is one example discussed later in the book—also brought new markets for once-elite fresh foods.

So then what happened to the prestige value of freshness? Now we're used to perishables that last months and travel half the globe. Today it's infinitely easier to buy a single burger than an entire steer. Yet in certain ways the ubiquity of more-or-less fresh foods has simply driven the standards higher. The most status goes to whoever can find and afford the absolute freshest product, however fresh is currently defined. Seafood inspires the most extreme spending, whether on first-of-the-season Copper River salmon, sushi made from ultra-premium bluefin tuna, or a giant grouper scooped out of a tank in a Shanghai banquet hall. Across the fresh food supply, though, few labels can top the prestige value of the locally grown.

For most of human history, of course, perishable foods were by definition local. They traveled far only if they could go "on the hoof" (like cattle) or at least be kept alive and breathing. In the late nineteenth century, refrigeration, steamships, and railroads together pushed the frontiers of the fresh food supply across continents and oceans. In rural New England as well as in parts of Western Europe, change came brutally fast. Farmers planted orchards and bought cows only to find markets flooded with produce from lower-cost hinterlands. In cities, though, consumers saw little of this pain. Instead they saw the bounty hauled in from far away. They read menus of restaurants like the New York chain Schrafft's, which in the 1930s listed the mileage traveled by its exotic produce. The fresh oranges, grapefruit, and strawberries in its fruit cocktail had cumulatively covered 7,800 miles en route to Manhattan, while the makings of a vegetable salad together racked up 22,250 miles.[9] At the time, such menu options seemed worth boasting about. They demonstrated technology's conquest of borders, distance, and seasons; they offered customers fresh foods from the places they grew best.

Doubts about this sunny view of global freshness came and went. The experience of World War Two made dependence on faraway food look risky, especially in countries such as Britain. By the 1970s, at least some consumers worried that the people producing their foreign fruits and vegetables might have been subject to abusive conditions; by the late 1980s many feared that foreign produce might be tainted with dangerous bacteria. Yet only at the very end of the twentieth century did local food begin to seem like the cure for global food's accumulated ills.

Some of the strongest popular support for food "relocalization" developed in Britain, where pro-global food policies were blamed for devastating epidemics of mad cow and foot-and-mouth diseases. Local food activists tallied up "food miles" not, like Schrafft's, to thrill consumers, but rather to show how imported peaches and peapods contributed to resource depletion, global warming, the disappearance of family farms, and culinary decline.[10] In the United States, "locavores" swore off food outside a certain radius. Farmers markets proliferated, as did local food options at restaurants, school and university cafeterias, and even some mainstream supermarkets. At the same time, bookstores filled up with local food memoirs and manifestos. Everyone from chefs to philosophers to novelists weighed in on the virtues of the locally made, the homegrown, even the hunted and gathered.[11]

The romance of the local has reinvigorated small-scale agriculture in parts of the country where it once looked doomed. The farming regions surrounding foodie-packed cities like New York and San Francisco are fending off sprawl with microgreens and gourmet butter. But the local food movement does not yet spell good news for all the world's small farmers. Consider Burkina Faso, an extremely poor country in West Africa where I lived for a while in the 1990s. Each winter it sends planes full of fine-grade green beans *(haricots verts)* to France, its former colonial power. A century ago, colonial officers forced the peasants to grow beans, mainly so they could eat like they did back in France. Today, farmers are forced to continue doing so by both the inequities of the global economy and the poverty of the local one. Burkina Faso's cotton sells in a world market long flooded by American harvests. Its local marketplaces are full of subsidized American wheat and European powdered milk. They are also full of people who can only afford to spend pennies per day on fresh vegetables, hardly enough to cover the costs of producing them. So many farmers, with encouragement from American and European aid agencies, instead depend on the kind of export trade that local food activists love to hate.

Those farmers in Burkina Faso would agree that much about their livelihood does not make sense. Growing picture-perfect green beans for export is hard work, and risky, too. Given how quickly the beans go bad, it takes only one late flight to ruin a season's earnings. But the farmers have children to feed and send to school. They can't wait around for a trade that makes sense. The global market's demand for year-round freshness means, for them, a chance to earn a better-than-nothing income—something that local food activists often overlook.

Burkina Faso is one of several countries visited in the following chapters. The international itinerary of the refrigerator tour shows how many technologies besides refrigeration, from steamships to shrink-wrap to fish sedatives, have helped to create today's global fresh food supply. It also shows how global demand for fresh foods has shaped livelihoods and landscapes all over the world. Yet in certain respects the history of modern freshness is a very American story. The United States might be the birthplace of Twinkies and prefab hamburger patties, but it's also a country where the pursuit of freshness as an ideal has produced all kinds of technological and commercial innovations—some of which have shaped how not only Americans but also the rest of the world eats.

Among the first results of this pursuit was a nationwide cold chain. The United States became what one historian called the "first refrigerated society," not because Americans invented refrigeration itself (they didn't), but rather because they embraced it most readily. European engineers saw Americans' chilled railcars and municipal cold stores as evidence of a pragmatic national character. But geography also mattered. North America's vast distances and extreme temperatures encouraged all kinds of food producers and merchants to invest in the preservative powers of first ice and then refrigeration. As more and more of their products appeared in urban markets throughout the year, it became easier to convince consumers that they, too, needed an icebox at home.[12]

By preserving freshness, refrigeration brought progress measurable in many ways: the exploding farm output of states such as California; the increasing variety of fruits and vegetables sold by big city greengrocers; the falling prices of chilled eggs and beef; and eventually the growing stature and survival rates of children raised on more and safer milk. It was just one of many technologies that made life altogether easier and less hazardous, at least for middle- and upper-class Americans. It also inspired boasts about the utopian achievement of the United States: a seasonless food supply. "The agricultural feast of harvest time and the resultant waste; the famine of fresh farm products in winter, and its corresponding high prices—are no more," proclaimed one American shipping manager at the first international conference on refrigeration in 1908; "the golden mean has been reached."[13]

Yet this growing material comfort also left many people vaguely *un*comfortable about where progress was taking them. In *No Place of Grace* Jackson Lears emphasizes that this "anti-modern impulse" was neither universal nor uniquely American. But many influential members of American society—writers, artists, businesspeople, and even scientists—shared apprehensions about the New World's breathlessly fast transformation. They romanticized pioneer days and rural living more generally. They saw contemporary urban existence as both too complex and not challenging enough, as somehow cut off from real life. Many warned that society was becoming "soft," and some predicted decline. "Once let the human race be cut off from personal contact with the soil," wrote a pathologist in *The North American Review* in 1888, "once let the conventionalities and artificial restrictions of so-called civilization interfere with the healthful simplicity of nature, and decay is certain."[14]

Hunger for a supposedly more natural, authentic, pre-modern life pervaded many aspects of American culture, from landscape painting to outdoor sports.[15] It also influenced turn-of-the-century views about the value of freshness in food. During an era when nutritional opinion otherwise changed dramatically, the belief that modern life demanded *fresher* foods remained a constant theme. Partly this reflected the assumption that city-dwellers' stomachs, like their muscles, had gone soft; they would get sick on the salted and semi-spoiled meat that had sustained past generations. Similarly, "pure milk" campaigners argued that city children needed the freshest milk even more than country children, simply because their urban upbringing left them more vulnerable to bacteria of all kinds.

In addition, many fresh foods appeared to contain the qualities that modern Americans needed to maintain busy work and social lives. Fad diets based on raw and otherwise "vital" foods grew popular even before the discovery of vitamins in the mid-1910s. By the mid-1920s the notion of therapeutic freshness had gone mainstream. Consumers learned that fresh fruits, vegetables, and dairy products would protect against the ills of an overly processed, "too modern" diet, and keep them young, slim, smart, "regular," and energetic.[16] Ads for iceberg lettuce, among others, told them that a diet rich in fresh produce was essential, not just to their own health, prosperity, and happiness, but also to the strength of the nation: "In your body millions of little workmen—known to Science as cells—transform the food you eat into teeth, hair, bones, organs, glands. Yes, and into beauty, vitality and character. If you are forcing these cells, through an unbalanced diet, to make 'bricks without straw,' you are growing old before your time. You are tossing into the discard years of that virility which enables men and women to attract, achieve, win in love and business as well as in war."[17]

Then as now, the search for beauty and vitality led some Americans back to the land. Most did not want to become ordinary "dirt farmers" in the hardscrabble prairie states; their vision of pastoral bliss did not look like Nebraska. Many instead headed to California, described by its boosters as a land of health, wealth, and gorgeous scenery. California's mild climate also supported farming that appealed to nonfarmers, both aesthetically and financially. Urban refugees became egg ranchers in Petaluma, lettuce growers in Salinas, and orange growers in Riverside. Once they were settled on the land, romantic visions of living in nature gave way to the need to make it productive. Growers quickly adopted the latest technologies—pesticides, commercial feeds, and fertilizers—as well as marketing techniques. The founders of Sunkist, themselves mostly transplants from East Coast enterprises, proved especially successful at using the state's natural beauty to promote their own fruit.[18]

Sunkist and other California growers' associations also sought help from advertisers such as J. Walter Thompson (now known as JWT). Like many of the era's manufacturers, the growers faced an oversupply crisis by the end of the nineteenth century. But theirs was more acute, simply because they did not have the option of slowing down the plant. Their land was often deeply mortgaged, and they could not afford to tear up orchards or scale back planting. They could not stop ripening. They could not store their produce for more than a few days or perhaps weeks. So they had to make people eat more fruit. Advertisers found fruit's raw, unpackaged nature no obstacle to their craft; with clever words and colorful art, they could spin the orange as a new, improved, and branded product, fresh in every sense.

Advertisers' job, of course, was to put a fresh spin on even the most inert and unglamorous goods, from oat flakes to drain cleaner. More broadly, advertisers had to portray buying as a form of consumption as vital, natural, and satisfying as eating or sex, but also part of a distinctly contemporary "American tempo." "Besides being new," the J. Walter Thompson Company told its designers, an advertisement "must be virile, snappy, magnetic."[19] It had to sell a specific product along with a broader ideal of the good, full, modern life as one always full of modern new goods. In this sense, fresh foods were easy to advertise, regardless of vitamin content. They always seemed alive; they always had to be bought anew. And they could almost always be portrayed as solutions to modern consumers' problems.

These problems went beyond concerns about soft and sluggish digestive systems. The American advertising industry exercised a powerful influence over an emerging consumer culture by paying attention to everything consumers doubted, feared, and disliked about it. In a society full of new arrivals, advertisers offered advice about how to entertain and eat for success. In a Puritan culture wary of gluttony, they stressed how fresh foods could improve productivity. In gray, crowded cities, Sunkist billboards and grocers' windows offered colorful distractions. In magazines full of stories about a fast-changing world, refrigerator ads showed consumers how to keep up, and keep safe.[20]

In short, advertising sold both goods and reassurance, as it still does. It promoted newness for its own sake and as protection against modernity's potential harm to social, physical, and spiritual well being. Fresh foods embodied this paradoxical message better than most goods. Different kinds of perishables appealed for different reasons, as the following chapters should make clear. But all were marketed as pure, natural, wholesome, and vulnerable. Thus the need for refrigerators, which were portrayed as godlike protectors of family health and social status—as long as consumers had the latest models.

Refrigerator ads urged consumers to accept both technology's power over the perishable and technology's own perishable nature. They played on consumers' fears of decay while promoting the need for planned obsolescence. More broadly, they showed consumers how a diet rich in foods that did not last could, with the help of refrigeration, bring lasting improvements to their own lives. It was a message that appealed to Americans' broader optimism about the capacity of machines not simply to master nature but also to contain and improve it; to make nature easier and more pleasant to live with.

In other words, refrigeration counted among the technologies that promised to preserve, however indirectly, Americans' access to the "fresh, green landscape" that first greeted the Pilgrims. Leo Marx's *Machine in the Garden* traces this pastoral theme through American literature from Hawthorne through *The Great Gatsby*. Whether or not ordinary Americans read Emerson or Thoreau, many saw the ads portraying a domestic version of the pastoral ideal. Thanks to a cold chain that linked farms, ranches, and oceans to household kitchens, they could enjoy nature's fruits with little effort or worry. Refrigeration promised the garden in a machine.[21]

The selling of refrigerated freshness did not always go smoothly, even once most consumers had a Frigidaire or at least an icebox at home. Increasingly, producers of fresh food found that they had to compete not just for a share of the consumer's dollar but also for her time. They had to deliver freshness in an ever more convenient (but still seemingly natural) package. And even as fresh foods in general became more affordable to middle-class consumers, marketing continually cultivated desire for new luxuries. These remained frustratingly beyond the means of many Americans, like the Chicagoans described by the writer Edna Ferber in the short story "Maymeys from Cuba":

> Just off State Street there is a fruiterer and importer who ought to be arrested for cruelty. His window is the most fascinating and the most heartless in Chicago. A line of open-mouthed, wide-eyed gazers is always to be found before it. Despair, wonder, envy, and rebellion smolder in the eyes of those gazers . . . It is a work of art, that window; a breeder of anarchism, a destroyer of contentment, a second feast of Tantalus. It boasts peaches, dewy and golden, when peaches have no right to be; plethoric, purple bunches of English hothouse grapes are there to taunt the ten-dollar-a-week clerk whose sick wife should be in the hospital; strawberries glow therein when shortcake is a last summer's memory, and forced cucumbers remind us that

we are taking ours in the form of dill pickles. There is, perhaps, a choice head of cauliflower, so exquisite in its ivory and green perfection as to be fit for a bride's bouquet; there are apples so flawless that if the garden of Eden grew any as perfect it is small wonder that Eve fell for them . . . Oh, that window is no place for the hungry, the dissatisfied, or the man out of a job.[22]

Freshness has long reminded people of their place, for better or worse. It reminds farmers and fishermen how far they are from the most profitable markets, just as it reminds consumers how far they've come—or still need to go—to be able to eat whatever fresh foods they choose. For well-off consumers in well-off countries, choosing has become one of the hardest parts of eating. Is it best to buy organic? Local? Free-range? Omega-3–enhanced? Yet most of the world's consumers do not face the so-called omnivore's dilemma. Such choice depends on having enough money as well as access to a myriad of modern technologies. Even seemingly antimodern consumer movements—the raw milk "underground," community-supported agriculture, locavores—rely on highways (paved and informational) that can't be taken for granted everywhere.

Such movements share more than infrastructure with the supermarkets and fast food chains they oppose. They also share a faith in freshness as a quality inherently worth promoting, whether on food's packaging or as a reason to avoid all packaged food. We've come to see freshness as a quality that exists independent of all the history, technology, and human handling that deliver it to our plates—a quality that, ironically, transcends time and space precisely because it is so sensitive to both.

one

Refrigeration

COLD REVOLUTION

Primitive man spent practically all his time getting, caring for and preparing food. In a real sense, the aim of human progress has been to make these processes ever easier and easier. The less time we are forced to spend thinking about food, the more we have for higher things, so called. The modern refrigerator, which makes it vastly easier to care for food, may help to produce love songs—we hope.

—Gove Hambidge, "This Age of Refrigeration," *Ladies' Home Journal* (August 1929): 103

Unlike some kitchen appliances, the refrigerator doesn't cry out for attention. Sure, some fridges talk or broadcast TV, and some boast fancy brand names. But the best refrigerators are like clean windows: entirely forgettable. They let us open and shut them, gaze at their interiors, leave them for extended periods—all while scarcely noticing they're there. We don't have to think about their role in our lives until they fail us. For several years after college I thought a lot about refrigerators, because everywhere I lived they failed me. In one apartment after another, I trusted my groceries to the type of seen-better-days appliances that, as I learned from one repairman, landlords like to leave in their rental units. They leaked and buzzed and groaned. Small glaciers accumulated in the freezers, but they didn't keep food very cold. Leftovers didn't last, lettuce wilted, and milk spoiled before its sell-by date.

The dates stamped on cartons of milk, eggs, and other perishables all assume, of course, that the refrigerator is working not just at home but throughout the entire cold chain. They assume that farms and trawlers, packers and bottlers, shippers and supermarkets all have the machines, energy, and know-how needed to keep our food cool, whether it's traveling twenty miles or five thousand. Admittedly, it's a fragile system. A few days without power can ruin an entire city's food supply. But we take for granted that refrigeration serves a good purpose, in and beyond the home. It allows us to think about "higher things" than the care of our food.

Yet early refrigeration was not an easy sell, especially outside the United States. Its backers found it hard to understand why anyone would oppose such a useful technology. After all, refrigeration let people eat seasonable, perishable foods from wherever they wanted, whenever they wanted. It had the potential to eradicate waste, stimulate production, and improve health. It promised an altogether more *rational* food supply, at least in theory. In practice, refrigeration undermined not just farmers' and merchants' local markets but also traditional understandings of how food quality related to time, season, and place. It threw into question the known physics of freshness. Refrigeration's conserving powers, in other words, threatened radical changes all across the food chain.

The Refrigerated Society

Humans used cold to slow spoilage long before they understood how it worked. They learned what materials kept perishables coolest and which seasons were safest for long-distance commerce. In some regions, they also stored and traded the coolants. The ancient Chinese stored ice, and the Romans kept snow in covered pits. At least as far back as the sixteenth century, pack animals hauled ice from the Alps to the Mediterranean, and from the Andes to Lima. But this frozen water was a luxury good, used more to chill drinks than to preserve food.[1] Ice first became a mass-market commodity in the northeastern United States in the first half of the nineteenth century, at a time when the region's rivers and lakes regularly froze solid in winter. The founder of the New England ice trade, the Boston entrepreneur Frederic Tudor, aimed to ship harvested ice to tropical colonies. He'd heard it could be done; the English had apparently shipped ice cream at least once to Trinidad, "packed in pots of sand from Europe."[2] He thought ice would be even more appreciated, since it could chill everything from cocktails to villas.

The third son of the noted Boston lawyer William Tudor, Frederic was a compulsive weather watcher but not an otherwise cautious man.[3] In late 1805, a few months after he got the idea to sell ice to the West Indies, the twenty-three-year-old sent his two brothers to Martinique to drum up business. He followed soon after, bringing 130 tons of frozen cargo, harvested from a lake near his family's farm. Upon arrival he found that his brothers had failed to find the promised buyers. It soon became clear that the French colony's residents had no idea what to do with ice. Few bought it, and fewer still knew how to store it. Some submerged it in tubs of water; others packed it in salt. The results were disappointing, to say the least. As Tudor's precious cargo melted on the dock, he made batches of ice cream, mainly to prove that it could be done.

Tudor lost thousands of dollars, but he remained convinced that he could make money from New England ice. Eventually he did. In the 1810s and 1820s, he climbed out of debt and built markets for ice stretching from Norfolk to New Orleans to Havana. After the Martinique fiasco, Tudor made sure that his ice shipped only to ports with well-insulated storehouses. In 1833, Rio de Janeiro and Calcutta received their first shipments. British expatriates in India became especially good customers. The *Calcutta Courier* ranked

Tudor among the "benefactors of mankind [such as] the importer of the potato into Europe."[4]

Back at home, ice harvesting became much faster and cheaper after one of Tudor's suppliers, Nathaniel Wyeth, invented the horse-drawn ice cutter in 1825. Wyeth used the "ice plough" to carve uniform blocks of ice out of Fresh Pond in Cambridge, Massachusetts, where his family owned a resort hotel. Previously a winter skating rink for Harvard students (and now the city's main water source), by 1837 the pond had become the site of an ice works employing 137 men, 105 horses, and one bull.[5] Tudor, meanwhile, had become known as the "The Ice King of the World."[6]

Encouraged by Tudor's success and Wyeth's labor-saving invention, new ice companies proliferated in the northern states. Some of them competed for pieces of the southern U.S. and Caribbean trades. Many more focused on supplying local markets. From New York to Cincinnati to Baltimore, ice-harvesting industries attracted seasonal workers, increased waterside property values, and packed so much ice in sawdust, wood shaving, and marsh hay that prices for these once-useless materials shot upward.[7] Prices for ice, meanwhile, dropped from five or six cents per pound in 1827 to anywhere from half a cent to three cents by the early 1830s. Ice also became much more convenient, as many city dealers offered daily home delivery. In Boston, fifteen-pound blocks of ice delivered daily by horse-drawn carriage cost families two dollars a month, or eight dollars for six months.[8]

Some of this ice was chopped up for cold desserts or drinks, including the ice water that temperance societies promoted as a "healthful beverage."[9] But much of it went into the iceboxes of well-off city-dwellers, who used it to store their dairy products, fish, meat, and even fruits and vegetables. Although the icebox was far from a standard fixture in every home, by 1838 the *New York Mirror* considered it "an article of necessity" on par with the din-

Ice from Fresh Pond in Cambridge, Mass., once cooled cocktails in Calcutta and Havana. (*Ballou's Pictorial,* March 17, 1855, p. 172)

ing table. Its appeal seems to have benefited, at least in some circles, from the fashion of serving salads and other foods considered
lighter, more sophisticated, and more French than colonial fare.[10]
That said, the icebox itself was not a sophisticated machine. A
metal-lined butter-storage tub patented by the Maryland engineer
Thomas Moore in 1803 remained the prototype for most home iceboxes even in 1840. Food stayed colder than it would have in a
cellar or a well (which most city-dwellers didn't have anyway) but
not necessarily more appetizing. Most early iceboxes stored ice at
floor level, which prevented the steady melting and cool air circulation necessary for effective refrigeration.[11] As a result, their interiors quickly grew moldy and rank, and foods absorbed one another's odors. No wonder Miss Leslie's 1840 *House Book* recommended
that families buy two iceboxes, one for dairy products and another
for meat.[12]

Soon the domestic ice demand outstripped the overseas trade.
New York City's annual consumption increased from 12,000 tons
in 1843 to 65,000 tons in 1847 and 100,000 tons in 1856. Boston's
consumption leapt from 6,000 to 27,000 to 85,000 tons during the
same period.[13] During California's gold rush, the demand for ice
in San Francisco was so great that a few of the city's businessmen
founded the American Russian Commercial Company and began
harvesting ice in Alaska's Kodiak Harbor.[14] A crackpot idea half a
century earlier, the frozen water trade now spanned the globe.

Cold on Demand

In 1855 the magazine *De Bow's Review* called ice "an American
institution . . . as good as oil to the wheel. It sets the whole human
machinery in pleasant action, turns the wheels of commerce, and
propels the energetic business engine." The ice habits of Americans
also testified to the country's democratic character, the magazine
claimed. For while ice was a rare luxury in most of Europe—"con-

fined to the wine cellars of the rich"—in America its use appeared "as widely extended among the people as the heat is, and at a trifling individual cost."[15]

The winter that produced this all-American commodity was not, of course, so widespread. Most years, northern freezes provided ample supplies of ice for cities in the South. But when the winters were too warm or too short, poor harvests led to high prices and, as ice came to seem like a necessity, even fears of summertime "ice famines." This was one reason the Florida physician John Gorrie invented a machine that could make ice regardless of the weather. His primary goal was to chill cities, not food. Like many mid-nineteenth–century medical experts, Gorrie thought that too much tropical heat led to mental and physical degeneration, as well as to the spread of diseases such as malaria. But since his schemes for citywide air conditioning seemed unlikely to attract investors, he designed a machine that he thought would have commercial appeal as well as humanitarian benefit. Essentially it used engine-compressed air to absorb enough heat from containers of water to freeze them. A few New Orleans businessmen expressed interest in Gorrie's invention, and *Scientific American,* in an 1849 review, described it as "a beautiful and comprehensive system."

The magazine conceded that the machine was still in the design stage, and "perhaps unavoidably very imperfect in plan and execution." Certainly its size (enough to fill a small garage) made it less than ideal for use in a home or a small shop. The magazine also put to rest rumors that Gorrie's machine froze water instantly. Still, the reviewer concluded, if countries in the tropics could manufacture their own ice, public health and productivity might improve enough to "modify the existing relations of the inter-tropical regions to the rest of the world.[16] In other words, manufactured ice would beat not just the heat but also the economic backwardness that nineteenth-century thinkers often blamed on tropical torpor.

Despite such predictions, Gorrie lost his financial backers, and he died in 1853 before his machine ever made it to market. The idea of winter-free ice-making, however, had by then captured the imagination of scientists and inventors in many countries.[17] France's Ferdinand Carré designed one of the first freezing machines to rely on the absorption of ammonia to drive the cooling process. Although not exactly a precision technology, the unit was much smaller and simpler than Gorrie's compressed air machine. And unlike the Florida physician, Carré found a market in New Orleans, thanks to the Civil War's disruption of the coastal ice trade. By 1865 the ice-starved city had smuggled in three of his machines. By 1868 New Orleans was chilling its seafood and mint juleps with the Louisiana Ice Manufacturing Company's factory-frozen city water.[18]

Ice plants went up throughout the U.S. South during the second half of the nineteenth century; by 1889 Texas alone had fifty-three. The industry grew more slowly in the North, at least until the public began to question the safety of the region's "natural" sources. Some Hudson River suppliers, for example, harvested ice downstream from where Albany and Troy dumped their sewage.[19] This practice didn't result in any known epidemics in New York City, probably because freezing usually kills bacteria such as typhoid. Still, at a time of growing awareness about how germs work, it's not surprising that many consumers didn't like the idea that they might be using frozen waste to cool their kitchen iceboxes or perhaps even their cocktails.[20] Ice manufacturers encouraged their disgust and boasted that they themselves used only the purest distilled water. In response, ice harvesters ran ads suggesting that "artificial" ice lacked the vital qualities needed to keep foods cold. "There is no life in it," claimed one such ad, "and like the lazy hired man it will not work."[21]

Especially in the South, ice inspired dreams of easy money. After all, the main raw material was either cheap or free, and the market appeared insatiable. Households, grocers, farmers, food shippers—everybody wanted more ice. Or so the companies selling ice-making equipment suggested. Production exploded, and by 1890 ice gluts had replaced famines in the South. Prices plunged accordingly. Similar oversupply crises hit the North as manufacturers moved into the harvesters' territory.

Many ice producers went bankrupt; others stepped up their marketing. The journal *Ice and Refrigeration* offered regular advice on advertising tactics. Step one was a frontal attack on seasonal buying habits. As the journal put it in 1910, "the large bulk of domestic consumers . . . do not begin the use of ice until the sun burns holes in the pavement, and only discontinue the use of ice with the first frost." Indeed, the journal found that only 10 percent of households in many states bought ice in the winter.[22] But the solution, according to *Ice and Refrigeration*'s advertising experts, was "quite simple": "By changing one thought in the minds of the American woman—that is all. Take the word 'extravagance' from the thought 'ice is an extravagance in the winter' and replace it with the word 'economy' . . . and you have sown the seed for a crop of profits three hundred days in the year instead of one hundred days of profit and two hundred of loss."[23] The best advertising didn't just sell ice; rather, it marketed an ice-enhanced lifestyle, with recipes included: "The housewife should be instructed in the real economy of keeping her ice box well filled . . . Then there is a whole lot that can be said about the luxury (and it is a real luxury) of a big roomy refrigerator. One that will hold the things that have to go in it without tipping over the milk or breaking a lot of dishes every time anything is put in or taken out. There is no end to the delicious iced dishes and drinks that can be worked into good copy by a clever advertising man."[24]

At the turn of the twentieth century, many producers of perishable goods looked to the "clever advertising man" to help boost consumption. In the case of ice, consumption soared. By 1914, ice sales in Chicago, Baltimore, and Boston had increased fivefold since 1880, and in New Orleans nearly thirteenfold. Between 1879 and

1919, sales of iceboxes rose from less than $1.8 million to more than $26 million.[25] And yet more than half of U.S. urban households and most rural households still didn't buy ice at the end of the 1910s.[26] There was plenty more selling to do.

Fridge Fears

Outside the United States, the selling went slowly. Early on, Frederic Tudor had rightly assumed that Europeans living in the tropics would learn to appreciate ice, even if the locals did not.[27] The British especially considered cool houses and drinks necessary for good health in hot climates. But in Europe itself, American ice didn't find much of a market. This was partly due to competition from Norway, which supplied breweries and other commercial ice users from the mid-nineteenth century onward. More fundamentally, most Europeans saw little need for ice at home. They liked *gelato* and *crème glacé,* but not iced drinks. Even the British drank their beer warm. The Scandinavians were accustomed to preserving food with ice, but they certainly didn't need to buy it from North America. And southern Europeans, it seemed, just didn't store much fresh food.

When U.S. icebox manufacturers surveyed American consuls in the 1890s about potential markets in their European host cities, the news from France and Italy wasn't good. "In the great cities of Marseilles and Bordeaux butchering is done every day in winter and twice a day in summer, and the meat is cooked within a few hours after killing," reported the consul in southern France. "The mass of the population use no ice, but purchase their supplies of food twice a day, consuming the total purchase at once, making no effort to preserve anything." A Genoa-based diplomat wrote that "the Italian people do not consider ice necessary at any season in order to enjoy good health or a good appetite; many of them believe ice injures rather than soothes the palate and stomach. Economy is practiced here to such an extent that fully ninety-seven families out of every one hundred purchase only sufficient food for daily wants. Nothing remains over for the morrow—not even bread or vegetables."[28]

These consular reports were not entirely accurate. In France and Italy, as in most other traditionally agricultural societies, people had long preserved food, especially in rural areas. They couldn't afford not to. They turned pork into salami and sausage, and milk into cheese. They dried fish and fruit, pickled vegetables, and stored eggs in glass-water. By the late nineteenth century, they home-canned. None of these preserved products filled the same place in the diet as newly harvested food. In addition, apples, pears, and grapes were kept for the winter months in specially designed *fruiteries*. But generally only the wealthiest households enjoyed these expensive novelties. Neither truly fresh nor really preserved, they didn't fit the budgets or eating habits of ordinary people.

The same could be said of the early American icebox. The Old World eating and shopping habits of the societies the consuls described didn't require the protections and convenience that it supposedly offered. Some fast-spoiling foods that Americans most commonly put in their iceboxes, such as uncultured milk, Europeans consumed relatively rarely. Others, such as fresh meat and fish, they prepared quickly and purchased less often in hot weather. Most other fresh foods kept fine for the few hours or perhaps days it took to finish them off. Why store them in a cold smelly box? Doing so certainly didn't improve their taste.

Differences in the geography of food shopping also shaped people's ideas about the merits of the icebox. As H. D. Renner notes in *The Origin of Food Habits*, cities in Britain and the United States traditionally concentrated commerce in the "High Street" or shopping center, whereas cities in continental Europe layered apartments atop shops. In Europe, freshness was never far. "It makes a vast difference to the housewife whether she has simply to walk downstairs to get all she wants for cooking . . . or whether there is a 5, 10 or 20 minutes' walk to the shopping centre. The housewife on the Continent, in the apartment house built in the nineteenth century or earlier, could almost do her shopping for each meal separately . . . The shops were able to keep their provisions fresh and what the housewife bought for immediate use was in the freshest condition."[29]

Consumers who bought fresh food daily expected shopkeepers to do the same. So not surprisingly, societies that saw little need for home refrigeration also questioned its commercial uses. In France, the public distrusted cold storage for at least two reasons. First, it enabled unscrupulous merchants to hoard goods and otherwise manipulate the market. American consumers shared this concern up to a point; when food prices skyrocketed in 1910–1911, many believed that speculators were using cold storage to create artificial scarcity. The French, however, suspected even short-term, small-scale cold storage, because it spared merchants from having to bargain over or liquidate stocks at the day's end. The power they gained over food also gave them new power over their customers. This was the second source of distrust: if merchants cold-stored their perishables, how could the public know which, if any, were truly fresh?

Such concerns framed the laws governing commerce at Les Halles, the central wholesale market in Paris. Except under special conditions, merchants there could not store their stocks overnight. Any produce that did go into the cool (but not chilled) basement storage chambers had to be inventoried and inspected by the market police, and then sold separately from the presumably fresher goods. All these measures aimed to ensure fair prices and high-quality fresh produce, both top priorities for the retail grocers and restaurant chefs who frequented Les Halles. For many merchants, their customers' trust and loyal patronage trumped any of the touted advantages of "artificial cold."[30]

Opposition to cold storage stretched beyond the market itself, as the Parisian fruit wholesaler Omer Decugis discovered. He began selling fruits from his native southern France in 1850, right around the time that express (but nonrefrigerated) trains began carrying fresh produce. By 1880, Decugis had established himself as one of the capital's most successful fruit wholesalers, with a large store just outside Les Halles. A pioneer in long-distance fruit shipping, he probably saw storage as a way to profit from post–peak season high prices. So Decugis equipped his store with what was apparently France's first commercial refrigeration chamber. But when his clients found out that the company's "fresh" fruit had actually been subjected to storage, they were outraged. To regain their trust, he had his *frigo* destroyed in a public square. Decugis's business survived the incident; five generations later the family still trades in fresh produce, now on a global scale. But Omer Decugis waited twenty-three years before rebuilding his cold storage chamber, and even then he was still the first fruit wholesaler in Paris to operate one.[31]

In retrospect, the public's antipathy toward cold storage might seem ridiculous. Certainly members of France's small refrigeration industry thought so; they called this syndrome "frigoriphobie."[32] But the few farmers and merchants who did use cold storage must have ended up with some pretty frightful food. Writing in the trade journal *L'Industrie Frigorifique* in 1903, one merchant marveled at how fish kept at 1 or 2 degrees Celsius (around 34–36 degrees Fahrenheit) remained "in the same condition as when it was just caught" for up to a month.[33] These days the consensus is that most fish doesn't last more than a few days at above-freezing temperatures. Many delicate and warm-climate fruits, by contrast, can't stand extended cold. They may look good on the outside, but their insides turn to mush. Yet at the turn of the century, the few French growers using cold storage (like their more numerous American counterparts) reported keeping peaches and strawberries at 1 degree Celsius for a month, and tomatoes for three months.[34]

French refrigeration experts had little patience for those who questioned the technology's benefits. When the newspaper *Le Matin* ran a letter in 1911 calling for the mandatory labeling of cold-stored fruit "because it is almost always uneatable and always harmful to the health," the trade journal *Revue Générale de Froid* mocked the letter-writer's "phobia." If refrigerated fruits were specially labeled, the journal predicted, "we are confident that they would soon command a premium."[35] Another newspaper letter-writer argued that fruit producers should turn their surpluses into preserves rather than "indulge the caprices of 'snobs' who bought out-of-season cold storage fruit." *Industrie Frigorifique* dismissed this idea as an example of the "profound ignorance" that slowed France's progress, and not just in refrigeration: "If the Californians had followed the advice of our wise compatriot, they would never have succeeded in creating their immense orchards . . . rather than make jam, let us imitate the Americans."[36]

"Foods That Will Win the War"

World War One marked a turning point in refrigeration history. In Europe, high demand for red meat weakened political opposition to chilled and frozen imports and spurred state investment in refrigerated steamships and cold storage. Frozen beef became a staple of British troops and even, by some accounts, their meat of choice. On the other side of the Channel, France cut back sharply on meat imports from South America and Africa once its own cattle population had recovered. But by then cold storage was more widespread and less mysterious, and more butchers and consumers knew how to handle meat subject to its influence. French importers learned to put on the market only frozen meat "in absolutely fresh" condition—an idea that no longer seemed so oxymoronic. Government-issued pamphlets on how to prepare "frigo" meat added to its respectability, as did the growing (if never large) number of shops that specialized in frozen cuts.[37]

In the United States, wartime food conservation taught the pub-

Be Patriotic
sign your country's
pledge to save the food

U.S. FOOD ADMINISTRATION

During World War One, stopping spoilage became a civic duty. (U.S. Food Administration, 1918)

lic to appreciate perishables *besides* meat. The government's Food Administration (FA), directed by Herbert Hoover, called on civilians to pledge to one wheatless day and one meatless day per week, plus less butter and sugar. In place of these staples, the FA pushed the fresh foods unsuitable for shipping to the front: milk, fruits, vegetables, and fish.[38] Government-approved cookbooks such as *Foods That Will Win the War and How to Cook Them* told the housewife that by training her family to accept meatless dishes, she'd be "doing her bit toward making the world safe for democracy."[39] Nearly twelve million households—about half the total U.S. population—took the pledge.

The FA also crusaded against food waste and extravagance. It promoted backyard gardening and farm-direct purchases of eggs and meat, since these goods cost less and lasted longer than produce from the grocery store. Women's magazines, which regularly printed FA propaganda, offered tips on selecting, storing, and using up fresh foods. Milk had to be scalded; lettuce belonged in a brown bag in the icebox. Leftovers could go atop salads, and scraps into soups.[40] Many of the FA's recipes—for casseroles and stews, loaves and puddings and pies—hardly emphasized the freshness of their ingredients.[41] Still, by calling on Americans to grow their own, "eat what would spoil," and throw nothing away, Hoover's agency sent a clear message: fresh foods belonged in every patriotic kitchen. Ideally, so did refrigeration.

This message didn't reach everybody. Working-class households, many of them recent immigrants, either couldn't afford or didn't want iceboxes, and they didn't appreciate social workers telling them how to eat.[42] Middle- and upper-class housewives, on the other hand, rallied to the FA's cause. They turned out for the food-conservation lessons held in downtown department stores, signed the pledge cards, and followed the meal plans. And by the war's end, a Department of Agriculture survey indicated that these women and their families were, in fact, consuming more fresh produce and milk—and less red meat—than they had before.[43]

Patriotism wasn't the only force for change in middle-class eating habits. Indirectly, the Food Administration's conservation measures helped popularize the ideas of researchers such as E. V. McCollum, one of the founders of vitamin science. McCollum's work helped to overturn a decades-old scientific belief that fresh fruits and vegetables added little to the diet besides color, variety, and cost. It also addressed concerns about the dangers of "overnutrition" and over-processed food, especially to the country's growing population of sedentary office workers. McCollum and other advocates of what became known as the "Newer Nutrition" emphasized that even some of the most calorie-poor kinds of fresh produce were worth eating for their vitamins. Along with eggs and dairy products, they would keep Americans' bodies and minds at peak performance and also protect against the deficiencies of an otherwise increasingly processed food supply.[44]

The Food Administration itself rarely referred to the health benefits of consuming fresh milk and produce, even though it sponsored E. V. McCollum's research. Hoover did not want to dilute patriotism with appeals to self-interest; nor did he want to endorse the era's vegetarian and raw food movements, which he considered foolish fads.[45] Moreover, not even researchers understood how vitamins actually worked. They only knew that they existed in the foods that prevented diseases such as scurvy and pellagra.

For cookbook writers and advice columnists, this was information enough. The cookbook *Everyday Foods in Wartime* emphasized that a "protective" diet would offer benefits even when peace came: "when war conditions make the free consumption of meat unpatriotic, it is reassuring to think that we really can get along without meat very well if we know how . . . Cabbage, peas, lettuce, dandelion greens, beet tops, turnip tops and other 'greens' are well worth including in our bill of fare for their iron alone."[46] Both during and after the war, the *Ladies' Home Journal* editor Christine Frederick urged readers not to skimp on fresh produce. "As a nation Americans eat far too little fruit," she complained in 1919, noting that many housewives still considered it an extravagance. "The cost of a fruit diet should be balanced by less expenditures for drugs and physicians' fees!"[47] A 1920 article in *Good Housekeeping*, titled "What Shall We Eat to Be Well?" went further, suggesting that humanity's postwar survival depended on a healthy dose of freshness: "Half the world is undernourished, physically run down. This means that vast populations are lacking in that bodily resistance which bars out disease. As never before the world is liable to illness . . . The memory of the recent influenza epidemic is still fresh enough to give us pause. The world needs food, America along with it. Particularly it needs the right food. These discoveries of 'protective' foods come at exactly the right moment."[48]

The discoveries certainly came at the right moment not just for the producers of these foods—dairy farmers, fresh fruit and vegetable growers—but also for the ice and refrigeration industries. In the 1920s, all found ways to push the protectiveness of their products. California's Sunkist growers set the precedent, with ads extolling the cure-all qualities of citric acid. Lettuce ads portrayed the iceberg as nutrient-packed ammunition against aging, fatigue, and fat.[49] For their part, ice and refrigeration companies simply pointed out that protective foodstuffs had to be, well, protected. "The need for fresh vegetables and fruits in the diet now admits no question," a Frigidaire recipe book said. "The preservation qualities of Frigidaire make it possible to preserve these important foods in ideal condition." McCollum himself recommended buying the biggest icebox possible and keeping it well filled, "for it never pays to buy less ice than the maximum which [it] will hold."[50] Indeed, it could be perilous, as an ice industry ad in *Ladies' Home Journal* warned:

Your baby! That priceless bit of humanity, so helpless, so dependent—how you hover over him with anxiety when he is peevish and fretful. Nine times out of ten, the trouble is due to some lack in his diet . . . He needs milk to make him strong. He needs prunes to keep his bowels active. He needs cod liver oil to prevent rickets. He needs orange or tomato juice for growth-building vitamins . . . All of these foods are perishable, extremely so. And unless they are FRESH they are harmful to the baby—even dangerous . . . If mothers only knew the extent to which insufficient ice is responsible for the ills of babyhood, they would never run the risk, just to save a few cents daily.[51]

Baby's milk was just the beginning. Like McCollum and other "newer" nutritionists, the refrigeration industry warned consumers that they couldn't eat like their forefathers. Sedentary city life, they were told, made for soft stomachs. A refrigeration salesman, writing in 1927, advised his peers to play up this point in their publicity materials:

> [The consumer] doesn't understand that food may look all right and smell all right, but still cause him a lot of trouble if he eats it . . . Your grandfather could eat a piece of tainted meat and go out after lunch and plough over an acre, and the physical exercise that he got made it possible for him to digest food that you and I, sitting at a desk before and after lunch, can't eat . . . As we become more and more desk workers, the necessity for guarding diet becomes more and more paramount. I don't believe we can throw off the poisons of bad food today because we don't take the physical exercise to do it.[52]

During the interwar era, advertising, home economics courses, government agencies, and the writings of scientists like McCollum all helped popularize ideas about the value (and perils) of fresh foods.[53] Advances in production and distribution, meanwhile, made such foods increasingly affordable and attractive. Not least, changes in American lifestyles and beauty norms helped boost demand for products that, like iceberg lettuce, appeared both "slimming" and simple to prepare. Altogether, these factors help explain why the average American household at the end of the 1920s ate more fresh fruit, green vegetables, and dairy products than the previous generation, and considerably less red meat (from 72.4 pounds per capita in 1899 to 55.3 pounds in 1930) and grain.[54] While the new American appetite for freshness was hardly consistent—sales of canned goods also leapt in the 1920s—it both encouraged and benefited from the spread of refrigeration.

The Ice Man Goeth

Compared with the fast-changing American diet, the icebox evolved at a glacial pace. If anything, the mass-produced models of the early twentieth century worked less well than their predecessors.[55] They were cheap but poorly insulated, sometimes with nothing more than a few sheets of paper between the wood and the walls. Cold air leaked out but didn't circulate well, contributing to "off" tastes and smells. Icebox interiors needed regular, thorough scrubbing. Worst of all, they needed ice, a coolant more perishable than any of the foods it was supposed to preserve. A 1921 article in *House Beautiful* described the tedium of icebox upkeep:

> All day long, at almost every house, the ice-man has been driving up each day, or every other day, and lugging in a cake of ice. Somebody has had to meet him and let him in. Somebody has had to lift things off the dwindling ice of yesterday to make room for the new supply. Somebody has had to wipe up the wet spot where the ice-man set the cake while he was waiting. Somebody has had to clean up the dirt that he has tracked in. Somebody has had to pull out the pan each day from underneath and empty out the water that has dripped down from the ice compartment. Somebody has had to wipe up the water spilled in doing it. Somebody has had to keep smelling around the ice-box, day by day, to see when it began to get foul and needed scouring. Somebody has had to keep the ice in mind and telephone when the iceman forgot to come, or came too late. Then, somebody has had to pay the sizable bill.[56]

The article neglected to mention that the ice sometimes contained splinters, dirt, or insects. This no doubt reinforced consumers' belief that their ice bills were *too* sizable. This belief persisted even when periodic gluts drove dealers into bankruptcy. Consumers didn't appreciate how much it cost to store ice; they just saw

frozen water, sloppily delivered.[57] The iceman was "a national joke
. . . apt to be a rough, uncouth individual whose route across the
kitchen floor was marked by dirty footprints and puddles of water." He was also apt to bring less ice than his customers paid for,
provided he even showed up.[58]

By the 1910s, the icebox was an appliance behind its time. It
begged for obsolescence. And in fact inventors had been trying
since at least the late 1880s to develop an iceless alternative, with
little success. They faced at least three challenges. First, size and
noise: the refrigerating systems used in breweries, slaughterhouses,
and steamships weighed anywhere from five to two hundred tons—
not including the several-ton steam engines that kept them running.
Even when engineers figured out how to make such a system (not
including the food compartment) small enough to fit in a house, it
usually wouldn't fit in one room. So the machinery had to be installed in the basement, and the coolant piped up to the "iceless"
box in the kitchen. Even from this distance, one engineer noted of
his own experimental model, it could "be heard in the major part
of the house, and if I were not interested in the results, I would not
tolerate it."[59] Second, the technology had to become much, much
cheaper. While mass-produced iceboxes cost about $30, the earliest
mechanical fridges went for around $1,000. In the early 1920s they
still cost more than $450, plus another $50 for electricity.[60]

Third and most important, refrigeration had to be made safer
and easier to use. The big commercial systems demanded constant
supervision and frequent maintenance, and even then it was not
uncommon for them to catch fire, explode, or leak toxic gases.[61]
An ice plant inferno at the 1893 World's Columbian Exposition—
where the public witnessed several firemen burning to death—was
one of the earlier and more spectacular disasters. But even in 1920,
insurers considered industries using ammonia-based refrigeration
to be among their riskiest clients.[62] News stories of cold storage
warehouses going up in flames obviously didn't reassure consum-

ers. Nor did the reports that early home fridges needed repairs about once every three months.[63]

The companies that finally overcame these challenges had little background in refrigeration but plenty of money and skill to devote to its improvement. General Electric (GE), for example, first ventured into the field in 1911, when it began manufacturing a machine invented by the French abbot Marcel Audriffren several years earlier. Audriffren designed his system to cool wine rather than food, but GE saw potential in both its hermetically sealed design and its use of electricity rather than gas. As the historian Ruth Cowan points out, gas-powered "absorption" refrigeration technology offered at least one major advantage over the electric "compression" alternative: it didn't need a motor, and thus it lent itself to smaller, simpler, and quieter designs. But GE's biggest customers, the electric companies, had nothing to gain from a gas appliance. So GE threw its fortunes into the development of an electric model. Its two biggest rivals, Kelvinator—founded by two former General Motors (GM) executives in 1916—and Frigidaire (founded in 1916 and later bought by GM) followed suit.[64]

After several years, millions of dollars, and a few commercial disappointments, GE hit gold with its Monitor Top, released in 1927. By then the majority of urban households had electricity (up from 10 percent of all households in 1910), so the potential market was much larger than it had been in the 1910s.[65] Mass produced in a plant built just for that purpose, the Monitor Top boasted all-in-one construction—the motor, about the size of a large pumpkin, sat on top of the refrigerator compartment. Although it cost more than an ordinary icebox, a $1 million advertising campaign convinced consumers that the Monitor Top was worth it. In the weeks before it introduced its all-steel model in 1929, GE sponsored radio programs, parades, and parties nationwide. Newspaper ads, giant neon signs, and dancing puppets in shop windows all celebrated the housewife's liberation from ice.[66]

Publicity for "iceless" refrigerators in the late 1920s promised cleaner, more consistent cold—and no more waiting around for the iceman. (*Ice and Refrigeration,* July 1926, p. 157)

Needless to say, ice dealers did not join the festivities. In fact they were already on the counteroffensive, as a poem published in *Ice and Refrigeration* shortly after the release of the Monitor Top suggested:

"To Ice!"

Tinkle tinkle, little chunk,
How I wonder if it's bunk—
All these stories that I hear,
That you've something now to fear.
Ice machines are springing up,
Thick as fleas upon a pup,
Loud their chorus—"Watch us grow,
When we come in, ice must go."

"Tell me then," my chilly friend,
"Do you see in them, your end?"

"I AM ICE!" came proud reply,
"On me the millions still rely.
I am faithful, safe and sure,
Qualities that must endure.
Foods entrusted to my care,
I keep fresh beyond compare,
Simple, saving, pure and cold,
I work without being told.
Little I take, much I give
WHO DARES SAY I SHALL NOT LIVE?"[67]

The ice industry's survival strategy went well beyond poems. Icebox makers began to build better-insulated boxes, and icemen began to wear uniforms. In some cities, dealers promoted the construction of "outside icing" portals on houses, so that no one had to be home to receive the delivery. At the same time the National Association of Ice Industries responded to the electric refrigerator companies' "pernicious propaganda" with its own publicity.[68] It created a Household Refrigeration Bureau (HRB) and hired as its director Dr. Mary Pennington, the former head of the Department of Agriculture's Food Research Laboratory. Pennington, a respected scientist and well-known defender of cold storage (sometimes referred to as "the Ice Queen"), churned out pamphlets with titles such as "The Romance of Ice" and "Cold is the Absence of Heat." Some stressed the value of refrigeration in general, targeting the many households that had never been in the habit of buying ice. Food safety and economy were common themes, as was the notion that purchased ice was simply more "up to date" than, say, ice cut from the pond on the family farm: "Farm housewives now are not willing to take chances on temperamental Jack Frost . . . The farm ice house and the farm spring house are rapidly giving way to the regular purchase of symmetrical cuts or whole cakes of manufactured ice to be obtained in the nearby town or at the country 'cash and carry' station or even at the roadside gasoline supply and whisked home in the flivver to maintain freshness in Baby's milk and in Biddy's new laid egg."[69]

Other pamphlets stressed the superiority of ice. A well-insulated icebox, Pennington claimed, wasn't just cheaper to run than an electric fridge; it also did a better job of keeping meat moist and vegetables crisp and vitamin-rich.[70] Along with her assistant director, Mary Kingsley, Pennington was convinced that ice would sell itself once consumers appreciated its enduring value: "It isn't only the women of the households who need education but the men as well. When a man learns that a good refrigerator with sufficient insulation will cut his ice bills in half; save on the food bills because of less spoilage and waste; his business sense is aroused, and he is an ardent convert to the cause of 'Better Refrigerators Cooled by Ice.' Neither will he listen readily to the salesman for the electrical refrigerator."[71]

By the end of the 1920s, the HRB had distributed more than a million copies of each pamphlet. Intending them for home economics teachers as well as for housewives, Pennington emphasized that her advice was "impersonal and impartial" and not to be confused with industry propaganda. "Our pamphlets go out in little gray, plain covers with blue lettering, lady-like, well-bred, nothing showy," Pennington said. Ice dealers' own ads, like the electric fridge makers' full-page spreads, were much less subtle. Both sides claimed to offer the cheaper, more convenient, and safer product. The ice industry's publicity compared ice and electricity bills, and pointed out that "ice does not get out of order." Whereas the manufacturers of the electric fridges touted the high-tech kitchen, the ice industry portrayed technology as trouble. "No improvement has been made upon the convenience of ice," said one ad. "[It] entails no costly machinery, no charge for electric current, no repair bills, no noise. There are no fuses to blow nor belts to break—

nothing to worry about when you are out of the city during the week-end."[72]

The ice industry's ads contrasted the questionable safety of chemical refrigerants with the natural qualities of their own product. They described ice as not just a superior preservative but also a sanitizer. "Just as falling snow purifies the atmosphere," one ad asserted, "so melting ICE purifies air in the refrigerator." The electric fridge makers' ads, on the other hand, told readers that safe refrigeration was by definition stable—and ice obviously was not. As one ad put it, "You spend money for ice that melts and trickles down the drainpipe. Up and down—rising, falling ice-box temperatures. High or fluctuating temperatures invite bacterial growth and resultant food contamination—a serious menace to health."[73] General Electric boasted in 1929 that its own All-Steel Refrigerator maintained a temperature "several degrees below fifty . . . always!"[74] Kelvinator, meanwhile, promised to keep foods secure in its "zone of Kelvination" (below fifty and above forty degrees). At the time, this temperature was considered cutting-edge cold; these days the recommended temperature is around thirty-six degrees Fahrenheit.[75]

The electric fridge makers also described everything their appliances could do to food besides keep it fresh. Kelvinator's ads suggested that "Kelvination" worked magic on ordinary produce: "Kelvinated foods just fairly coax midsummer appetites. Taken from the cold frosty air of a Kelvinator-chilled refrigerator, they are irresistible. Think of sliced oranges, served ice-cold; of cantaloupe or grapefruit; chilled through and through; or of home-canned fruits, served cold in their rich juices. Think of the cream for your cereals cold and refreshing."[76]

With help from women's magazines, the electric fridge makers promoted "refrigeration cookery" as the new cool cuisine. Company cookbooks such as *Frigidaire Recipes* and Kelvinator's *New Delights from the Kitchen* devoted ample space to salads ("the uncooked salad greens must always be cold, crisp and dry") as well as to frozen desserts.[77] *Ladies' Home Journal* ran recipes for pigs' feet aspic and egg-and-asparagus molds, and urged readers to keep their fridges well stocked with olives, capers, pimentos, and other colorful garnishes for luncheon salads and "jellied things."[78]

Refrigeration cookery saved not only money and "food value" (many recipes used leftovers) but also time. Or rather, it allowed busy homemakers to reorganize their time so that cooking would not interfere with social engagements. As a General Electric cookbook reminded readers, "the hostess who entertains at a bridge party does not like to leave her guests very long in order to prepare refreshment." It offered make-ahead-and-chill menus for all kinds of occasions, from afternoon luncheons to children's birthday parties. Frigidaire proposed that even a dinner party could be "a simple matter, with salad ready to place on lettuce, sandwiches prepared ahead, a frozen dessert ready in a freezing tray and colored ice cubes for cooling the beverage."[79]

Nineteen-thirty marked a turning point in the kitchen cold war. The year before, methyl chloride leaks from electric home refrigerators killed seven Chicagoans. *Ladies' Home Journal* nonetheless ran a lengthy article on "the age of refrigeration," describing the mechanical units as "today's infant prodigies . . . You can get them now operated by electricity, by illuminating gas or by kerosene oil. The things are so completely nifty that to see one is to have a passionate desire for it. Already they are counted among the necessities for happiness." Consumers must have agreed, for in 1930 the industry's sales overtook those of icebox makers.[80] Aggressive and well-funded advertising helped (the ice industry's own publicity efforts faltered after Mary Pennington took a job with the poultry industry in 1931), but so too did cheaper electricity and ongoing innovation. A Frigidaire engineer's 1930 synthesis of Freon, a nontoxic, nonflammable refrigerant, ranked among the most important advances. Although later found to contribute to ozone depletion,

Freon eased consumers' fears about chemical-driven cooling. Since it needed less pressure than other refrigerants, it also opened the door to the development of smaller, lighter, and cheaper machines. Already the average price of an electric fridge had dropped to $275. By 1940 it was down to $154, and ownership exceeded 50 percent of American households.[81]

It took another generation before the refrigerator became a near-universal fixture in the American kitchen, alongside the sink and stove.[82] During this time, rising affluence, suburbanization, the teaching of "scientific housekeeping," and the marketing of kitchen appliances as labor-saving status symbols all helped to ensure that the United States remained the world's most thoroughly "refrigerated society." By mid-century, Americans could buy models so big that "bushels of fresh things" could fit inside, and so sophisticated that (according to the advertising) they kept foods from seven different climates in prime condition.[83]

Already by the end of the 1920s, however, the refrigerator stood out among appliances. It was part of a larger system that connected people and places in new ways, and by doing so it transformed what it meant to be a food consumer. *Ladies' Home Journal* described the home fridge as "the last link in the continuous chain of cold that constitutes the modern method of food handling: the careful housewife cooperates with a vast army of producers and purveyors. To insult good food by handling it without cold is today unthinkable."[84]

Refrigeration, in other words, was no mere convenience. In a 1931 article titled "The New Ice Age," the popular monthly *Golden Book Magazine* imagined what might happen without it: "If the stupendous system of food preservation and transportation which supports us were interfered with, even for a short time, our present daily existence would become unworkable. Cities with thousands of inhabitants would fade away. We would probably turn into beasts in our frantic struggles to reach the source of supply . . .

The Ladies' HOME JOURNAL   February, 1927

Frigidaire keeps food fresh
regardless of every change in weather

WINTER weather is never a safe substitute for refrigeration. Outdoor temperatures change from day to day. Icy cold days, *too cold* for proper food preservation, are often followed by days so mild that foods of all kinds may spoil very rapidly.

With Frigidaire Electric Refrigeration in your home you will be protected against every change in weather. Day and night, and day after day, your foods will be kept at low, even temperatures—fresh, pure and wholesome, retaining all of their original flavor. You will enjoy the better, more convenient, more dependable, and more economical refrigeration that is now being enjoyed by more than 250,000 Frigidaire users.

New low prices have made the *value* of Frigidaire greater than ever before. Metal cabinet models are priced as low as $225—and mechanical units for installation in the standard makes of ice boxes as low as $170. (All prices f.o.b. Dayton.) The General Motors deferred payment plan affords the most convenient and economical way to buy.

Visit the nearest Frigidaire Sales Office and see Frigidaire in actual operation. Or mail the coupon below for a copy of the Frigidaire Catalog.

Frigidaire
PRODUCT of GENERAL MOTORS

FRIGIDAIRE CORPORATION
Dept. T-52   Dayton, Ohio
Please send me a copy of the
Frigidaire Catalog.

Name...........................

Address........................

Refrigerator advertisements taught consumers that food could be kept in a naturally fresh state only if it was protected from nature. (*Ladies' Home Journal*, February 1927, p. 88)

It is not extravagant to say that our present form of civilization is dependent upon refrigeration."[85]

At the time, few questioned this dependency. Public health experts as well as chefs celebrated consumers' liberation from local food. "Refrigeration wipes out seasons and distances," *Ladies' Home Journal* enthused. "Duluth serves California or Florida orange juice for breakfast, Buffalo school children munch bananas from South America, Chicago dines on roast Long Island duckling. We grow perishable products in the regions best suited to them instead of being forced to stick close to the large markets."[86]

As food from far away became less exotic, shifts in seasonal supply became less pronounced. So did shifts in seasonal cooking and eating. In the recipe pages of women's magazines, "seasonality had virtually disappeared" by the end of the 1930s. What remained were holiday recipes, or dishes suited for hot or cold weather. By 1941, *American Cookery Magazine* had changed the title of its monthly "Seasonable and Tested Recipes" column to "Tested Recipes of the Month."[87]

Of course, restaurant menus and recipe pages don't always reflect how people actually ate or shopped. The price and quality of many fresh foods continued to vary over time and place, as they do today. New Yorkers could buy strawberries in January, but they weren't cheap, and they weren't necessarily good. Kansans could order "fresh" fish in restaurants, but smart diners probably didn't. Part of the decline in seasonal and local eating was a result of the rising consumption of foods that weren't remotely fresh—even if advertisements promised they were "at the peak of freshness" when canned or frozen.[88]

Together with advances in transportation, refrigeration revolutionized the geography of fresh food. Arguably, though, the most radical changes took place not on the map but in the minds of consumers. As the cold chain linked their kitchens with distant producers, the idea of the durable perishable no longer seemed as paradoxical as it once had—and still does, in some parts of the world. Consumers stopped expecting fresh food to be just-picked or just-caught or just-killed. Instead, they expected to find and keep it in the refrigerator. But this convenience hardly put an end to worries about food. On the contrary: an age of questioning freshness—its provenance, its worth, its power to make people feel happier and healthier—had only just begun.

two

Beef

MOBILE MEAT

The desire to export fresh meats was the father of all
ideas of refrigerating transportation, both by land and
sea.

—William G. Sickel, "Refrigeration on Ocean Steamships"
(1908)

In August 1999 the French sheep farmer José Bové drove his trac-
tor onto a McDonald's construction site in southwestern France.
Accompanied by several other farmers, Bové took hammer and
screwdriver to the half-built symbol of American fast food, disman-
tling large chunks of the building before the police intervened. In
addition to a spell in jail, the incident earned him national applause
and international fame. Bové later explained that he had nothing
against Americans or, for that matter, hamburgers. Rather, he opposed the globalization of corporate *malbouffe,* which threatened
markets for the produce of small-scale farmers worldwide. Roughly
translated as bad food or junk food, *malbouffe* also described, as
Bové put it, the "confused unease" that such food provoked.[1]

A century earlier, one of Bové's compatriots had been accused of
bringing France both kinds of *malbouffe.* Charles Tellier wanted
nothing more than to deliver American meat to his motherland.
The inventor of a refrigeration system that made it possible to ship

fresh beef halfway around the world, Tellier believed that cheap imports of this favored meat would boost workers' productivity, stimulate commerce and innovation, help to preserve native cattle stocks, and allow the poor to eat more like the rich. Early shipments of chilled American meat, however, met with skepticism, outrage, and the passage of prohibitive tariffs across much of Europe.

Charles Tellier reenters the story later in the chapter. For now it's enough to note that the transoceanic trade in "dead meat" fared much better in Britain than in France, for reasons both cultural and economic. The country built a fleet of refrigerated steamers, and its butcher shops filled up with meat slaughtered in Chicago, Sydney, and Buenos Aires. The Argentine capital became especially prosperous during the early days of globalized beef, as British and American meatpackers competed to build the biggest and most modern *frigorificos* (refrigerated packing plants), and cattle poured in from the fertile Pampas. As in Chicago, though, the big winners in Buenos Aires were the packers themselves. The idealistic Tellier did not anticipate the role of refrigeration in the birth of the "Beef Trust," nor the trust's ruthless treatment of workers, cattlemen, and independent butchers.

Upton Sinclair, of course, saw big beef in a very different light. In 1906 the young muckraker published *The Jungle,* a novel that detailed the atrocities of Chicago's meatpacking industry. Readers reacted not with "confused unease" so much as with gagging disgust. But even Sinclair, focused on the gore and misery of shop-floor work, did not appreciate the power that packers wielded over the idea of freshness. Not only did they convince consumers that fresh meat could come from far away; they also promised fresh opportunities for faraway regions. Through investment, employment, and commerce, the chilled meat industry promised to pull backward lands into the modern age. To be successful, the industry had to harness new technologies to older notions about beef, in particular, to the idea that it was a food of progress. At least in the cattle cultures of the West, beef was considered a social good, a strategic resource, a symbol of dominion.

Cattle, as Marx once noted, were the earliest form of mobile capital: wealth that could be bred, milked, fattened, and finally walked to a profitable slaughter. The companies that mastered their perishable meat became, in the food world, the first mobile capitalists. The packers took the Midwestern model of meat production with them to new lands, with worldwide consequences for how both animals and people ate. In other words, the story of meatpackers is not just the prehistory of a fast food nation; it is, as well, the story of how they took control of a vital fresh food and became themselves global powers, loyal to no nation.

Tough Flesh

Steak lovers know that fresh is not usually best when it comes to beef. Soft when just slaughtered, the meat becomes unchewable once cool. Over time it self-tenderizes, as the muscles' own enzymes (known as protease) break down tough connective tissue. Letting cattle carcasses hang for a few days is a traditional aging method still favored by beef connoisseurs. Unlike the industry's current practice of vacuum-packing beef in plastic soon after slaughter, "dry aging" also allows for evaporation, which concentrates the flavor.[2] Under ideal conditions—adequate fat content, steady humidity, and a temperature around thirty-five to thirty-eight degrees Fahrenheit—beef can age for several months. This process of controlled rot would turn most other meats deadly.

Beef's relative durability at cool temperatures made it a good candidate for early experiments in long-distance refrigerated shipping. In the centuries before refrigeration existed, however, well-aged beef was a seasonal and regional luxury. Pastoral peoples dealt with the meat's perishability in various ways. In Africa, nomadic cattle herders killed their animals rarely and ate them very quickly, often at feasts that celebrated the cattle themselves. Europe's early cattle cultures also feasted on fresh beef, reinforcing the status of the cattle owners while solving the problem of storage.[3] By medieval times, curing and pickling helped to slow spoilage, while laws subjected butchers to inspections and penalties for selling bad meat.

Such laws were the last and least-effective defense against spoilage. The availability of fresh beef depended above all on the stamina of the cattle themselves. Traditionally cities around the beef-eating world depended on supplies that arrived on the hoof, traveling from wherever pastures were rich and abundant. In early modern West Europe, such pastures grew scarce as settlement and grain farming spread. Increasingly, cattle trekked from the Continent's sparsely populated edges: Denmark, Ukraine, Hungary, and even the Russian steppes. On the British Isles, Welsh cattle supplied London. Herds forded rivers, veered around war zones, and on the longest routes stopped over for the winter. Even under the best conditions the animals arrived thin and weary and had to be fattened up before slaughter. Despite its slow pace, the on-the-hoof trade proved lucrative for cattle owners and merchants. Between the fifteenth and seventeenth centuries, proceeds from such sales built grand homes for the Danish and Hungarian nobility, and cattle became one of Europe's biggest trades in terms of value.[4]

Within Europe, of course, some peoples appreciated beef more than others. By the sixteenth century, England's rural aristocrats were famous for eating huge quantities. Even their servants probably ate more red meat than many mainland Europeans. Long after Celtic cults of bull worship gave way to Christianity, the animal's meat still represented land-based wealth and all the power that came with it.[5] Nobles displayed that power, among other places, at banquet tables, where social rank determined which cuts of beef a diner received. Only military men ate nearly as much red meat as nobles, in their case to bring strength rather than status. Of the more than two hundred pounds of beef rationed each year to a British sailor, some would have been salted or pickled. But since a steady diet of preserved meats was considered demoralizing if not unhealthful, cattle routinely trailed armies and sailed on marine expeditions.

In the 1620s cattle also sailed to colonial North America with the Pilgrims. Along with imported sheep and hogs, they adapted especially well to the environment in New England, at least once the settlers planted grasses from old England.[6] Despite losses to wolves and harsh winters, the region's herds soon fed a coastal trade that stretched to the West Indies. Relatively little of this bounty could be enjoyed fresh. New Englanders slaughtered their cattle and hogs in the fall, when the animals were fattest, and immediately packed most of the meat in barrels of salt. Most households ran through their salt beef by spring or summer and then counted on meat from their other farm animals—as well as whatever they fished, caught, or bought—to tide them over for the next several months. Except in the homes of wealthy families and in the largest towns, then, fresh beef was a strictly seasonal food.[7]

Colonial North America's livestock were mere specks on a map compared with the herds that spread from Chile to northern Mexico in just two centuries. Like the Puritans, Spanish colonists brought cattle over for more than food. The animals cleared forests, hauled ore, powered sugar mills, and eventually became hides and tallow. Beef-fat candles lit the silver mines of eighteenth-century Mexico.[8] Not least, Europeans in both North and Latin America used cattle to establish order and ownership over lands where the concept of private property didn't exist. As it happens, many of the animals themselves went wild; cattle reached Texas before the Spanish did. Unfenced herds ranged far beyond where their meat could feed anything but wild dogs. Skinned carcasses were left to rot. If Latin America's vast herds represented impressive wealth, their flesh was effectively worthless.

Power Food

In the mid-nineteenth century, science reinforced the customary belief that red meat made men stronger. The most influential research came out of the laboratory of the German chemist Justus von Liebig, whose work with the newly invented microscope seemed to show that digestion converted meat directly into blood and then tissue. Meat, he argued, was "flesh-forming," whereas other foods simply generated bursts of heat and energy.[9] This theory was debunked by the century's end, but in the meantime it helped to convince a generation of experts and politicians that meat was a top public-health and even national-security issue. It also helped to sell new preserved beef products such as Liebig's Extract, a bouillon made in Uruguay and endorsed by the enterprising scientist. Its German manufacturer claimed that a pound of extract contained the "essence" of thirty-six pounds of beef, making it ideal for armies, explorers, and convalescents. Florence Nightingale gave it rave reviews, and aggressive marketing eventually pushed it into shops on every continent.[10]

While Liebig's extract was convenient, it hardly satisfied hunger for the real thing. In much of West Europe, demand for meat was rising, and so were expectations. No longer content with salted provisions and the dairy products known as "white meats," working-class families were getting more of their protein from the butcher's shop. Partly this reflected gradual improvements in buying power, as wages rose relative to the cost of staples such as wheat and sugar. Partly it reflected the aspirations borne of city life; more than their counterparts in the countryside, working-class city dwellers saw how the other half ate. They noticed that well-off shoppers bought steaks and roasts, not salt pork and stew bones. They came to feel that Sunday dinner was not complete without a good-sized hunk of fresh beef—even if, as Friedrich Engels observed in *The Condition of the Working Class in England,* the meat was actually a "half-decayed" slab bought just before the Saturday closing time, when butchers liquidated their stocks.[11]

British officials were especially concerned about consumers' changing tastes. While cured meats could be imported from any number of countries, on-the-hoof supplies could not. Despite clearing millions of acres in Ireland for pasture, Britain was running out of space for large livestock. It was also increasingly wary of live animal imports, especially after cattle plague from mainland Europe reached London's livestock markets in 1865.[12]

As beef prices rose, experts conferred about "the meat question" —namely, where and how to get more of it. Maintaining the country's access to a bountiful supply of beef, in particular, was more than a matter of taste or even national pride. Many believed that it was also necessary for maintaining the country's status as a world power. Alongside advances in cattle breeding and feeding, then, came a rash of new food-preservation methods suited for long-distance transport. Many aimed to replace ice, which was too heavy and perishable for voyages across the equator.[13] One of the arbiters of these new methods was the Royal Society of Arts, an organization founded to encourage technological and social progress. In the 1860s the society's Food Committee offered prize money for advances in preservation, which it then reported in the society's weekly journal. Over plates of tinned lobster and tongue, dried cod from Spain, and "compressed" vegetables from Holland, the committee's members discussed matters of taste, cost, and morality. At least a few of them hoped for an invention that would render obsolete the traditional on-the-hoof trade. As Rear Admiral Charles Elliot wrote in 1861, it was a "barbarous" and inefficient way to procure fresh beef: "No man of observation can visit one of our great cattle markets in large cities, or attentively examine the conditions of the beasts during the voyage from Ireland, or in closely-packed cars on long journeys, without being struck by the humiliating juxtaposition between the great and growing advancement in the decencies and conveniences of life and revolting proofs of unnecessary violence, too often of savage cruelty, and always of deterioration of the meat, and superfluous expense, which are the results of our present modes of fulfilling this great requirement of our existence."[14]

The shipping of live cattle by rail and steamship had only just begun. Soon many of the cattle that rode the rails from Kansas to Boston continued by boat to Britain. Despite high mortality rates, the trade persisted because it delivered recognizably fresh meat—which was more than society members could say about certain cheap but barely palatable salted products coming out of Latin America.[15] One correspondent described the Argentine salt beef known as *tasajo* as "rancid beyond my conceptions of any foregoing rancidity . . . it filled the house with its wretched odor, and made my friends sick. Even the dog looked suspiciously at it . . . it would be admirable food wherewith to feed convicts in prison. I think that if anything would deter them from crime it would be the dread of such feeding."[16]

Somewhat less offensive were beef and mutton first soaked in calcium sulfite and then packed in barrels of melted butter. Shipped from Australia, the meat did not win over British consumers, but it found an outlet in "charity meals and soup kitchens." In Uruguay packers injected an "anti-putrescent solution" of salt, saltpeter, vinegar, and spices into the veins of cattle (presumably after they were dead). The result was judged "quite edible" when fresh, but then defeated its purpose by quickly spoiling. The British Meat Preserver Company proposed a powder that, when mixed with water, "restored" already tainted flesh.[17] And the chemist G. C. Steet suggested keeping meat in containers full of "artificial atmosphere."

Like many of the contributors to the society's food-preservation contest, Steet emphasized both national-security concerns (Britain's "gallant defenders," he said, deserved better than salted rations) and the need to think far beyond the nation:

> While in our densely populated towns and districts we can scarcely supply sufficient cheap food for the bodily sustenance of our laboring population who chiefly require good animal food . . . and while at the same time there exists in other regions of the earth a superabundance of animal food not only fit for sustenance but of excellent quality, it should be the duty of every philanthropist—to say nothing of mercantile and social interests—to make use of the abundance of one part of the earth for the supply of the necessities of another part, and thus act up to the precept of our great Exemplar, who required his followers to gather up the fragments, that nothing should be lost.[18]

Ultimately the "mercantile interests" did achieve this goal, though not because of any Biblical injunctions against waste. And for all the attention that the society gave to newfangled preservation techniques, the companies that came to control much of the world's meat supply built their first fortunes on ordinary ice. They innovated in management and marketing, not science. To appreciate what they contributed to the history of freshness, it helps to learn the story of the little-known inventor behind the technology they all came to use: the refrigerated steamship.

A Modern Invention

"All my life I've had a weakness for ammonia." So began Charles Tellier's memoir about the inventor's long battle to bring refrigerated beef to France. Although titled *The Story of a Modern Invention,* it's really the story of an inventor born in the wrong place at the wrong time.[19] The French *père du froid* ("father of cold") spent most of his adult life trying to sell his vision of a "rational" global food supply to countrymen who saw it as madness. He wanted to turn fresh beef into a commodity as stable as wheat and then import it from regions where it cost almost nothing. Given the thriftiness of French consumers, he thought they'd welcome good cheap beef, since food was their single biggest household expense, and meat the most expensive food. He thought they would embrace an invention that promised, as he put it, *la vie a bon marché*—the good life for less. Even French farmers and butchers, he thought, would appreciate the spoilage-stopping powers of refrigeration. Instead they saw its power to spoil cattle prices and carefully guarded markets.

Unlike many of the men who went into the beef business in the Anglophone world, Tellier had no family money; his father had lost his textile mill during the 1848 Revolution. Tellier's early engineering work, already slow, was interrupted by a jail spell over a patent dispute. Although he protested the charges, his biographers pointed out that the ammonia-fueled refrigeration system he developed in 1860 was not based on any new discoveries. It just applied the vacuum-compression process to a new purpose: cooling air rather than freezing water.[20] This advance nonetheless had great commercial potential because it produced a more stable, portable, and drier cold than ice—all qualities especially important for storing and shipping raw beef.

But not just beef. Tellier's first customer was Justin Menier, one of France's largest chocolate makers and, as Tellier fondly described him, "a man of progress."[21] Menier just wanted a cold storage room, but that same year Tellier met three Uruguayan men with a much grander plan. Their country had immense herds of cattle, they said, but exported only low-value products such as hides and beef extract. They thought French consumers would appreciate fresh Uruguayan beef, especially since they typically preferred leaner meats than the British did. They were so confident that they wanted to buy a steamship that Tellier would then equip with refrigeration. Tellier leapt at the offer, agreeing first to ship fresh meat in the opposite direction, to prove that it could last the distance. Unfortunately, it didn't. Twenty-three days out to sea, the refrigerator aboard the *City of Rio de Janeiro* broke down and the meat, in good shape before the mishap, had to be thrown overboard.

A few years later, war with Germany cost France at least three million heads of cattle. With beef scarce and costly, the government imported canned ham, admitting that the market for this unpopular product would be "inevitably limited."[22] Tellier saw opportunity. When promoting his South American fresh beef scheme, he emphasized that cured meats were hard to digest and "should never be daily nourishment, especially not in the cities where our bodies need more delicate treatment than in the countryside." He also insisted that frozen meat, which Australia had just begun exporting to Britain, would never suit French tastes. But French investors proved a skeptical audience: "No one believed it could succeed . . . some even told me that I would poison my crew with spoiled meat." To boost his credibility, Tellier sought the endorsement of the prestigious Academy of Sciences. Its members demanded proof that his invention worked, and he provided it. In 1874, Louis Pasteur, France's father of germ theory, declared that beef stored for fifty-one days at near zero degrees Celsius was indeed edible and wholesome.[23]

Even with the academy's stamp of approval, no bank would put up the million francs Tellier needed. "They treated me like someone hallucinating," he recalled. "No one would seriously listen." Undiscouraged, Tellier took his company public. He wrote up a classified ad promising such good returns that a Parisian newspaper initially rejected it, suspecting a con job: "5000 francs of profit for 1000 francs investment. Honorable and sure business, funds deposited at the Bank of France. Send to the post office, addressed to the initials C.T."

Despite (or maybe because of) the ad's lack of details, responses

Charles Tellier, France's long-suffering "father of cold." (Courtesy Science, Industry, and Business Library, The New York Public Library, Astor, Lenox, and Tilden Foundations)

poured in at a rate of fifteen to twenty a day. Within a few months, Tellier was shopping for steamers in Britain. He settled on the *Éboé,* a packet boat that had previously plied the rivers of colonial West Africa. It was "pretty and solid," Tellier noted, but smelled of rancid palm oil, its previous cargo. By the autumn of 1876, the *Éboé* (renamed *Le Frigorifique*) was cleaned, refrigerated, packed with an experimental load of meat, and set to sail from the port city of Rouen. It even received the blessing of the city's archbishop—a deviation from church convention justified, Tellier was told, by the voyage's scientific importance.

Tellier and the *Frigorifique* left Rouen on September 20. He had with him a couple of officials from the chamber of commerce, a science reporter, an artist, a few crewmembers, and his dog Ox. Four days into the journey, they hit a terrific storm, and for the next three days no one ate. Tellier sipped cognac. Even Ox looked bad: "She regarded me with sad eyes, as though asking, 'where have you taken us?'" The crew's mechanics were too sick to stand, so Tellier had to check the holds of chilled meat himself.

After the storm they docked in Lisbon to rest and make repairs. Tellier met with ambassadors and the king of Portugal; the chilled cargo was inspected and declared perfectly preserved. When this news reached Paris, Tellier's shareholders concluded that the expedition had served its purpose. Refrigeration worked; now they wanted to cash in. Some sold off their shares at five times the original price. Others saw no need for the voyage to continue. Tellier, an awkward and easily flustered man, returned to Paris to calm matters, eventually resigning his directorship. The steamer continued south without him; he came later with the French merchant marines.

The *Frigorifique*'s fortunes improved once it crossed the equator. Arriving in Montevideo on December 23 and in Buenos Aires on Christmas Day, the steamer caused a sensation in both cities. The dignitaries and journalists who attended on-board banquets reported that the 105-day-old steak was "excellent, bloody, savory," and as good as any beef sold locally. The real cause for celebration, though, was the machinery, not the meat. The South American press described the *Frigorifique*'s arrival as an event of monumental proportions. Argentina's daily *La Nación* called it a moment "awaited more anxiously than the mythical hero Theseus' return to

Athens," where he was to announce the death of Minotaur. Tellier, it seemed, had slain the monster of isolation. "The big problem that has preoccupied thinkers of both hemispheres for nearly two centuries can now be considered definitively resolved," declared Buenos Aires's newspaper *El Nacional.* "Hurray," effused the city's *La Liberté,* "Hurray a thousand times for the revolutions of science and capital. The dawn of a new day rises for La Plata."[24]

For the region's beef industry, it did mark a new day. Although South America did not export much fresh beef for several years—frozen proved much more reliable—Tellier showed that it was possible. Soon foreign-financed refrigerated packing plants *(frigorificos)* went up in the major port cities, and the meat of millions of cattle went northward. But little of that meat traveled to France, and none of it earned any money for Tellier. After the *Frigorifique*'s return voyage (during which the refrigeration malfunctioned, spoiling the South American meat aboard) it was a popular attraction at the 1878 Paris International Exposition. Unlike Tellier, the shareholders who had taken over his company showed no interest in building a real commercial enterprise. "The task was miserably beyond them," he lamented. "They only know how to be backseat drivers."[25]

Maybe they'd begun to have second thoughts about importing beef into a country where many farmers and butchers fiercely opposed it. If farmers feared lower cattle prices, butchers at Paris's giant La Villette slaughterhouse feared extinction. For them, the *Frigorifique* was a Trojan Horse: beneath cheap chilled beef lay the dangerous idea that fresh meat not only could come from anywhere but also could be slaughtered by anybody.[26] Well-organized and connected, the butchers' union helped to convince the French government to impose increasingly stiff tariffs on foreign meats.[27] One shipping company (Les Chargeurs Reunis) continued to import frozen South American beef and mutton despite the tariffs, but in 1900 the government began to require exporters to ship meat with organs attached. Supposedly a measure to help health inspec-

A French pamphlet on frozen imported meat offered advice on "how to choose it, how to cook it, how to prepare it." (*Le Frigo*, 1919)

tors detect disease, it rendered the trade completely unprofitable. It wasn't exactly thriving anyway. Few butchers would handle imported "dead meat" and even fewer had cold storage. Widespread confusion about the difference between frozen and chilled products—both became known simply as *le frigo*—didn't help matters any.[28]

Frigo meat became the stuff of urban legend. A survey conducted in Paris in 1912 turned up "complete ignorance . . . for some, this refrigerated meat is simply preserved in cans . . . others imagine that it is a quarter of an animal encased in a block of ice. A few even assured us that this meat is preserved by injecting it with liquid air!"[29] Some city council members claimed that French workers simply could not digest the same meat that British workers ate so readily.[30] In response, supporters of the *frigo* trade pointed to studies suggesting that French workers, deprived of the cheap red meat enjoyed by their Anglo-Saxon peers, were turning to red wine instead.[31]

It took skyrocketing inflation and food riots to convince the French government to allow limited imports of *frigo* meat in 1912. By then Tellier was an old and discouraged man. The following year the country's Legion of Honor recognized his achievement with a medal, as well as a banquet featuring refrigerated foods from all over the globe. It must have been the best meal the eighty-five-year-old engineer had eaten for a while, because he was dirt poor. When he died in Paris a few months later, it was rumored that he had starved to death.[32]

Dead Meat for the Masses

Of all the inventors and entrepreneurs who tried shipping fresh or frozen meat in the late nineteenth century, Charles Tellier shows better than any that the challenge was far from merely technical.[33] In France, his "modern invention" clashed with the food habits and political forces of a predominantly rural society. Most French people either grew at least some of their own food or had direct access to the farm produce of family members and friends. Imported *frigo* meat not only failed their standards of freshness, especially in its early days, but also threatened producers and tradespeople whose political clout far outweighed that of consumer groups.[34]

The United States and Britain, by contrast, were rapidly becoming mass-market societies in the late nineteenth century. The growth of the refrigerated meat trade did more than coincide with this broader development; it pushed it forward. The success of the Chicago meatpackers hinged on management and marketing techniques that, as Albert Chandler showed in *The Visible Hand,* eventually revolutionized how the country and the world did business. More directly, the packers convinced the meat-eating public that the work of turning "animal into edible" could be a big and distant business. Put somewhat differently: they convinced consumers, not only that fresh beef could come from far away, but also that their main relationship to meat—and indeed, to all once-living foods— was *as* consumers.[35] Today this might seem like an obvious point. But packers had to campaign to win over skeptics, with monumental consequences. Their efforts helped to sever ties that fettered the expansion of the meatpacking industry: between cities and their pastured hinterlands, between shoppers and their neighborhood butchers, and between people who bought the meat and those who dressed it in faraway slaughterhouses.

Karl Marx called this process "commodity fetishism"; the packers would have called it commonsense marketing. Either way, it worked. The outrage unleashed by *The Jungle,* centered more on filthy meat than on oppressed workers, testifies to how successfully the packers reengineered the moral economy of meat. For them, cleaning up filth was easy enough. And despite official investigations and popular unease about their near-monopoly power, the companies avoided penalty by delivering what consumers had learned to value: the cheapest fresh beef on the market.

To achieve that goal, Chicago packers used ice to eliminate

Ownership of most of the nation's refrigerator cars helped the big packers push down shipping rates, and eventually push local butchers out of business. (Courtesy Smithsonian Institution, National Museum of American History, Pullman Photographic Collection)

wasted time, space, and animal matter. Traditionally packhouses sat idle during the warm months, and workers had to be hired anew each year. The discards after butchering fouled the city's waterways.[36] The on-the-hoof trade was slow and consumed the same flesh that cattle merchants intended to sell. Shipping live animals was faster, but they still had to be fed and watered en route, and even then deaths and "shrinkage" (cattle typically lost 10 to 15 percent of their body weight) cut into profits. Moreover, it was simply inefficient to send entire animals to market when only about 55 percent of their carcasses became sellable meat.

In the late 1850s Chicago's slaughterhouses began using ice to extend the pork-packing season, and also to preserve organs and other by-products. The packer George Hammond, though, saw its potential for rail transport.[37] His wasn't an original idea. By the 1840s, the Western Railroad of Massachusetts had developed an insulated railcar for shipping lobster, oysters, wild game, lemons, and other "nice delicacies" (apparently little came of this plan). By the 1860s the Illinois Central Railroad used ice-chilled boxcars on its intrastate "strawberry express" trains.[38] But Hammond, already a well-established shipper of cured meats, developed the first more-or-less regular Chicago-Boston trade in chilled beef. The railcar patented in 1867 proved barely up to the job. It had to be re-iced daily, and it cooled unevenly. Some parts of the meat arrived discolored, and others simply spoiled. Half-carcasses suspended from the car rafters "swung like pendulums" when the trains rounded corners, causing more than a few wrecks.[39]

Hammond built his trade slowly and attracted little attention. Then in 1875 an ambitious young competitor arrived from New England. Gustavus Swift, the son of a successful wholesale butcher, came to Chicago to buy and sell cattle. "A born expansionist," according to his son, he soon decided that the real money lay in dressed beef. It helped that he already had some family money, plus useful New England connections. The Boston engineer Andrew Chase, for example, built him a better refrigerator car, one that stored ice on top (allowing for more even cooling as well as for easier re-icing). When no railroad company would manufacture the cars, Swift borrowed money from Boston bankers and acquaintances in the New England meat business and commissioned the Michigan Car Company to build him a fleet.[40] Then he tapped these connections again to develop his market. He and his younger brother Edwin went from town to town, inviting wholesale butchers to become distributors of Chicago-dressed meat. While plenty refused (more on them later), Swift's business grew fast enough to convince the city's other big packers, including Philip Armour and Nelson Morris, to follow his lead.

The packers' early rail shipments coincided with the rise of a transatlantic chilled beef trade. The New York livestock dealer T. C. Eastman made the first successful delivery to Liverpool in 1875. For refrigeration he relied on tanks full of melting ice—a cooling method much cruder than Charles Tellier's ammonia-based system, but acceptable for the two-week voyage.[41] The meat itself caused a sensation. The morning it arrived, customers at one shop were so eager to buy it that a policeman had to maintain order. Queen Victoria herself tasted some, declaring it "very good."[42] The editor of *The Farmer* agreed: "the meat has stood every test . . . and both City and West End speak favorably of it. Not only is it fresh, but it has that quality which housewives know as 'old killed,' so often wanting in our home-killed meat."[43]

Encouraged by early windfall profits, packers and meat wholesalers stepped up shipments to both the East Coast and British markets. Within two years, most were losing money. It became clear that if chilled beef was to compete against beef delivered on-the-hoof, both the meat and the marketing needed work. In Britain, consumers' curiosity was replaced by disdain for the Texas Longhorn's ultralean meat. As the Scottish journalist James MacDonald wrote in *Food from the Far West,* "the beef which grows upon the bones of the long-horned Spaniards away down in Texas . . . is not beef that finds acceptance in the homes of beef-eating Britons." Not only were the animals largely wild; their meat also suffered from the Texan "peculiarity . . . [of] allowing the steers to run about and pick up food as best they can."[44]

British businessmen read accounts like MacDonald's with more than idle curiosity. Encouraged by descriptions of free land, free cattle, and "sporting" lifestyles, they poured millions of pounds into new Western cattle companies in the early 1880s.[45] Many followed their money west, settling on ranches between Texas and Montana. Most lasted only a few years, but their breeding of Longhorns with much stockier British varieties helped raise the value of the West's meat. Ranchers also began to sell their range herds to specialized "feeder" farmers in Iowa and Illinois, where turning crops into fat-marbled beef became a profitable form of Midwestern agriculture. With their acreage divided between cornfields and cattle enclosures, feeder farmers founded the modern feedlot industry.[46]

Consumers' ideas about how fresh beef should look posed a different set of challenges. In Britain, scarce cold storage made speedy sales imperative. Long before the chilled beef actually spoiled, it turned soft and sweaty, making it appear suspiciously *un*fresh. Poor handling by British wholesalers didn't help matters any.[47] Gustavus Swift dealt with such problems in his usual direct fashion. He pitched his product in person to London's meat wholesalers and checked regularly on their selling style. "Father was forever turning up at Smithfield market in the gray London dawn with his cheery insistence on having his beef cut properly," recalled his son.[48] It took twenty trips to London and back before he was satisfied, at least temporarily. Eventually Swift and the other Chicago packers opened up their own "American meat shops" around Britain.[49]

Even when the beef arrived at market in perfectly good shape, it looked different from the traditional butcher shop offerings. As the trade journal *Ice and Refrigeration* noted, American consumers expected beef to be bloody red. This "relic of barbarism," it said, "must be 'educated out' of the public mind": "This bright color is always in evidence that the meat has been but recently killed, and is not thoroughly cured . . . the beef is always bright in color, but invariably tough and stringy, no matter what was the character of the 'critter' . . . it is only an ignorant prejudice—born of the habit of seeing on the block the bright colored but half seasoned beef, killed in a country slaughter house, always offensively dirty."[50]

It wasn't just surface appearance that disturbed consumers. To many it seemed unnatural to eat any flesh that had not arrived at market alive. Swift's son described how "the hidebound, rockbound conservatives of New England" initially reacted to Chicago-dressed meat: "Eat meat dressed a thousand miles away? No Yankee had ever been served a steak which originated more than a few miles from the stove that cooked it, no sir, not if he knew it! To people accustomed to having a slaughterhouse just outside the limits of every town, the very idea of Chicago-dressed beef was repugnant. The meat was actually fresher in condition if not in time. It had been produced in cleanliness instead of in a filthy small-town shambles. The cattle were in better condition when slaughtered. But all this made little difference. Prejudice is founded on feelings, not on knowledge."[51]

Many independent butchers did their best to reinforce such feelings. In both the United States and Britain, they saw Chicago-dressed meat as a threat to their entire profession. Selling someone else's factory product was not the same as practicing a craft, and Chicago's packers intended to leave them no choice. Swift invited wholesale butchers to join his company as distributors because he wanted experienced handlers of fresh meat. If they refused, he set up his own depots and ran them out of business—and then, sometimes, gave them a job. As the packers moved into one city after another, butchers organized to oppose them. They posted signs outside their shops: "No Chicago dressed meat sold here," and threatened boycotts of any business that did sell it. In 1887 they formed the Butchers' National Protective Association, which pledged to protect consumers against the "diseased, tainted, or otherwise unwholesome meat" pouring out of Chicago's packhouses.[52]

The government did not seriously investigate the charges of unwholesomeness until after publication of *The Jungle* in 1906. In the meantime, the Chicago packers beat down prejudice and butchers' resistance with unbeatable prices. Whether they sold their meat out of their own retail shops or, as in smaller towns, directly out of railcars, they entered markets by selling at or below cost. On occasion, it was rumored, they gave meat away.[53] Almost everywhere, consumers tried it, liked it, and left local butchers with little choice but to sell the Chicago meat themselves. These tactics rarely failed. Even though the packers usually raised their prices once they had established themselves in a particular market, they could still undersell the local competition. By the end of the 1880s the only major cities with many independent butchers left were New York and San Francisco. One had a large kosher market; the other a public committed, as one journalist put it, to "transacting their own business in their own way."[54]

The packers' marketing methods went beyond fire-sale prices. Swift especially recognized that consumers could be seduced by display. Whereas butchers traditionally stored their carcasses out of view and cut them to order, Swift insisted that shops selling his meat present customers with an attractive array, from the choicest loins to the cheapest chuck. This approach boosted the overall image of meat "dressed a thousand miles away," encouraged impulse purchases, and speeded up sales of less popular cuts.[55]

If such retailing strategies now seem elementary, they were not simple. The very capacity to offer shoppers an appealing, competitively priced fresh-meat selection depended on the packers' less visible power over people, nature, and markets. This power extended well beyond Chicago itself, where hundreds of thousands of recent immigrants depended on packhouse jobs. Efficient use of packhouse labor and machines required destroying the seasonality of production and controlling the climate of distribution. So the packers built immense lakeside ice-harvesting and storage facilities in Wisconsin and northern Illinois (one of Armour's icehouses alone held 175,000 tons) as well as "icing stations" alongside the rail lines. Telegraphed in advance of a train's arrival, station workers mounted platforms to load half a ton of ice into each passing car.

In addition, each packer built a network of "branch houses" to manage local distribution, advertising, and accounting. They ran "peddler cars" full of fresh meat into remote rural areas, luring farm families away from their salt pork and loggers from their

Once chilled by ice, Chicago's slaughterhouses could store and sell almost every part of the animal. ("Market Room," 1892. Courtesy Chicago History Museum)

barreled beef. They developed national and international markets for every conceivable by-product—from margarine to hairpins—and used the profits to keep prices down on their main product, fresh meat. Not least, the packers invested in refrigerator cars, or "reefers." Swift built his original fleet of 10 cars up to 5,900 by 1903; Armour by that time owned 13,600 cars.[56] Together with six other packing companies, they owned 25,000 cars, or more than 90 percent of the national total.[57] On a day-to-day basis, their near-monopoly of the cold chain ensured that the railroads took their meat whenever and wherever it needed to go. Over the long term, it destroyed the railroads' own control over shipping rates. This was a major victory for the packers, because many of the rail companies had invested heavily in the live cattle trade and were determined to protect it by charging a high premium for dressed meat.

For much of the 1880s, the railroads and the packers argued over what constituted a "fair" or "rational" rate structure now that the refrigerator car had confused the old measures of time and space. Both sides were stubborn because the stakes were high.[58] On one level, they were fighting over pennies per pound, significant only because the volume of traffic was so great. On another level, all parties realized that this struggle would decide the future of livelihoods and trades. So local butchers sided with the cattle shippers, and the intensely competitive packers learned to cooperate.

The railroads' resolve collapsed first. Once the Grand Trunk Line agreed to carry dressed meat for cheap, it had so much business that the other companies soon followed. A rate war ensued, and by late 1888 Chicago-to-Boston shipping costs had dropped to six cents per 100 pounds of meat, down more than 90 percent from two years earlier.[59] By the time rates stabilized, Chicago's "Big Four" packers—Swift, Armour, Hammond, and Morris—had secured their conquest of the national beef supply. Whether or not consumers followed the controversy, they saw the outcome at their neighborhood butchers' shops, where the men behind the counter had become "mere cutters and retail handlers" of Chicago-branded products. Before long, their tastes acclimated. "People have been gradually educated up to the point where they appreciate the enhanced qualities of refrigerated meat," the industry boasted in 1895, and now preferred it to "meats handled in the old manner."[60]

Whether they preferred meat handled by only four companies was another matter. During government investigations into what became popularly known as the Beef Trust, the packers often argued that their product's perishability required trust-like behavior. If they did not control and coordinate all aspects of the supply chain, one of the Armour brothers testified, the losses from spoilage would hurt everybody; "we should just be slaughtered in no time."[61] Only a highly concentrated industry, in other words, could consistently deliver the kind of fresh meat that consumers had come to expect. But such an industry performed other services as well. "There was a time when many parts of cattle were wasted, and the health of the city injured by the refuse," Philip Armour said in an interview, "Now, by adopting the best known methods, nothing is wasted, and buttons, fertilizer, glue and other things are made cheaper and better for the world in general."[62] It sounded like Charles Tellier's dream: the good life for less.

Of course, the Chicago packers' methods for making fresh beef gave the world many by-products besides cheap buttons. The meat-packing industry's drive to produce fatter beef both faster and more cheaply has, over time, contributed to modern ills ranging from soil erosion to the spread of lethal cattle-borne diseases.[63] In the early days, though, the sheer scale and efficiency of the packers' operations impressed overseas thinkers and businesspeople as much as it did the Union Stockyards' many foreign tourists.[64] One of France's leading refrigeration experts suggested in 1903 that "all the market's current inconveniences," including high meat prices and low farmer earnings, could be fixed by following the example of the "Packing House Americans." He used the English term, for no French word quite described what the packers had built.[65] But it was plain to see. The prosperity of the United States and Britain seemed to prove that the benefits of Big Beef outweighed the possible costs. Produced and distributed efficiently, this fortifying food did not just reward progress; it was its most potent fuel.

Fresh Pastures

Progress eventually pushed the packers beyond and then out of Chicago. Within the United States, the search for locations offering cheaper labor and distribution costs began early. By the 1890s, Swift and Armour had plants scattered across the West. Less than a hundred years later the entire industry had shifted westward, closer to the cattle and farther from organized labor. But it was at the very beginning of the twentieth century that the packers first looked abroad for raw materials as well as for markets. At home, a booming, increasingly affluent population was rapidly eating up both cheap land and cheap cattle. As the frontier disappeared, so did the United States' exportable beef surpluses. At the same time, Britain's harbors had begun to fill up with steamers loaded with meat from Argentina, Australia, and New Zealand, all lightly populated countries with immense cattle herds. Although the quality of these products did not yet match that of North America's corn-fed exports, their prices were unbeatable. Given the size of the British market—it bought 60 percent of the meat traded globally—Chicago's packers obviously did not want to lose it to competitors from the southern hemisphere.[66] So they went south themselves.

Argentina was an obvious first stop. With an estimated thirty million cattle and only six million people, it boasted more beef per person than almost any other country on earth. Its million square miles also included some of the world's best grazing land. "Nothing in Kansas or Nebraska can compare with the Pampas of Argentina for flatness," reported a U.S. agricultural official in 1908. "It is an ideal pasture country; cattle thrive summer and winter on no feed but grass, and very little shelter is required."[67] It was also not too remote; chilled beef could last the three- to four-week journey from Buenos Aires to Britain, which was not the case with Australia's meat. Not least, Argentina's ruling elite (who included many ranch owners, or *estancieros*) had traditionally welcomed any enterprises that added value to the country's huge and half-wild herds. A few small British companies had exported the infamous salt beef since the late eighteenth century, while members of the Argentine Rural Society had imported British breeding stock since the 1860s. In 1877, Argentina celebrated the successful (if unrepeated) arrival of Charles Tellier's ship *Frigorifique*.

But much had changed since Tellier's voyage. Hundreds of thousands of European immigrants had arrived, expanding Argentina's labor force and helping the government in its mission to "pacify" the countryside. British-financed railroads now carried crops and cattle from the Pampas to port cities. Tens of thousands of live animals were shipped each year to Britain, suffering appalling mortality rates. Most important for the packers' purposes, British and Argentine firms had built several *frigorificos*.[68] Most produced frozen meat, which proved much easier to ship than fresh but commanded a much lower price. Slow freezing left the meat gray and ice-crusted, and better-off British consumers avoided it entirely.

Argentina's meat exports exploded just after the turn of the century, helped by high demand (large quantities fed British troops in the Anglo-Boer war) and improved technology. Faster cooling reduced ice damage in frozen beef, while sterilized, dehumidified cargo holds delivered mold-free chilled beef. If everything went right, the meat could arrive in Britain looking as fresh as when it had left the *frigorifico*. Although corn-fed North American meat still commanded higher prices, even Philip Armour predicted that Argentina would soon become "a formidable competitor."[69] British investors agreed, especially after their own country, hit by a foot-and-mouth epidemic, banned imports of South America's live cattle in 1900. They poured millions into the construction of new *frigorificos,* and by 1906 British-owned firms controlled three-quarters of Argentina's meat export trade.

Their domination did not last long. A year later, the *Buenos Aires Standard* announced the arrival of "the American Beef Trust." Swift and Company had bought a *frigorifico*. In 1908 the National Company (representing Swift, Armour, and Morris) bought another. Argentines gave them a much more mixed reception than they had Charles Tellier. On one hand, they'd felt the ripple effects of *The Jungle,* which was widely read in Britain. There the typical consumer did not distinguish between North and South American meat, the *Financial Times* said, but rather "imagined that he had swallowed the garbage from cesspools of an entire continent." Argentina's beef exports suffered accordingly. In addition, the country's own media warned people about the American packers' power to control cattle prices. The *Review of the River Plate* predicted that "as soon as these firms are established they will start a combine and estancieros will be unable to offer their cattle to the best buyer. There will be no best buyer."[70]

On the other hand, *La Prensa* saw cause for optimism: "the trust is the foundation of American prosperity, the explanation of the rapid industrial development of the United States . . . it is the genuine product of American civilization and since our conditions are similar we must take it for our model. If the meat trust wants to come here and the existing firms refuse to sell out, the trust will build new plants—modern, better equipped, and better operated."[71]

Gustavus Swift knew how to show Argentina's cattlemen that he meant business—and good business for them. At a 1910 livestock show, jaws dropped when he bid $11,500 a head for five champion steers. Like the other American packers, he made clear that he wanted only top-grade, alfalfa-fattened cattle ("chillers") but would also pay top prices for them. As in the American West, the combination of U.S. packers' colossal purchasing power and the British market's high standards for quality encouraged both improved breeding and the emergence of specialized fattener farms.[72] In the port cities, it brought the predicted investments in bigger, more modern *frigoríficos*, and thousands of new jobs.

The U.S. packers' arrival also intensified competition among the *frigoríficos*. The Argentine and British firms had previously relied on "friendly chats" to decide who would ship meat abroad and in what quantities. It ensured reliable access to scarce cargo space as well as stable prices in Britain. But the U.S. packers were not interested in chatting until they could be guaranteed of the outcome. Employing old tactics on a new scale, they launched a price war in 1910–1911. They bought more cattle, chartered more steamers, and flooded the British market with fresh meat. Argentine ranchers didn't complain; as cattle prices soared, they called the U.S. packers "saviors" of the country's livestock industry. British consumers were equally pleased by the rock-bottom beef prices. The British and Argentine-owned *frigorificos,* however, suffered huge losses. The price war ended with an agreement to limit exports and give each company a share, which would be periodically negotiated. As intended, the U.S. packers locked in their gains, which amounted to two-thirds of Argentina's fresh beef exports.[73]

A few Argentine politicians warned that the U.S. packers' massive exports might drive up domestic beef prices—a disastrous prospect given the country's famously carnivorous appetites.[74] But this argument did not go far in a parliament dominated by ranchers, who instead pointed to the newfound prosperity of Argentina: in three decades, it had gone from a backwater to the ninth richest country in the world.[75] Buenos Aires boasted a $10 million opera house, better water and sewer systems than those of most European cities, and, despite a booming population, a crime rate only a tenth of Chicago's.[76] It would be a mistake, said one parliament member, to restrict an industry that brought "so much glory and so much wealth to the republic."[77] Certainly it had brought wealth to ranchers like him, many of whom lived like the English landed gentry whose bulls they imported. They built mansions in the countryside yet spent most of their time between Buenos Aires, Paris, and London.[78]

The ranchers' glory days ended shortly after World War One. As international demand for their meat collapsed, they realized just how little clout they commanded vis-à-vis the likes of Swift and Armour. The packers effectively controlled both cattle prices and grades; they decided what counted as a "chiller" versus a "freezer." The state's first legislative attempt to prop up cattle prices in 1923 went nowhere. The packers simply stopped buying livestock for three weeks—they had plenty of frozen meat in storage—until ranchers begged to have the law reversed. The incident reflected, as one historian later observed, "a circumstance which has become increasingly familiar, the rise of business organizations with greater power than governments."[79]

Over the next two decades, Argentina's meatpacking industry contended with British and U.S. import bans, prompted by fears about foot-and-mouth disease; a new British policy to favor "Empire beef" from Australia and Canada; and plummeting Depression-era prices.[80] In the postwar era, praise for the country's "magnificent development" faded. Instead scholars puzzled over how Argentina became, as one delicately put it, "a curious model of economic and social retrogression."[81] Yet while instability discouraged investment in prized herds and packhouses, Argentines' appetite for beef did not waver. Consuming an average of 240 pounds per person per year (more than twice the U.S. average), they ate their way to a once-unimaginable beef shortage by the early 1970s. Later, beef prices remained so central to the overall cost of living that politicians sometimes took drastic measures to keep them low. Faced with spiraling inflation in 2006, President Nelson Kirchner found it easier to ban beef exports than to convince consumers to eat less of it. The prospect was unthinkable, some said; they would sooner starve. Even though beef no longer brought glory, it sustained a sense of national well being.[82]

Chilly Empires

Well before Argentina's meatpacking industry went into decline, the American packers found their monopoly undermined by a British company that, in certain ways, beat them at their own game. The corporate empire built by William and Edmund Vestey stretched into more countries and controlled more of the fresh beef supply chain than any of the American companies. From an early date, it was also more zealously committed to milking money from waste. In 1876 the Vestey brothers' father, a Liverpool produce importer and wholesaler, sent William to the United States to help out with the family business. He ended up in Chicago, where he turned the big packers' trimmings into canned beef for the British market. Thirteen years later he was thirty, rich, and bored by a brief stint of retirement. On a trip to Argentina he fell into his next enterprise: the freezing and export of partridge, a British delicacy that the Argentines rarely ate. In a matter of months the brothers ran short of cold storage space in Liverpool, at which point they built a new warehouse out of scrap timber.

The Vesteys' partridge storeroom became the first property of the Union Cold Storage Company, which in turn became the biggest such company in the world. As British demand for all kinds of fresh produce exploded at the century's end, the brothers built warehouses in Riga and Moscow, Hankow and Shanghai. These served as outposts for a massive trade in fresh and frozen eggs, as well as whatever other perishables the Vesteys thought they could sell back in Britain. Unreliable shipments from China convinced them that they needed their own refrigerated transport; out of this decision came the Blue Star Shipping Line, founded in 1911 with two secondhand steamers. Over the next few years they moved back into meat, investing in retail butcher shops in Britain as well as freezing plants in Australia, New Zealand, and Argentina. In Buenos Aires they bought and sold slaughterhouses "as though they were so many lots of steers."[83] Their crown jewel, the Anglos-Buenos Aires, became for a while the world's largest packhouse.

By 1921 the Vesteys were Britain's biggest meat retailers, with more than 2,300 butcher shops. They owned cold storage or meatpacking enterprises on every continent. In a few countries, such as Venezuela and Australia, they bought huge tracts of ranchland—a degree of "upstream" integration that not even Gustavus Swift attempted. The Vesteys' quest for cheap beef took them as far as

Madagascar, at the time a French colony. More famous these days for monkeys than for meat, colonial Madagascar was considered a carnivore's treasure island. No one knew how many long-horned zebu grazed its highlands (estimates varied between five and ten million), but it was clear that they far outnumbered people. Some thought the island had more bovines per person than even Argentina. And compared with Argentines, the Malagasies ate little beef. They built up massive herds to show status and pay bridewealth, saving slaughter for special occasions. Yet their cattle worship did not preclude cattle sales. The island's pre-colonial empire exported animals to neighboring islands and hides as far as the United States. Most important from the meatpackers' perspective, fresh beef in the markets cost only a tenth as much as in France.[84]

In the early colonial period, French officials salivated at the prospect of exploiting the island's bounty. An 1896 account described it as "a country where the livestock, the game, the food grains, in a word, all that is necessary for life, grows profusely . . . all the elements that our men of state could use for the renewal of our forces and for the prosperity, vitality and survival of our race."[85] The abundant livestock, in particular, suggested parallels with British colonies that already supplied the motherland with large quantities of frozen beef and mutton. "The great island of Madagascar," one official wrote in 1903, "could become the Australia of France . . . The cattle-raisers could send us incalculable quantities without depleting their stocks. And since a good bull costs hardly 50 francs there, you can easily imagine how much profit could be made."[86] The problem for many years was that France did not want *frigo* meat even if it came from one of its own colonies. The earliest, French-owned packing plants turned out canned beef for the army, which the troops scornfully called *singe*, or "monkey."[87] But the Vesteys timed their entry well. Their Madagascar packhouse began operating shortly before World War One, just as the French government was loosening its restrictions on *frigo* imports. Run by the Vesteys' Paris-based subsidiary, the Compagnie Frigorifique Générale (CGF), the packhouse came up to speed quickly. During the war it slaughtered some 25,000 animals a year, accounting for half the island's meat exports. Given the distance, it froze rather than chilled its beef, but a three-year contract with France's Ministry of War ensured steady high demand.[88]

France continued to import Madagascar's *frigo* meat while it rebuilt its domestic herds, then stopped abruptly in 1921. Undiscouraged, CGF found new buyers in Egypt, Italy, even Britain.[89] It also rebuilt its plant, turning it into "a true industrial city," complete with housing, a library, two tennis courts, and a Chicago-style disassembly line. A 1927 French study noticed "appreciable prosperity" in the region where it drew its thousand-person workforce, and praised its *systeme Taylor,* a reference to the workplace efficiency principles developed by the American engineer Frederick Taylor. "Just like in the great American packing plants, the work gives the maximum results with perfect regularity, thanks to the division and specialization of tasks. This parceling-up of the work also suits perfectly the Malagasy labor force. The worker, staying in place and repeating always the same movements, acquires great dexterity. He cannot dawdle because the assembly line rolls constantly."[90] The company also started up ranches, modeled after the Argentine *estancia.* The breeding of imported Charrollais bulls with zebu cows produced impressive results, and government-sponsored livestock fairs reported large turnouts.[91] Altogether, the colony's governor-general said, the beef export business was "a factory that's just getting organized."[92]

The CGF factory operated in Madagascar until the end of the 1950s. Except during a brief export boom in World War II, the island's "grand future" as a meat exporter dimmed steadily, and eventually not even the Vesteys saw any way to make money there. It lost its comparative advantage as a land of dirt-cheap beef for a number of reasons. Heavy colonial livestock taxes discouraged the raising of large, mature animals, while frequent cattle raiding (a traditional Malagasy rite of manhood turned commercial and violent during the colonial era) undermined the security of company ranches. Not least, Malagasies themselves began eating more beef, especially in the rapidly growing cities. As in Argentina and the American Far West, Madagascar ceased to be a land where cattle cost next to nothing.[93] For the meatpackers, Treasure Island was no more.

The Spirit of Capitalism Made Flesh

In 1912 James Critchell, the British author of *A History of the Frozen Meat Trade,* concluded that the vast distances between regions of supply and demand posed "no stumbling block to the industry, and no disadvantage to the consumer." Cheap refrigerated freight had made the planet small, and the world's meat-eaters reaped the benefits. But he wondered about the future. "The struggle for existence as the population of the world increases will of necessity be so keen that a stimulating meat diet will be essential; a worldwide demand is certain. Whence are supplies to come? Not from the consumers' countries; there will hardly be elbow-room by-and-by for men and women in these hives of industry, let alone pastures spacious enough to feed cattle and sheep. The 251 vessels that now ply across the ocean from the South freighted with meat will, when that period comes, have doubled—trebled—who can say?"[94]

Critchell could not have imagined the vessels that now ply the oceans in all directions, carrying not just chilled and frozen beef but also the corn, soy, bonemeal, hormones, antibiotics, and genetic material that go into the making of the "world steer." Although the sociologist Steven Sanderson coined this term to describe the product of new livestock industries in countries such as Brazil, now the world's biggest beef exporter and home to the world's biggest beef packer, the world steer is not itself a uniquely

Latin American species.[95] Contrary to Critchell's predictions, the global international feed-grain trade allows even Japan, a country with very little elbowroom, to raise a good chunk of its own beef—though the country also buys fresh Kobe-style marbled beef from Kansas. Upton Sinclair's term for the Chicago Beef Trust—"the spirit of capitalism made flesh"—could describe beef itself. The hardy meat of an adaptable animal, it can be produced or sent fresh almost anywhere that the market demands it.[96]

Neither the Chicago packers nor the Vesteys can take credit for the world steer. In the late twentieth century, Britain's richest family saw its overextended empire unravel in a series of popular movements against its ruthless labor policies and control of indigenous lands. In the meantime, the Vestey Company's decades-long, multibillion dollar tax-evasion scheme was finally exposed and prosecuted.[97] The Chicago packers declined in a more conventional fashion. They simply did not react quickly enough to the rise of supermarket and fast food chains—retailers who wanted beef that was not only fresher and cheaper but also more conveniently packaged. By 1962 the Big Four's share of the U.S. beef supply stood at 38 percent, down from 78 percent less than three decades earlier.[98]

In their place, a new and equally elite club of meatpackers upped the efficiency ante. Its founding member was Iowa Beef Packers (IBP), started by two former Swift employees in 1960. IBP's first innovation was location: it built its plant right next to a feedlot in Dennison, Iowa, where meat traveled via interstate rather than railcar, and where big-city unions held no sway. Its one-story packhouse also used conveyor belts rather than chains and hooks to move carcasses. This both sped up and deskilled the disassembly line, allowing the company to employ the cheapest and most replaceable labor. In addition, IBP's plant was much more thoroughly refrigerated than the old Chicago houses, which only cold-stored their final products. By turning on the chill directly after the kill, IBP added to the lifespan of its meat while reducing shrinkage.

Last but not least, IBP dispensed with awkwardly shaped hunks of carcass in favor of pre-cut, vacuum-packed boxed beef. This slashed IBP's shipping costs, suited retailers' demands for convenience and longer shelf life, and opened up higher-value markets both at home and abroad. Like Swift's refrigerator car, boxed beef offered the upstart Iowa packer such huge advantages that it dominated the market before its competitors had time to respond. Within twenty years IBP had become the nation's largest meatpacker, and many of its cost-saving methods—including hard-line labor practices—had become industry norms.[99]

In their push to produce beef as cheaply and globally as possible, new generation packers like IBP have much in common with the fast food companies who are also their biggest customers. The fare offered by these companies might now seem the exact opposite of "fresh" in the farmers-market sense of the term (though the Wendy's chain makes a point of advertising that its own burgers are "fresh, never frozen"). It seems this way only because, at least in the industrialized world, we are so far distant from the days when most people's red meat was salted, spoiled, or just scarce.

In *Fast Food Nation* Eric Schlosser points out the irony "that a business so dedicated to conformity was founded by iconoclasts and self-made men."[100] But an even more profound irony defines the industry that supplies McDonald's with its meat. Now infamous for its mistreatment of workers, it owes much to a man who believed that his "modern invention" would improve workers' health and happiness. Charles Tellier also thought that by bringing fresh beef to the masses, the refrigerated steamship would forge a virtuous circle between consumers wanting ever-better meat and distant producers who would prosper by supplying it. In retrospect his vision might appear hopelessly naïve. But it was genuine and not Tellier's alone. To make the elite's meat available to all, anywhere and anytime, once seemed like a dream worth realizing.

three

Eggs

SHELL GAMES

An egg is full of original sin from the moment it is laid
and asks only for a little leisure in a warm place to
indulge in all its proclivities for wickedness.

—M. E. Pennington, "The Egg and Poultry Demonstration
Car" (1914)

Although its plain exterior doesn't show it, the egg is a food of
extremes. Among the raw foods we keep in our refrigerators, none
are more fragile than the egg, yet few last as long. Few foods are
sunnier when truly fresh, and none more noxious when they've
truly gone bad. And while today's eggs don't look much different
from those sold a century ago, their white and brown shells hide a
history of radical and controversial technological change. In fact,
of all the perishables first touched by commercial refrigeration, no
food proved a harder sell than the cold-stored egg. Even in the
fridge-friendly United States, popular distrust of the "storage" egg
endured for decades.

As refrigeration experts saw it, the basic problem was consumer
ignorance. People needed to give up the outdated idea that the only
good egg was a local and recently laid egg. But as the public saw it,
the basic problem was uncertainty. The egg's unrevealing nature became an unappealing mystery once merchants began chilling mas-

sive stocks. The perceived danger lay not so much in cold storage itself as in the way it could be used to cheat people out of wholesome and fairly priced provisions. In this sense, the story of the egg is part of a much bigger saga about how refrigeration upset the moral as well as the physical order of the fresh food market.

Ever since such markets existed, the threat of imminent spoilage limited the bargaining power of sellers, whether they were farmers or merchants. Sellers risked losing their goods if they searched too long or too far for better prices than what their local customers offered, just as they risked losing those customers if they overcharged or otherwise deceived them. Refrigeration gave sellers much more leeway on both fronts. In the case of eggs, it provided storage for the good ones as well as a scapegoat for the bad ones. But as merchants used the technology to hide their own duplicity, they also cast doubt on an entire class of commodities. Consumers wanting honest eggs steered clear of the cold storage label.

Eventually the chicken resolved the moral hazard of the egg. The premium attached to genuine, newly laid freshness pushed the focus of innovation from the warehouse to the henhouse, where farmers bred and fed their flocks for ever-higher output. The impressionable hen, it turned out, responded even to minor home makeovers. Now we can trust that our eggs are fairly fresh, relative to what people used to eat. How much else we can trust about them is another matter.

The Everlasting Egg

The most incredible part of the egg is its inedible shell. Made of almost pure calcium carbonate—the same tough mineral compound found in bone and marble—the eggshell forms inside the oviduct in only about half a day.[1] The egg enters the world with a thin coat or "bloom" of albumen, which seals the shell's microscopic pores against bacteria and fungal spores. A nesting hen's warm, moist feathers eventually rub away this coating, ensuring that the fertile eggs breathe, and the rest quickly go stale.[2] For this reason farm wives knew to gather eggs frequently and not wash them until just before use. Modern chicken farms do wash their eggs, soon after they drop from the hens' cages onto nonstop conveyor belts. A quick dousing of mineral oil replaces the albumen bloom.

"Stale" describes an old egg pretty well because, like a loaf of bread, an egg ages mostly by drying out. In a newly laid egg, the white fills up the shell and holds the yolk in place. As it loses moisture and carbon dioxide, the white shrinks and its proteins rearrange themselves. The result is a runnier egg, which you'll discover if you try to poach or fry it sunny side up.[3] It's not as pretty as a newly laid egg, and it might not be free of bacteria (more on those later). But if its shell is sound, that egg can survive for months, unspoiled, in your refrigerator. Indeed, even by the time they arrive in the supermarket, many "fresh" eggs today have been sitting in their cartons for three weeks or more.[4]

A century ago, people commonly ate eggs that were only a few hours old, simply because so many of them lived on or near farms. But they also routinely ate eggs in November that had been laid in May. Often they didn't have much choice. That's because the egg was once a springtime crop, much like asparagus and sweet peas. Responding to changes in temperature and day-length, hens in the northern hemisphere laid most of their eggs between April and June. Egg quality as well as quantity dropped off in the summer, as the hens rested and molted. "Broody" hen varieties stopped laying altogether, choosing instead to sit on the eggs they'd already laid. Except for the rare "everlasting layers"—varieties that could lay through the winter if they were well fed and housed—most hens began laying again only in late winter, when their pituitary glands produced more of the hormones needed for egg maturation.[5]

From an evolutionary perspective, this seasonality made perfect sense, because chicks hatched in fall or winter would be least likely to survive. Culturally, the reappearance of eggs in a hen's nest signaled "as surely as buds swelling on trees" that winter would soon end.[6] It also marked the beginning of a season when people celebrated Easter and other spring holidays by eating, painting, rolling, and otherwise reveling in eggs. Not only did they symbolize rebirth and fertility; eggs were also "sweeter" and far more abundant than they'd be at any other time of the year. In the mid-nineteenth century, for example, New York City received seventy-two times more eggs in May than it did in January. Prices varied accordingly; depending on the month and region, off-season eggs might easily cost three times as much as spring eggs.[7]

In America as in much of the egg-eating world, newly laid eggs were a rare luxury during the fall and winter. Yet rather than give up eggs altogether during those seasons, people found ways to store them for months or even years. Not to be confused with *preserved* eggs—that is, the pickled kind found in English pubs, or China's "ancient" salt-preserved varieties—home-stored eggs were by some accounts "just as good" as fresh. At the least, according to the 1850 guidebook *American Poultry Yard,* stored eggs ensured that holiday hostesses did not have to "disappoint the little folks of their Christmas plum pudding or the ladies of their 'egg nog.'" For keeping eggs just a few weeks, *The Book of Poultry* advised putting them "in a cool but not very cold place—about 50 to 60 degrees is best—and with the large end down." For longer periods, people commonly put their eggs in jars filled with a sodium silicate solution known as glass-water.[8] Other storage techniques, such as those suggested in the popular book *Dr. Chase's Recipes,* took more work but apparently yielded worthwhile results:

Eggs to Keep from September to April, as Good as Fresh. This is from J. B. Strathnairn, who says: "I take a tub of any size and put a layer of common salt about an inch deep in the bottom; then grease the eggs with butter (of course salted butter) and place them in the salt with the small end down . . . then fill the vacancies with salt . . . then cover the top with salt, and put them where they will not freeze. I have kept eggs in this manner from September until April as good as fresh. The grease on the shell keeps the salt from penetrating, thereby keeping the eggs fresh, while the saving qualities of the salt keep them from becoming putrid. This recipe is both cheap and good, as the salt can be fed to cattle afterward."

Eggs—To Keep Two Years Perfectly Good. This is from Emily Audinwood: "I have tried several experiments but find none work so well as the following . . . Two pounds of coarse salt boiled 10 minutes in 1 gal of rain water; pour off into an earthen jar. When nearly cool, stir in 5 tablespoons of quick lime; let it stand till next day; then put in the eggs and keep them tightly covered until wanted for use."

Eggs, Preserving Six Months, Equal to Fresh. A writer in the *English Mechanic* says "In the year 1871–2, I preserved eggs so perfectly that after six months they were mistaken when brought to the table for fresh laid eggs, and I believe they would have kept equally good for a twelve-month. My mode of preservation was to varnish the eggs as soon after they were laid as possible . . . I pack the eggs in dry bran."[9]

Many of the families who stored eggs got them from their own chickens. Unlike cattle ranching, poultry farming remained a "people's business" in turn-of-the-century America. The vast majority of farm households kept at least a few hens; most flocks numbered between ten and one hundred birds.[10] They didn't live very differently from their jungle-dwelling ancestors; they ate grain and grass and barnyard grit, and during warm weather they roosted in the trees. Hens sat on their own eggs, which they sometimes laid in the grass.[11]

Women usually tended the family flocks. Along with butter, sales of eggs brought in "pin money," at least for a few months of the year. Poultry experts thought it only natural that women handled the chickens, "owing to the fact that they seem to understand the temperament of the sitting hen better than men."[12] More likely, the tedious work of gathering and packing eggs didn't appeal to many men, especially since it peaked at a time when local markets were glutted and prices discouragingly low. As one female poultry extension agent explained in 1890s Wisconsin, men lacked the "patience and gentleness, as well as eternal vigilance" that hens demanded. A successful egg business, she said, "does not depend entirely on any one thing, but on the many little things which men dislike so much to do, and which make women more especially fitted for the work."[13] Men did raise chickens when and where they could do so on a large scale, such as in the breeding business.

Before the spread of rail transport and refrigeration, relatively few eggs traveled far between farm and market. Rapidly industrializing London was exceptional among cities, in that it was already importing eggs from France and Ireland in the 1840s. U.S. cities got most of their eggs and chicken meat from their immediate hinterlands, making New York, Virginia, and Pennsylvania the biggest poultry-producing states in the years just before the Civil War. Commercial chicken farms also cropped up in New Jersey's Vineland district.[14] In more rural areas, meanwhile, many farmers just took their eggs to the local country store, where they could exchange them for groceries and other dry goods.

Expanding transportation networks cracked open these local trades. By the 1860s, eggs sold in New York came from as far away as Minnesota; by the mid-1870s they were coming from southern states (Mississippi, Tennessee, and Georgia) as well as from Canada and Mexico. Eggs also traversed oceans, either across naturally chilly northern routes or, after the 1870s, in refrigerated steamships. As with beef, Britain's supply networks quickly stretched across the globe. By the end of the nineteenth century, eggs from

A chicken-farming family in Petaluma, Calif., once known as the "egg basket of the world." Circa 1910. (Courtesy of Petaluma Historical Library and Museum)

Normandy had lost market share to cheaper and better-chilled eggs from Denmark, Siberia, the United States, and even Australia and South Africa.[15] Ports in California and the Pacific Northwest, meanwhile, received shipments of eggs from China.[16]

Access to distant urban markets quickly encouraged regional and local specialization. In the United States, towns in Illinois and Ohio competed with New Jersey's Vineland district for shares of the New York egg market. But no town saw a bigger egg boom than Petaluma, California. With a river, a railroad, and San Francisco all nearby, Petaluma boasted an ideal location and year-round mild weather. Petalumans began specializing in poultry farming in the 1880s, and soon the town became a magnet for East Coast urbanites, many of them Jewish émigrés seeking better lives as family farmers. By the 1910s Petaluma was known not only as one of California's wealthiest towns but also as "the egg basket of the world."[17] It shipped 100 million eggs a year to markets up and down the West Coast, as well as to Alaska and New York. In 1918, Petaluma reportedly produced enough eggs to circle the earth.[18]

By that point some eggs did in fact traverse large parts of the globe. Well before such large-scale production became profitable, though, cities needed somewhere to keep eggs during the long offseason. Cold storage warehouses transformed the landscape of American egg production as dramatically as railroads did. They also turned fresh eggs into political hot potatoes.

Cold Storage Controversies

The earliest cold storage warehouses first went up in the 1860s in Indianapolis, Cleveland, Boston, and New York. Chilled by ice and insulated with sawdust, they primarily stored fruit. As demand for perishables increased, cold storage became an industry in its own right. Temperature control improved first with better insulation and the mixing of salt into the ice, and then in the 1890s with the gradual replacement of ice by mechanical refrigeration.[19]

By 1904, the United States had more than six hundred cold storage warehouses, most of them in urban settings and many of them several stories tall. With a combined capacity of more than 102 million cubic feet, they held staggering quantities of food: Boston's Quincy Market Cold Storage Company had room for 150 million eggs at a time. In addition, meatpackers, breweries, dairies, and fruit wholesalers were quickly building millions more cubic feet of refrigerated storage space, also mostly in urban areas. No other country—not even import-dependent Britain—had comparable facilities.[20] European refrigeration engineers marveled at the speed with which the United States had built a nationwide network of "industrial cold," and at the variety of fresh, seasonal foods stored and sold year-round in East Coast cities.[21]

In the cities themselves, however, Americans weren't so keen on the storage part. Like consumers elsewhere in the world, they distrusted claims that refrigerated foods could stay tasty and wholesome for weeks or even months longer than their natural lifespan. They also suspected the motives of anyone who stored large quantities of food in order to sell it during periods of scarcity. Although these concerns did not slow the early twentieth-century warehouse-building boom, they did spark investigations and calls for regulation at both the state and the federal level.

People had good reason to doubt early cold storage, simply because it often did not work very well. Ice-chilled warehouses were difficult to keep at a stable temperature and humidity. They also lacked adequate air circulation, so goods might freeze in one corner and spoil in another.[22] While mechanical refrigeration facilitated temperature control, many merchants and warehousemen did not know which temperatures were appropriate for different foodstuffs. Nor did they know how to keep them from developing a telltale "storage taste." Merchants bore the cost of obviously spoiled food, such as moldy meat, but sometimes they just marked down prices. This happened at Boston's Quincy Market in 1900, when eggs there were discovered to have a "fruity flavor," most likely absorbed from fruit in the next chamber. The company reassured shoppers that this was "not a novel occurrence" and that the eggs were still fine for cooking.[23] The storage goods that sowed the most distrust, however, were those that consumers discovered were *not* fine only after they bought them. Rotten eggs, mushy fruit, frozen beef that spoiled as it thawed: stories of these unfortunate purchases circulated long after the technology of refrigeration had greatly improved.

So did theories about the health dangers of cold storage. Early on, rumors centered on refrigerated beef, which was alternately linked to cholera, cancer, and appendicitis.[24] Again, the newness of the technology meant that even the experts didn't really know what happened to food that spent months under its influence. They contributed to the kind of hearsay that one letter writer to the *New York Times* reported in 1906: "The writer has just returned from a seaside resort, where he found among the visitors as well as the residents annoying illness was the rule rather than the exception . . . the two physicians I consulted expressed the opinion that while weather conditions were responsible for some of these troubles, they thought cold storage, artificially preserved, and low-grade articles of food were responsible for their share of it. This bears out the opinion of the writer."[25]

In 1906, shortly after the passage of the Pure Food and Drug Act, the Department of Agriculture's Food Chemistry Bureau undertook one of the first systematic epidemiological investigations of cold-stored foodstuffs. The findings were mixed. Dr. Harvey Wiley, the bureau's chief and a self-proclaimed "champion" of cold storage, reported that some foods, namely meat and fruit, actually benefited from up to three months in a warehouse. Milk, cream, and eggs, by contrast, all deteriorated "immediately" in storage. A few years later, expert opinion shifted. A Massachusetts study declared that properly stored foods, including eggs, were perfectly safe and wholesome—and could stay that way for nine months or longer. The *Journal of the American Medical Association* agreed: "the charge that cold storage in general is detrimental to public health is refuted by an impartial examination of this subject in hygienic aspects."[26]

But what happened when intermediaries used cold storage to hide *improper* trade practices? Here the public had, if anything, more reason to worry. In the early years, merchants routinely stored goods that were already going bad, such as meat or fish that had spent all day in an exposed market stall. Cases of eggs went directly from the railcar to the warehouse, without candling to check for rot or staleness.[27] Retailers then retrieved them in the "blind faith," as one Department of Agriculture report put it, "that some magic property of cold storage had made good eggs of all the bad eggs, or else accepted the situation as one for which they were not responsible and which they could not remedy."[28]

At first, the warehousemen saw no reason to prevent such practices. In fact, by offering their clients easy credit, they encouraged them to buy and store eggs and other foodstuffs on a much larger scale. And if some of that food turned out to be less than fresh, the warehousemen could point out that it was their job only to keep other people's provisions, not to ensure that they were any good.[29] Not surprisingly, this "buyer beware" attitude didn't sit well with consumers. Cold storage eggs seemed especially dubious because they revealed nothing to the naked eye. As one food chemist put it, "It is a well-known fact . . . that a cold storage egg guards the secret of its age as jealously as some men and many women."[30] So even though eggs marketed as "freshly laid" did not come with any quality guarantees either (and in September or December it was highly possible that they'd been sitting in a barn for a while), such eggs continued to cost much more than their cold-stored counterparts—say, fifty cents a dozen versus thirty-five, at the height of winter. Meanwhile, women's magazines warned against preparing cold storage eggs *as* eggs: "Of course I shouldn't think of trying to boil or even scramble them," wrote one *Good Housekeeping* columnist. "But they help out in cooking."[31]

Before long, cold storage companies realized that they needed to set some standards—requiring, for example, that all eggs be candled before entering their warehouses. Yet such measures could not prevent retailers from taking advantage of the price differential to sell all their best eggs (including the cold-stored ones) as "fresh," and all their inferior eggs (including those fresh from the farm) as cold-stored. Warehousemen's efforts at quality control also didn't satisfy those who felt that the very practice of large-scale, long-term food storage reeked of unscrupulous speculation.

Here again, it's not hard to see why people got such ideas. Just as grain traders had long taken advantage of seasonal harvests to buy cheap and sell dear, industrial cold storage made it feasible to do the same with eggs. Combined with easy credit, cold storage also encouraged egg "gambling" and the rise of informal futures markets. In early spring, "gamblers" in New York and other major cities placed orders for hundreds, thousands, or even tens of thousands of cases of not-yet-laid eggs. Even if they didn't beat the market at the time of purchase, they could very likely sell those eggs at a much higher price six months later.[32]

Anyone who stored spring eggs in order to sell them for a profit in the winter was, by definition, speculating. But were they hoarding? Were they driving up prices? Proponents of cold storage contended that, on the contrary, the practice made off-season eggs much more affordable. By reducing the likelihood of ruinous spring gluts, they said, cold storage also encouraged farmers to produce more eggs, thus increasing overall supply.[33]

Or at least that was the argument. But when food prices skyrocketed in 1909–1910, the giant food-filled warehouses looked awfully suspicious. It did not help matters when newspapers reported that members of the "Beef Trust"—the meatpackers Armour, Swift, and Morris—were storing millions of eggs in their facilities in Chicago and other midwestern cities. The first such reports appeared in 1902 and suggested that the meatpackers' eggs stocks foretold a giant "food combine."[34] Within a few years the press routinely referred to "the egg trust," but also noted that early springs and winter warm spells—which brought on unexpected egg-laying—could undermine its speculative profits. "Hens as Trust-Buster," ran one headline in spring 1905. "Hens Happy—Smashed Egg Trust's Shell Game," said another in January 1906.[35]

When prices began to rise rapidly in 1909, talk of a refrigerated "food trust" once again filled the press. No one seemed quite sure who was in it or how much food they controlled, partly because companies did not have to report this information. The mountains of perishables going into public cold storage houses did indicate that the meatpackers were not the only suspects. And the role of refrigeration in this trust appeared obvious.[36] As a *Washington Post* letter writer, signed "Householder," put it in 1909:

> I see editorials now and then commenting on the high prices of food . . . the explanation, however, is very simple. It is not because food is scarce. It is because there is a food trust . . . I don't think it is an organized concern, but to all intents and purposes, it might as well be. I understand that in our own markets, right here in town, the dealers get together and decide on prices before they open their stalls. Cold storage is responsible for the prices they charge . . . They have thousands and thousands of eggs, but keep them in the refrigerator, in order to keep prices high.[37]

Soon politicians and Progressive reformers called for government action. The Chicago city council had voted down one of the first attempts at cold storage regulation in 1906, after "energetic" opposition from warehousemen and their customers.[38] From 1909 onward, several other cities and states introduced their own cold storage bills. Most of these bills proposed to limit legal storage time to anywhere from one month to one year. Most also called for warehouse licensing and inspections, and the labeling of all cold-stored foods.[39] In New Jersey, the push for legislation began with a price-fixing investigation. In 1910 a grand jury indicted six companies (including the meatpackers Armour and Swift) "for conspiring to increase the price of foodstuffs." The jury did not buy the argument that cold storage evened out scarcity and glut, instead noting that "when eggs were plentiful they were cornered and kept from the market at times when it was natural that they should be cheap."[40]

In turn, warehousemen blamed the press for turning cold storage into the favorite cause of populist politicians. Editorials in *Ice and Refrigeration*, the industry trade journal, expressed a mix of anger and bewilderment: "A consideration of the agitation which has led to the legislation, discloses a curious condition of the public mind. It has the symptoms of an unreal nervous and mental disease. . . . It is not based upon knowledge, fact, or investigation, and has no sponsors nor advocates except the sensational newspapers and the politicians who are looking for an issue."[41]

The press did run some inflammatory headlines about "storage horrors." Tapping into two reliable sources of reader outrage, the *New York World* suggested that consumers were being simultaneously poisoned and cheated: "Thousands of Tons of Food Unfit to Eat Foisted on Public by Freezer Owners," raged one headline. "Bad Eggs, Poisoned Poultry, Deadly Fish, Unwholesome Butter, and Decaying Vegetables Kept to Get Benefit of High Prices," said another. The newspapers also fueled suspicions of speculative hoarding by publishing figures on the quantity of warehoused provisions that, as the New York *Evening Mail* claimed in January 1910, could "feed the population of the United States for a month." These reported reserves included the meat of fourteen million cattle, twenty-five million dollars worth of fish, and more than one hundred million dollars worth of other perishables—including, no doubt, lots of eggs.[42]

Ice and Refrigeration urged its readers to combat these "sensational" charges with more positive publicity. Informative advertising and recipe booklets, argued one columnist, would help calm the "agitation" of consumers who "do not understand the function of a cold storage warehouse, and do not understand anything but the simple fact that here is a big building . . . containing food that they want to get hold of." In Chicago, city officials and reporters were

This ad for cold storage eggs stressed the honesty of the seller as much as the "fresh-tasting" quality of his stocks. (*Ice and Refrigeration* 40, 1911, p. 30)

invited to a luncheon featuring exclusively cold storage ingredients. The *Chicago Tribune* announced the event with the headline "To Dine on Embalmed Food."[43]

The warehousemen and their major clients also stepped up their lobbying, especially after the U.S. Senate began considering a federal cold storage bill in the spring of 1910. If nothing else, they wanted a say in shaping whatever legislation they couldn't defeat entirely. After all, as one Pennsylvania congressmen warned, "the *laissez faire* attitude is no longer tenable in the face of impending struggle, and a negative position . . . only serves to whet the appetites of those who are determined to take a fall out of the cold storage interests."[44]

The Senate's cold storage bill emerged from an investigation into the high cost of living, led by the Massachusetts senator Henry Cabot Lodge. But here as in the press, questions of fairness and freshness were rarely far apart. In hearings held by the Committee on Manufactures, meatpackers reported on how cold storage affected meat pricing, and chemists from the U.S. Department of Agriculture (USDA) told how it affected meat texture and taste.[45] Physicians discussed digestibility, and lawyers brought lists of consumers willing to go on record as having eaten and even enjoyed cold-stored foods.[46] A fish dealer offered to bring in fresh and frozen samples "prepared by a first-class chef," claiming that the committee members would be unable to taste the difference. To one senator, this boast only proved the need for government oversight, because it illustrated "the ease with which the consumer can be deceived with respect to food he is taking into his system."[47]

To protect consumers from the dangers and deceits of cold storage, some senators called for legislation limiting storage time to six or even three months. But refrigeration experts argued that such timeframes were ridiculously unscientific, for they didn't account for differences among types of food or types of storage. Six months, for example, was far too long to keep peaches and tomatoes but

As food prices climbed in 1909–1910, *Puck Magazine* portrayed cold storage as a weapon of highway robbery, used by speculators against farmers and consumers.

not long enough to ensure a year-round supply of high-quality spring eggs.

Among the opponents of restrictive time limits was Dr. Mary Pennington, the director of the USDA's Food Research Laboratory and a woman once called the "voice of conscience in the refrigeration world."[48] She was also one of the Western world's foremost authorities on chicken and egg hygiene. Like many refrigeration advocates, she blamed the negative image of cold-stored goods—especially eggs—on ideas leftover from the days when "the nearby farm was the source of supply and when 'freshness' was measured by the number of hours that elapsed between the gathering of the produce and its delivery to the consumer. The fewer the hours, then, the better the goods, because the farmer had no facilities for preventing decay . . . Because the consumer has insisted that he must have produce 'right fresh from the country' the vendor has imposed upon his ignorance by pretending to give it to him. In reality the vendor cannot obtain such goods as his customer demands, hence the falsehoods that are the stock in trade of every retailer."[49]

If consumers and congressmen wanted honest trade, in other words, they had to accept that refrigeration had altered the physics of freshness; no longer did it depend on time or distance. Whether or not the senators agreed with this view, it ultimately convinced them that setting time limits on the cold storage of an entire nation's food supply was just too complicated. After months of hearings, the bill died in committee.

Controversy surrounding the marketing of cold storage eggs, however, remained very much alive. In December 1912, women's groups took matters into their own hands—literally. In Philadelphia and later in New York and Chicago, they organized citywide egg sales, attracting crowds of customers as well as national news coverage. "Women Solve Problem of the High Living Cost," a *Los Angeles Times* headline announced, "Philadelphia Society Dames Break Corner in Eggs and Demonstrate to World How to Smash

Trusts While Puzzle-Brained Statesmen Are Holding Long-Winded Investigations and Conferences."

In Philadelphia, the media probably showed up partly to see the "society dames" who, according to the *Los Angeles Times,* were "famous beauties and matrons . . . known all over the continent." For the city's housewives, the biggest draw was the price: twenty-four cents a dozen, or five to fifteen cents less than what they'd pay elsewhere for cold storage eggs. And these were, the ladies emphasized, eggs from storage—but all "were carefully inspected . . . so that there will be no fear of antiques."[50]

The egg sale organizers had no illusions about getting fresh-laid eggs in December, which Chicago's Clean Food Club noted were "rarer than strawberries in January." And they had no qualms about advertising their stocks as "fine Aprils." Their gripe lay with the speculators who restricted the supply of such eggs, and the retailers who clearly profited from the pretense that they were fresh. As the leader of New York's Housewives' League explained, "The clerk tells (the housewife) he does not think she would like the storage eggs, and advises her to take fresh eggs. She pays 60 cents for the supposed fresh eggs, which in reality are storage eggs. Nine times out of ten when a woman believes she is getting fresh eggs she is getting storage eggs."[51]

The women's egg sales succeeded in pushing prices down until spring and encouraged politicians to move forward with legislation. By 1915 eleven states, including California, Indiana, New Jersey, and New York, had passed cold storage acts. Combined with the easing of inflation and the outbreak of World War One, these laws helped calm the "hysteric attacks" on the industry. By 1915, the warehousemen's association was grateful that "the intervention of foreign strife has given the press so much to write about for the past year that cold storage topics have slept peacefully.[52]

Still, campaigns to promote cold storage eggs were far from over. The USDA led the charge in rural areas, where the typical egg's journey from nest to market was neither quick nor climate-controlled. Poultry extension agents traversed the country on a specially equipped demonstration railcar on what Mary Pennington described as an "egg-saving mission." In 1913–1914 the car traveled 7,000 miles and visited 117 towns, mostly in the Midwest. Farmers, egg dealers, and even schoolchildren were invited aboard to learn about the merits of prompt cooling. At the same time, the USDA officials acknowledged that most farm families couldn't afford to buy a refrigerator or even an icebox for their eggs. So they were advised to keep their eggs "on the cooler places on the farm" and deliver them promptly to their local dealer.[53]

In the cities, meanwhile, the cold storage industry urged egg dealers to make the most of the new mandatory labeling. *Ice and Refrigeration* praised a Cleveland dealer's newspaper ads, which promised that his wintertime cold storage eggs were "as good and sweet tasting eggs as you ever bought at this time of year—not a bit lime-y." The advertisements of the Quincy Market Cold Storage Company told consumers that the storage label was "the mark of a Good Egg. Look for this sign in your grocer's window."[54]

Promoters of cold-stored eggs also borrowed ideas from the Chicago meatpackers, who had used cut-rate pricing to win over consumers to chilled "dead meat" three decades earlier. Reporting on its public relations campaigns in 1915, the warehousemen's association said that "so far as general education goes, the only article which seems to require particular stress is 'eggs' . . . The practical problem of proving to a doubting housekeeper that good refrigerated April and May eggs at 30 cents per dozen are better than summer or early fall so-called fresh eggs, at 60 cents or more per dozen, is a problem which can only be solved by practical demonstration. Special sales of cold storage eggs at practically cost prices have been conducted in several large cities . . . These sales created a marked change in public sentiment wherever conducted."[55]

Apparently progress on this front was slow, because in the mid-

1920s *Ice and Refrigeration* was still complaining about the "many erroneous ideas" that American housewives had about cold storage eggs. And experts like Mary Pennington (by then head of the Household Refrigeration Bureau) were still trying to make them see the light. At a January 1923 "health exposition" in New York City, for example, the cold storage exhibit featured a giant refrigerator, an educational movie, and the well-known home economist Anna Barrows, making omelets out of nine-month-old storage eggs. "The clear, thick whites, firm yolks, and quick whipping, seemed to surprise many of the observers," noted Pennington. "Some timid folk, who seemed to think they had never eaten cold storage foods, had to be encouraged to 'try some.'"[56]

The public's uncertainty was understandable. Because eggs were not individually labeled, merchants could still easily switch them between cartons in order to sell as many as possible at the higher, "freshly laid" price. As the California State Board of Health reported in 1925 (fifteen years after the state passed its own cold storage law),

> There has been a temptation, irresistible by many dealers, to candle out the better grades of cold storage stock and sell them to retailers as fresh eggs. Retailers have deliberately purchased the better grades of eggs, cold storage stock, and sold them to their trade for fresh eggs. This sort of so-called "bootlegging" in eggs has grown to such proportions that it is estimated that 75 percent of the cold storage pack has been "bootlegged" . . . Most of the eggs offered the consumer under the name "Cold Storage" have been inferior ones, remaining after "bootleg" stock had been selected. This practice has served to enhance the prejudice against "cold storage eggs," because the consumer has had no opportunity to use the good ones under that name.[57]

So despite free omelets and illustrated ad campaigns, consumers remained unconvinced that nine-month-old storage eggs could taste, as Pennington insisted in one public address, "not only good but really delectable."[58] Eggs marketed as "freshly laid"—whether they really were or not—continued to command a premium. In response to intractable egg "bootlegging," some states eliminated the cold storage label altogether and required retailers to sell eggs by grade instead. Because grades rate eggs according to visible quality indicators (including the size of the internal air cell) this policy neatly erased age as a measure of freshness. National egg standards, adopted by the USDA in 1925, also ignored chronological age. The geographical measure of freshness endured longer, mainly because farmers' organizations demanded protection in the form of state-level "fresh egg" laws. In Florida, for example, *only* Florida eggs could be sold as "fresh"; eggs from anywhere else had to be labeled "shipped." In the Northeast, geographical origin labels (for example, "New York whites") encouraged consumers to buy local.[59] Ultimately, though, the most revolutionary changes in the meaning of egg freshness—changes that ironically helped to destroy both cold storage and local egg trades—didn't occur in any state house. The place of real foment was the henhouse.

Hens See the Light

From the farmers' point of view, cold storage initially appeared to enhance what one poultry manual called the "wealth-producing powers" of the hen.[60] By increasing annual egg consumption while alleviating springtime gluts, the storage industry helped to make chicken farming seem like a feasible full-time occupation, not just a so-called side business of farmwives. During the same years that refrigerated warehouses went up in cities across the United States, "chicken towns" such as Petaluma swelled with newcomers, and publications offering advice and encouragement *(The Call of the Hen; Productive Farm Poultry; The Golden Egg; Happy Hen; Industrious Hen;* and *Feathered World)* proliferated.[61] Between 1880 and 1920, the U.S. poultry population grew faster than that of any other kind of livestock. The number of eggs produced per person nationwide increased from around 9 dozen in 1880 to 17.3 dozen in 1910. The Petaluma Chamber of Commerce foresaw "no danger of overproduction," thanks to booming egg appetites among urban populations. As one of its pamphlets told prospective settlers in 1918, "the quicker you get some chickens the more pleasure and profit you will have."[62]

In fact, egg prices plunged a few years later, driving many small chicken farms under and demonstrating to the rest that cold storage alone couldn't guarantee stable earnings. After all, storage did nothing about the fact that, as one farmer reportedly put it in 1910, "the dum fowls won't lay when the price is high."[63] Efforts to convince the "dum fowls" to adopt more lucrative laying schedules weren't new. Mid-nineteenth-century poultry fanciers, for example, found that at least some hens, "if properly fed, and kept in a warm situation protected against the cold," could be convinced to keep laying into their normal seasons of rest. They also knew that promptly removing eggs from their nests—and raising them in an incubator, if they were fertile—helped to discourage hens' instinctive broodiness.[64] But at that point, most farm families were no more likely to tamper with their flocks' seasonal habits than they were to question the seasons themselves. Only at the end of the century did the "ever-lasting layer," once considered a gift of Nature, begin to look like a creature that could and should be engineered.

The high price of fresh eggs during the winter provided an obvious incentive, but not the only one. In addition, year-long egg-laying contests rewarded the heaviest off-season and overall layers.[65] They also generated useful information about the traits of the winning birds, such as the fact that they were typically early hatchers. Among breeds the Leghorn excelled, thanks to a weak maternal instinct and a strong metabolism. For poultry breeders and farmers, the big question was how to cultivate these traits within their own flocks.

For science, the big question in the early twentieth century was how poultry breeding could help them to solve the larger mysteries of heredity. To ensure the support of farmers as well as federal funding for this project, biologists at the Maine Agricultural Experiment Station defined their research in very practical terms: they sought to breed hens that laid the most when egg prices climbed highest.[66] They soon discovered that, contrary to accepted wisdom, chicks did not always inherit their mothers' laying abilities. A hen's lineage (and that of the breeding rooster) mattered much more than her own egg-laying performance. On the basis of these findings, practical manuals such as *The Hen at Work* advised farmers how to "breed and weed" their flocks, culling the poor performers.[67] Professional breeders developed varieties like the "Everlay" Brown Leghorns, which were advertised as having "the winter-laying habit bred into them. You can own a flock of these Real Money Makers at small cost."[68]

Whereas the Maine experiments centered on the nature of off-season laying, the poultry scientist Harold Lewis told farmers how to nurture this profitable trait, mainly by undermining the hen's own nurturing habits. As Lewis wrote in *Productive Poultry Husbandry,* "The hen is too valuable an egg machine to allow her to waste weeks and months in hatching eggs."[69] He explained:

> As soon as an individual hen has laid a number of eggs, her natural tendency is to try to incubate them. This instinct is much more pronounced during late winter and spring, and, if a continuous heavy production is desired, it is necessary to break up the broody habit as soon as possible . . . The best way . . . is to confine them from three to five days . . . in specially constructed coops with slatted bottoms, feeding them light rations of wheat, with plenty of water. . . . The desire to sit is thus more quickly discouraged.[70]

As Lewis's advice suggests, hens react to changes in diet just as they do to changes in light and temperature. This was old news to farmers. But early twentieth-century advances in nutrition, including the discovery of vitamins, opened a new era in scientific feeding. Poultry manuals described formulas and feeding plans that farmers could use to synchronize birds' life cycles with the egg market. A relatively sparse diet, for example, slowed down the maturation of chicks, so that they started and stopped laying later in the season. A few days of no food whatsoever could "force" an early molt, bringing hens back into full production before the springtime glut. And a rich, well-balanced winter diet—including buttermilk and leafy greens—could boost the off-season egg yield. To prepare the ideal rations, Lewis recommended that farmers equip their barns with caldrons, vegetable slicers, and bone cutters. Soon they could buy vitamin-fortified feeds and supplements at their local store. Farm journals were full of ads for products like Quaker's Ful-O-Pep Egg Mash and Purina Poultry Chow, which promised to boost egg production during "high price time."[71]

While careful breeding and feeding brought incremental improvements in off-season egg production, the most immediate results came at the flick of a switch—literally. Sometime around the end of the nineteenth century, at least a few chicken owners found that their hens couldn't tell the difference between sunlight and artificial light supplied by the newly invented light bulb.[72] Thus they could effectively be "tricked" into thinking that night was day (time to eat), and winter was spring (time to lay). Whether or not the authors of *The Chicken Book* are right in calling henhouse lighting the "single most important discovery in the history of the domesticated chicken," it certainly marked a turning point in the history of the egg. One "early adopter" of electric lighting, a Cambridge poultry farmer, reportedly described it as "the most definite control of production that we have. You can turn on a switch or turn it off."[73]

For farmers who wanted to sell fresh eggs when no one else had them, this kind of control was invaluable. Indeed, in 1919 *Scientific American* called it "neither more nor less than a miracle." The mag-

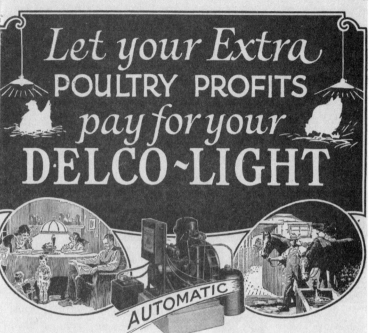

Let your Extra
POULTRY PROFITS
pay for your
DELCO-LIGHT
AUTOMATIC

EVERY poultryman, every farmer, or farm woman, who raises poultry, should write immediately for complete information as to how Delco-Light increases egg yield during the season of highest prices, increases poultry profits, pays for itself in a surprisingly short time.

There is absolute and positive proof, taken from literally hundreds of cases, which shows beyond any question that Delco-Light pays for itself over and again if it is used to increase daylight time—laying time—for poultry. The most exhaustive tests carried on by agricultural colleges, the practical experience of hundreds of poultrymen, prove the money-making abilities of Delco-Light.

And remember, too, that besides increasing poultry profits, Delco-Light will supply an abundance of clean, safe, brilliant electric light, and smooth, quiet, strong electric power for the farm home. Write immediately for complete information about the new low-priced automatic Delco-Light which starts, runs, and stops at the touch of a button. Find out about the easy terms of the General Motors Deferred Payment Plan. Mail the coupon today.

Delco-Light Company
Subsidiary of General Motors Corporation
Dept. F-6   Dayton, Ohio

Send Coupon To-day

Delco-Light Co., Dept. F-6   Dayton, Ohio.
Please send complete information about Delco-Light for the poultryman.

Name _____ Town _____ County _____ RFD ____ State _____

In the 1920s, henhouse lighting promised not just more wintertime eggs but also earnings high enough to electrify the entire farmhouse. (*Leghorn World,* October 1926) azine recounted the epiphany of one southern California farmer who "installed lights in his runs so that he could more conveniently display his stock to prospective buyers after dark, and soon was led to suspect that the lights, often on, made the hens lay 'better' . . . Now his technique is practiced all over California." A reader in Bemidji, Minnesota, saw even more dramatic results when he tried to warm up his barn with a 50-watt light bulb: "To our great astonishment our chickens began to lay eggs, and they have since literally turned night into day, cackling and behaving in liveliest fashion, and laying practically all their eggs in the night, sometimes 3 or 4 even after 10 pm."[74]

Despite such stories, the use of henhouse lighting spread slowly. This was partly because electricity itself spread slowly into rural areas; even in the mid-1930s it reached only about 10 percent of American farms. Gas or kerosene lanterns were the only alternatives, and they were inconvenient and fire-prone. But many farmers must also have been reluctant to try out such a risky and poorly understood technology. As one 1926 article in *Leghorn World* warned,

> Lighting, if it is properly used, is a most remarkable aid in increasing the egg production at the time when eggs are most valuable. It must be remembered, however, that slight mistakes . . . may have disastrous effects as birds are under more or less artificial conditions. When lights are used they must be used regularly. It is not at all practical to turn on the lights for two or three weeks and then suddenly change . . . it was found that discontinuing the use of lights suddenly dropped the egg production from about 30 per cent to almost zero over two weeks.[75]

In the mid-1920s poultry scientists still didn't know why lighting affected laying habits so dramatically. Nor did they know how far to push it. The traditional wisdom held that hens needed adequate rest, both daily and seasonally, and this belief was confirmed by reports of flocks that simply wore out after lights "forced" them to lay eggs all winter. "The vitality of the hens in the Northwest has been greatly reduced through forcing the birds to commence laying in the early fall," *The Hen Coop* said of Oregon poultry farms in 1922. "A bird cannot be kept off her perch from 4 am until 9 am without showing the effect of the daily grind for eggs."[76] So it was reasonable to think that too much artificial light would be akin to killing the goose that laid the golden egg. Moreover, even before scientists discovered the role of vitamin D in building strong shells and healthy chicks, it was common knowledge that hens needed regular access to sunshine and fresh air.

In the 1930s, scientific views on henhouse lighting began to shift. One series of studies at the Ohio Agricultural Experiment Station suggested that if a little extra light was good, then more was better. All-night light, in particular, was no longer considered "out of the question," but rather a convenient way to maximize hens' productivity. The station's scientists admitted in 1931 that their research, once again, simply confirmed a discovery first made by a farmer who was either lazy or careless:

> Many important contributions to science have come about by accident or by failure to observe customary practices; the present use of all-night light came about in this way. A poultryman, J. E. Morris, in southeastern Ohio initiated the practice when he disregarded the precautions generally observed in order to use natural gas for lighting the laying house. His difficulty was to turn the gas on and off automatically, as is the practice when electric light is used. The problem was solved by leaving the light on all night—surprisingly good results followed. This was in 1925 and since that time he and others in that vicinity have used all-night light with satisfactory results.[77]

The Ohio Station studies compared the wintertime eating and laying habits of Leghorns receiving no extra light, those receiving light beginning at 4 a.m., and those exposed to all-night light. Although none of the hens were very productive by modern standards, over a six-month period the birds with twenty-four-hour lighting laid twice as many eggs as those with no extra light (fifty-six versus twenty-eight; the hens with some extra light laid forty-eight eggs). The researchers noted that no birds kept up the heavy-laying for months on end, so leaving the henhouse lights on all winter would "usually mean less spring eggs." But whereas fresh eggs fetched low prices in the spring, fresh chicken meat sold well. The Ohio researchers thus suggested that selling the chickens would be a logical way to dispense with worn-out winter layers. Altogether, what made all-night lighting a worthwhile investment was not that it increased *overall* egg production but rather that it increased production during the most profitable period.[78]

For the many farms that kept barnyard flocks mostly for home use, electrifying the henhouse would have seemed like a ludicrous idea in the middle of the Depression. Simply keeping their flocks from starving was enough of a challenge.[79] For commercial poultry farmers, however, selling fresh eggs in winter seemed like a very smart survival strategy. After all, they were contending not just with the collapse in consumer demand but also with surging imports of cheap frozen eggs from China, which supplied bakers and food manufacturers.[80] So as New Deal electrification projects brought power to rural areas across the country, henhouse lighting became more commonplace, especially in major egg-producing regions. Poultry manuals and magazines provided advice on how to boost winter "henhouse morale" with the help of lamps, automatic on-off switches, and lighting schedules. Even farmers without electricity could light up their henhouses if they bought a Wincharger windmill, which cost $27.50 in 1941 and advertised immediate, fourfold increases in egg production.[81]

By that time, the January "rate of lay" had reached nearly 40 eggs per 100 layers. While this was still considerably less than the

June rate of 53.5 eggs, it marked an 80 percent increase in just fifteen years.[82] From the consumers' point of view, this was progress: fresh eggs in the winter were not nearly so rare nor costly as they once had been. From the point of view of certain egg dealers and storage firms, by contrast, this evolution spelled what *Business Week* called the "doom" of the cold storage egg trade: "Of the several factors which have forced the sales of storage eggs to a depressingly low point, the most important is the ever increasing supply of year-round fresh eggs. Millions of modernized hens, scientifically fed and housed, start laying three months earlier than their barnyard sisters. They come into production at a time when eggs are normally scarce and high-priced, and they lay twice as many eggs as those fed on grain and grasshoppers . . . Now there is only a limited demand for storage eggs, and grocers have to make a wide spread between them and fresh eggs."[83]

In response, egg dealers tried to convince the USDA that cold storage eggs should just be called *refrigerated* to eliminate their "unmerited stigma and unjustly unfavorable reputation."[84] The warehousemen's association, though, recognized that a new label could not save an obsolete commodity. "The shell egg is fading in importance," said one member at the association's fiftieth anniversary meeting in 1941. Instead, it was time to celebrate the nation's booming demand for frozen eggs.[85] Sold in cans or cartons, these were the kinds of stable, simple, uncontroversial commodities that warehousemen liked. After all, they didn't have to pretend to be "farm fresh."

The same applied to the eggs dried, powdered, and turned into military rations during World War Two. The war quickly ended the problem of oversupply; now the country needed just as many eggs as its hens could lay. Manufacturers of poultry supplies lost no opportunity to remind the nations' egg producers—the majority of them still women—of their patriotic duty. One maker of chicken feed ran ads in farm magazines portraying the enemy leaders as "faces on the henhouse door": "Crafty Hitler, cunning Hirohito, crazy Benito—let them be an everlasting reminder that we need eggs, eggs, and more eggs!"[86] Rural America obliged: after years of stagnant output in the 1930s, egg production increased 67 percent between 1938 and 1946. Consumption increased even more rapidly—in part owing to wartime meat rationing—from around 300 eggs per year per capita in the late 1930s to more than 400 at the war's end.[87]

As during the First World War, strong demand for eggs attracted newcomers to the business and new migrants to egg towns such as Petaluma.[88] By the early 1950s, however, family-run chicken farms found themselves competing against producers operating on a scale unimaginable in the pre-war era, when 400 birds still counted as a "commercial" flock.[89] These veritable "egg cities," whether run by corporations or by enterprising individuals—at this scale, mostly men—relied on machines to collect, grade, and pack the daily output of tens of thousands of layers.[90] As their economies of scale increased the all-important ratio of eggs laid per cost of feed consumed, they pushed tens of thousands of smaller producers out of the egg business within a decade. They also pushed forward pre-war developments in poultry breeding, feeding, and housing. Vitamin D–enriched feeds and ultraviolet light bulbs replaced sunshine, battery cages replaced free-range chicken houses, and the idea that hens needed exercise and fresh air gave way to the idea that continuous confinement was efficient, clean, and altogether modern.[91]

Admittedly, birds kept under these conditions were prone to bad behavior as well as to the spread of disease. But technology could fix these problems. Cutting off hens' beaks prevented them from pecking their cage neighbors to death; a regular diet of antibiotics averted epidemics while speeding up growth. The larger goal of such measures was to maximum the hens' productivity throughout their brief lives, with as little downtime as possible. And in this sense, they worked. Once hens had little or no exposure to seasonal variations in light, temperature, or feed, they obediently became full-time, year-round workers, producing around three hundred eggs a year. As the American Egg Board boasts, "Today's laying hen doesn't need to depend upon the fickle sun to tell her when laying time has arrived." Now it's *always* laying time. The USDA's charts of monthly egg-production rates, still jagged in the mid-1930s, had essentially flattened out by the mid-1970s.[92] Seasonal prices did the same, meaning that the financial incentive for long-term cold storage disappeared along with the very idea that eggs were freshest and best at Easter. The great egg warehouses were either torn down or turned into more valuable real estate, such as malls or condominiums.

High-Tech Trust?

By the early twenty-first century, the typical egg "production complex" was itself a formidable piece of property, with its eight or so hen "houses" holding more than 100,000 layers each. Companies such as Cal-Maine Farms—the largest egg producer in the United States, with a total flock of more than 20 million birds—operated complexes in rural areas across the country. But at a time when factory farming was often blamed for rural pollution, animal cruelty, and the spread of food-borne diseases (especially antibiotic-resistant forms of salmonella and other bacteria), the big egg companies no longer boasted about the scale and efficiency of their operations.[93]

Instead, they jumped into the rapidly growing specialty egg market. Although these eggs were much more expensive than conventional eggs, sales increased sevenfold between 1997 and 2003.[94] The options also multiplied. Consumers could choose between eggs that claimed to be some combination of drug-free, vegetarian, organic, Omega-3–enhanced, free-range cage-free, or "United Egg Producers certified." But with choice came confusion and doubt. Except for organic certification, the government did not enforce claims made on egg cartons. Cage-free hens might actually live in spaces as crowded as battery cages. Egg brands that sounded small and old-fashioned, such as "Farmhouse," actually belonged to mega-companies such as Cal-Maine. The United Egg Producers logo actually certified only that hens received food and water.[95]

Meanwhile, smaller companies struggled to prove that *their* eggs really came from happy and healthy flocks.[96] Pete and Gerry's of New Hampshire posted "chicken of the month" profiles on its website, along with a description of the hens' living arrangements: "Our hens are housed in cage free barns. They roam freely about the barns and have access to the outdoors when conditions are appropriate. The hens lay their eggs in nests that appear like small houses with red curtains. Our hens also have places to roost and to scratch inside the barns. When you enter the barns you are greeted by the happy sounds of our hens."

Even as egg producers reassured consumers about their hens' well being, however, egg freshness remained ambiguous. The expiration dates stamped on egg cartons did not reveal whether their contents were days or weeks old. Consumers contributed to the uncertainty by throwing out the cartons and promptly forgetting when they had bought the half-dozen on the refrigerator door shelf. But the more serious concern was about eggs that might be old even before they got home. Could you even trust those carton expiration dates? Not according to a 1998 *Dateline* report called "The Shell Game." Hidden camera footage showed workers at a large Pennsylvania egg farm taking expired eggs out of old cartons and putting them into new ones. The report warned that the practice was potentially lethal, given how bacteria proliferate over time. It was also not as uncommon as consumers would like to think.

Among the people watching that *Dateline* show was Bradley Parker, a twenty-eight-year-old investment banker who grew up spending summers on his great uncle's North Carolina egg farm. Parker knew just enough about egg handling to imagine how the "shell game" might be stopped—by imprinting each egg with an expiration date just after it was candled. He imagined that this service could be valuable, provided that the printing method was foolproof. Some European countries already dated individual eggs, but they used ink that could easily smear, splatter, or be scrubbed off. So Parker consulted engineers, tested out different approaches on his uncle's farm, and in 2001 launched Eggfusion, a company offering an entirely new way to keep track of each egg's age—and much more. Instead of ink, Eggfusion used laser beams.

The company's "laser marking systems" can be integrated into egg packhouse machinery. Besides providing tamperproof expiration dates, the systems collect and encode each egg with data about where and how it was produced (for example, organic or cage-free), storage temperature and humidity, and even the vaccination schedule of the flock from which it came. All this information is in turn fed into Eggfusion's "Freshness Network," so that it can be monitored and made available online. Technically, said Parker, "we can have accountability down to the bird," but that would require reorganizing the entire architecture of 100,000 bird laying houses. As it was, Parker recognized early on that producers would not likely pay hundreds of thousands of dollars to have their facilities retrofitted with sensors and laser guns. Instead, he proposed to offset the costs by selling advertising space on the eggshell—a medium, he pointed out, that reached into almost every American household.

The novelty of "eggvertising" won Parker's company a surge of publicity when CBS announced plans to laser-etch thirty-five million eggs with ads for TV premiers in fall 2006.[97] But eggvertising, Parker said, was "not what I want to be known for." Instead he wanted consumers, producers, retailers, and consumers to appreciate the value of the information that his technology provided. Consumers could rest easy knowing that no months-old eggs lurked on their door shelf. And if they wanted to know more about their origins, they could look up their eggs' codes on Myfreshegg.com, a site run by Parker's company. For their part, producers and retailers could gain a competitive edge just by demonstrating their commitment to freshness and safety. They could use the data Eggfusion collected to improve quality and cut costs and waste. And if their eggs were somehow either accidentally or intentionally contaminated—the Department of Homeland Security warned of packing plants' vulnerability to "agro-terrorism"—they could be immediately traced and recalled.[98]

One of the first companies to adopt the Eggfusion technology was Massachusetts-based Radlo Foods, which sold "Born Free" organic and cage-free eggs. It produced and purchased eggs in several states, packed and coded them in Pennsylvania, and shipped them as far away as Hong Kong. CEO David Radlo agreed with Parker that the technology appealed to different markets for different reasons. "Red state" consumers, he said, appreciated the security offered by traceable eggs, whereas blue state shoppers liked the assurance of freshness.[99] And if they bought premium-priced organic and cage-free eggs, they also liked knowing that this was in fact what they got. It's an open secret among egg dealers and supermarkets that some shoppers also play the shell game, for example, by switching Jumbo eggs into Medium cartons. That was one reason specialty egg companies, including Radlo Food, adopted sealable clear plastic cartons. Consumers could look but not touch.

In an interview with *Food Safety* magazine, Radlo emphasized that more work was needed on this front. "Human touch is where problems happen," he said. "Ideally we'd like to get to the point where the eggs drop onto a conveyor, where they are cleaned and dried, placed into cartons and sealed without being touched by anyone but the consumer who buys the product." It seems like an ironic ideal for eggs advertised as coming from "American Family Farms." But it is not a surprising one, given the egg's unrevealing nature and its history of dishonest handling. It is also not surprising given that there are only so many ways to add value to eggs without taking them out of the shells. And then, of course, they no longer count as fresh. Arguably the tamperproof, laser-coded, traceable egg is to be expected from an industry that has always had to reassure consumers about qualities they cannot see or even necessarily taste. Eggfusion's motto is "promoting freshness with every impression." But the company does not by itself get eggs to market any faster. Instead the coded shells are meant to make us trust the egg's innocence, a virtue as important as (even conflated with) its youth. They promote the impression of freshness, and that is ultimately what sells eggs.

four

Fruit

EPHEMERAL BEAUTY

These days, it's not enough to simply produce fruit; one must obtain fruit that is beautiful and, as much as possible, good-tasting. I say "as much as possible" because for certain fruits . . . good taste is not necessary—just beauty, gloss, and size. They are really more intended to dazzle the eyes than to satisfy the palate.

—Gustave Rivière, "Des moyens en usage pour accroître le volume des fruits et exalter leur coloris" (1894)

The Whole Foods store at Columbus Circle in Manhattan is a little unusual for an American supermarket. For one, it's in the basement of a glitzy vertical mall. An earnest, smiling employee welcomes shoppers as they descend into the cavernous store. A first-timer might briefly wonder why a supermarket needs a greeter. Maybe to keep up with the neighbors? No matter. Shoppers' attention shifts quickly at the bottom of the escalator where, just past the flowers, the fresh fruits beckon. The light flatters their curves and colors. Shapely pears, rouged peaches, dewy grapes; all beg to fill your shopping cart with their sweetness.

Actually, with the exception of the greeter you don't need to go to the Columbus Circle Whole Foods for this titillating experience. Most supermarkets these days put their fruit up front. When I asked produce managers at supermarkets around the country why they did this, they said things like, "the colors attract people," "it gets you to buy something not on your shopping list," and "it brings out a fresh image." And the fresher the image, the better a store does. Market research has shown that, all other things equal, consumers choose supermarkets where the fresh produce—and especially the fruit—looks best.[1]

But good looks obviously don't guarantee good taste. At one time or another we've all fallen for the beautiful and the bland. We've grown used to disappointing supermarket fruit—mushy waxed apples and giant watery berries, peaches that go straight from rock-hard to rotten—just as we've grown used to blaming the supermarkets for our disappointment. After all, it's their long-distance distribution systems that demand varieties as tough and durable as baseballs. It's their experts who have figured out that we shop for fresh fruit with crow's eyes, zeroing in on big, bright, shiny objects. What many people don't realize is that consumers have fallen for the beauty of fresh fruit for centuries—that is, for just as long as producers have tried to sell it.

This history doesn't absolve today's mass retailers of all blame for blandness. It does show that fruit's seductive powers have served different, though not entirely unrelated, purposes in different periods. For a long while beauty added value to fresh fruit simply because it was an ephemeral and thus exclusive quality. At the same time, to cultivate and own such beauty showed Taste, regardless of how it tasted. But once refrigeration and steam-powered transport began to bring city-dwellers fresh fruit from ever more distant hinterlands, beauty took on a more crassly commercial value. It became a way for growers to stand out in crowded markets, to stay ahead of tax and mortgage collectors, to stave off the paving of orchards—at least for a few years. In a world where other industries produced plenty of ugliness, growers pushed beauty as a mark of freshness. Consumers bought it. The results greet us today in the supermarket.

Eve's Science

"And the woman saw that the tree was good for food and pleasant to the eyes, and desired to gain knowledge." Adam and Eve were not the first to fall for the beauty of fruit. It is arguably the world's oldest marketing tool—far older than supermarkets, the Garden of Eden story, and indeed humankind itself. For most of their history, trees and other fruiting plants spread their seeds by convincing birds, mammals, and reptiles to carry them off, then drop or defecate them somewhere suitable for germination. To do this, plants developed fruits that not only protected their seeds but also, and more important, appealed to the species on hand, whether they were earth-bound turtles, color-blind but sharp-nosed bears, or color-sighted birds. In turn, animals' organs and instincts developed to maximize this sweet nourishment. Plant and animal, as food and feeder, co-evolved to make the most of each other's primal needs.[2]

Humans and other primates entered the garden late. Like other animals, we learned to identify ripeness by size and color and scent, and we learned to avoid fruit whose hard sour flesh protected immature seeds. But we also figured out that with fire and other tools, we could eat different parts of plants, as well as animals. These foods provided nutrients that fruit lacked, and many of them could be hunted or harvested during seasons when fruit was scarce. Grains in particular kept much longer than fruit. So despite fruit's accessible and willing nature—it is, after all, the only part of the plant that evolved to be eaten—most early agrarian societies relegated it to the margins of their diets. Fruit figured more importantly in the food supply of hunter-gatherers, especially in tropical and subtropical regions, but most didn't depend on it for the bulk of their calories.[3] In the Genesis story, it made sense for Eve to tempt Adam with an apple precisely because at the time fruit was a seasonal treat, not a staple. A wheat stalk probably wouldn't have had the same effect.

A couple of millennia later California's fruit growers begin to convince consumers that they should in fact eat fruit daily. Like poets and artists of past centuries, they portrayed fresh fruit as a symbol of gorgeous youth. Real fruit, though, appeals least when young. That's because its purpose shifts from protective to seductive only after its seeds have matured enough to germinate. Then it lets its defenses down, inviting the birds and bears and humans to carry it away. Ripening marks the last, most frenetic phase of a fruit's life, one that often continues after harvest. It's a time when enzymes break down starches into sugars, chlorophyll into colors, and hard pectin into soft. Cells hyperventilate, sucking up oxygen and spewing out carbon dioxide at an accelerated rate. Fragrant and hormone-fueled, ripening fruit asks to be devoured before it decays.[4]

Thus fruit signals when it's ripe and also when it's rotten. Whether it's truly fresh isn't so obvious—or always better. As long as their skins are intact, most fruits don't spoil as quickly as, say, milk or seafood, and they don't wither or wilt as fast as green vegetables. Some, like pears, need time after picking to fully ripen. In others, the qualities that consumers associate with fresh "taste"— namely, texture and moisture—don't depend on age alone.[5] They expect a fresh apple to be crisp, a fresh orange juicy. Fruits do dry out over time, but some varieties enjoy a big head start. A Granny Smith apple, for example, will have more crunch after six months than a Roma just off the tree.

The best fruit, then, isn't necessarily the absolute freshest. In that sense it's more like beef than fish or eggs or an ear of sweet corn. On the other hand, many fruits are more difficult to *keep* at their best than other perishables. This is partly because ripening is hard to stop once it starts. Refrigeration alone doesn't do the trick. That's one reason many branches of the fruit industry now rely on controlled atmosphere (CA) storage and packaging. CA technologies arrest ripening by keeping oxygen and temperature low and carbon dioxide levels relatively high. First widely adopted in the late 1940s, CA storage is most commonly used to keep apples crisp and to prevent "core rot" in varieties such as the Macintosh.[6] How well CA preserves flavor is debatable. Some apple growers claim that certain varieties get sweeter if they sit for a month or so in a cool place; few would say that about the many apples stored for six or eight months.[7]

Another reason refrigeration alone never solved the problem of fruit preservation is that many varieties, especially those native to the tropics or subtropics, tolerate cold poorly. Either they develop a form of scald, or their skins darken, or their flesh turns mealy. Contemporary growers and dealers typically store frigophobic fruits at only a few degrees below room temperature, and consumers are advised not to put them in the fridge for long, if at all. But in the early days of cold storage, researchers claimed that even peaches kept well for two months at near-freezing temperatures. These scientists must not have eaten their findings. An American fruit dealer, writing in 1892, reported more likely results after cold storing his peaches for three weeks: "They still looked pretty, but upon sampling them I was surprised to find that the juice had apparently all gone . . . leaving them as tasteless as raw potatoes."[8]

Fruit as Fashion

Even before the Old Testament blamed a pretty apple for humankind's fall from innocence, the Romans portrayed the fruit goddess Pomona as a dewy beauty, on par with her sister Flora, the goddess of flowers. They also propagated Pomona's gifts. The patricians grew apples, pears, pomegranates, and figs on their own farms, while the empire offered both new varieties and new lands for growing the old ones. England's oldest apples, for example, probably arrived during Roman times. Like much more ancient civilizations, the Romans relied on grafting to develop bigger, sweeter fruit, and on drying and fermentation to preserve their surpluses. In *De Agricultura* Cato the Elder instructs readers on how to make raisins and how to pickle pears in boiled wine.[9] The Romans also appreciated raw fruit, especially as the final course of a long and heavy banquet.

The Romans planted fruit trees across Western Europe. But the orchards declined along with the empire, their bitter fruit feeding the notion that apples especially were poisonous. Although people continued to eat fruit through the Middle Ages, suspicions about its wholesomeness were not entirely unfounded. Fruits sold in town marketplaces were often either unripe, overripe, or contaminated by the flies that swarmed around open cesspools. Not surprisingly, as wealthy urbanites began to adorn their tables with fruit in the late Middle Ages, the era's dieticians prescribed strict (though not necessarily effective) rules for avoiding digestive ills. Fruits classified by humoral theory as "cold"—peaches, plums, cherries, grapes, apricots, figs, and melons—were only supposed to be eaten before a meal, and apples, quince, and pears only at the end. *Le trésor de santé* recommended following each pear with a "good glass of old wine."[10]

By the late seventeenth century, fresh fruit had become *de rigueur* on elite tables, both as dessert and as display. In France, calling dessert *le fruit* was a sign of sophistication, regardless of what was served. Fresh fruit's rising popularity reflected partly what the French food historian Jean-Louis Flandrin called "the liberation of *gourmandise*" from medieval medical and religious norms. Those who could afford to eat for pleasure, ignoring centuries-old ideas about bodily humors and digestion, increasingly did so. And as more people ate raw fruit without getting sick, medical opinion gradually shifted, albeit not permanently.[11]

Fresh fruit was at least as fashionable to grow as it was to eat. Among the Parisian nobility and the upper bourgeoisie, it showed *le bon goût* to keep a walled orchard at one's chateau on the outskirts of the capital, and to treat friends to the most perfect fruits of each harvest. Louis XIV set an example at Versailles, where scores of elegant apple, pear, and peach espaliers trained the walls of the vast Jardin du Roi. His head gardener, Jean de La Quintinerie, became an international authority on fruit-tree pruning and garden design. He had orangeries built for the king's citrus collection, as well as climate-controlled fruit storehouses, ensuring that Louis and his guests could enjoy the garden's harvest year round.[12] That didn't always mean eating the fruit: pyramids of apples, pears, or oranges were, as La Quintinerie noted in 1690, "truly *à la mode*" as table décor.[13]

Enterprising Parisian landholders also contracted with tenant farmers to grow luxury fruit on a commercial scale. While different varieties went in and out of style, fruit fashion generally favored the delicate, the out-of-season, and the gorgeous. Under Louis XIV, most of the hardy apple varieties grown on peasant farms were dismissed as too common for a noble's garden or banquet table. More suitable were varieties such as the thin-skinned, pearly pale Calville de Blanc, or pears, which ranked among the king's favorite fruits. By the mid-eighteenth century, the peach had deposed the pear. The very fact that it was a semitropical fruit, difficult to grow in the climate of northern Europe except in well-protected gardens, added to its elite appeal. "No other tree climbs the walls so agreeably, or offers fruit so brilliant to the eye and pleasing to the palate," wrote Charles-Jean De Combles in his 1745 *Treatise on the Culture of Peaches*. He also observed that some varieties had merit despite their lack of flavor. Although the Pavie de Pompone was "worth little, when it comes to taste," he recommended planting a few trees because its "monstrous size and beautiful coloring" made it a perfect table ornament.[14]

French court society was not alone in its passion for fruit. In England, Ralph Austen's *Treatise of Fruit Trees* described orchard-keeping as a spiritual enterprise with earthly benefits. The front cover showed "Pleasures" and "Profits" shaking hands, while the text explained how fruit trees improved land value, impressed friends, and ensured their owner a good and enduring reputation. As he put it, "Planting fruit-trees is profitable to the Name of the Planter. Fruit trees are living, lasting monuments . . . One way to gaine, and keep a good name is by gifts . . . Now who can so easily give so great, so many and so acceptable gifts as the husbandman . . . while he lives he bestows young fruit trees, fruits and many acceptable gifts out of his Orchard, and being dead, his Orchards, Gardens and fruit-trees live, and flourish, and occasion a remembrance of his Name, for many ages after him."[15]

The big problem with showing off fruit to friends—never mind posterity—was that it just didn't last. One solution was to hire a botanical illustrator to capture the fleeting beauty of the fruit in a form suitable for framing and reproduction. In the days before photography, these illustrators possessed skills valued as much by science as by high society. Trained to sketch plants fast and precisely, they often accompanied botanists on their exploratory voyages. Back at home, they painted the fruits and flowers of the wealthy. The most esteemed illustrators' works appeared in lavish, coffee-table–style books such as George Brookshaw's *Pomona Brittanica,* William Hooker's *Pomona Londoninensis,* and Pierre Redoubté's *Un choix des plus belles fleurs . . . et des plus beaux fruits* ("A selection of the most beautiful flowers . . . and some of the most beautiful fruit"). As the last title suggests, a successful illustrator (Redoubté's clients included Marie Antoinette and Empress Josephine) knew how to make each painted fruit look better than freshplucked.[16] These images flattered their patrons while adding to the luster of fruit growing as a high-class hobby. Did they also shape the elite's standards of fruit beauty, just as portraiture and other art forms have long influenced standards of human beauty? It's not hard to imagine, given that reproductions of these images circulated widely, perhaps more widely than the fruit itself.

The Montreuil Mystique

Perishable, seasonal, fragile, and exotic, the peaches and pears grown in the private gardens and orchards outside Paris testified to the power, money, and cultivated tastes of the city's landed aristocracy. Yet some of the most coveted fresh fruit sold in eighteenth-century Paris was produced by the apparently illiterate peasants of Montreuil, on the city's northeastern edge. The abbot Jean-Roger Schabol wrote at length about the villagers' unique gardens. The Montreuillois enclosed them in walls of plaster—a material that absorbs heat much more effectively than brick—and oriented them all north-south, so as to capture the most sunlight. "These industrious inhabitants have found the secret," marveled Schabol, "for catching and keeping the sun rays in each corner . . . and for holding on to their heat." Indeed, both day and night the gardens were warmer than their surroundings by several degrees Celsius. In this microclimate Mediterranean fruits thrived. Peaches ripened a month before others on the market, when prices were still sky-high. In addition, the villagers trained their espaliers to stretch out across the east-facing walls like giant fans cradling each peach in a perpetual sheltered sunbath. This design produced not only unusually big and beautiful fruits but also more of them from each tree.[17]

The abbot's accounts of the Montreuil gardens set off a controversy among French horticulturalists, some of whom suspected that the Montreuillois had stolen their ideas from the king's garden at Versailles, where some of them once worked. But even those who doubted the Montreuillois's originality were impressed by their sheer output. As one observer noted, "It's a really interesting spectacle to look down from the surrounding hillsides on this immense multitude of gardens, carved up every which way by walls covered with trees and verdant vines. You think you're looking at a hive of bees . . . And what image better describes this colony of hardworking, intelligent people, to whom the Capital owes, in part, its joys at the table. Plums, pears, cherries, chasselas (grapes), apricots; all that can be sold for a profit they cultivate."[18]

This author neglected to mention Montreuil's peaches, which were considered among the best in France and later, some would argue, the world. Carefully brushed free of fuzz and tissue-wrapped, they sold by the piece for princely sums. Wealthy Parisians were their main customers and, with the help of hired gardeners, their main imitators. Peach espaliers *à la mode de Montreuil* became all the rage among what one author called "the rich and civilized." The fame of the village's peaches spread even to Britain, where the director of Kew Gardens, Joseph Banks, described Montreuil as the one place where "the true management of this delicious fruit can be studied and attained."[19] But the Montreuillois were notoriously secretive—realizing, no doubt, that their monopoly on the Parisian market for the high-end peach was as fragile as the fruit itself.

Fresh New World

In the newly independent United States, meanwhile, fresh table fruit was just beginning to find its way into city marketplaces. It took a long time, given the abundance and variety of fruits that grew in colonial North America. When the first settlers arrived, they discovered not just native persimmons, plums, grapes, and many kinds of berries, but also Indian gardens full of peaches and watermelons—fruits brought long before from Asia and Africa, respectively, by the Spanish and their slaves. They also discovered soil that was almost *too* rich for varieties accustomed to the long-tilled dirt of Old World gardens. One late seventeenth-century visitor, observing "limbs torn to pieces" in a New Jersey orchard, remarked that "the fruit trees in this country destroy themselves by the very weight of their fruit."[20]

From an early date, such orchards dotted the rural landscape of colonial North America. In Jamestown, landowners were even required by law to plant orchards, so as to reduce the colony's dependence on imported foodstuffs. In eastern New York, poor tenants tended the fruit trees of wealthy estate owners; further south, the same work fell to slaves. Even the burghers of New Amsterdam, in what is today lower Manhattan, maintained peach orchards (also, apparently, worked by slaves).[21] But whether the orchards were large or small, most of their fruit did not get eaten fresh—or at least not by humans. Instead, farmers distilled it into drinks more potable than the local water and more durable than the fruit itself. Cider was by far the most common beverage, but owners of larger orchards also made all kinds of brandy and fruit beer, and those with vineyards made wine. Besides alcohol, farming communities dried fruit to use in pies and dumplings and, in the case of the Pennsylvania Dutch, to ship overseas. For French settlers in eastern New York, apple butter was considered "a great delicacy," serving as a substitute for both sugar and real butter.[22]

In addition, many farm families kept at least a few apple or peach trees for the express purpose of fattening hogs. If today it's a rare and lucky pig that feeds on fresh-fallen orchard fruit, at the time it was an efficient way to turn a soon-to-rot surplus into a sturdy staple fuel, namely, salt pork. Plus, much of the fruit the pigs ate probably didn't taste or look very good to their owners. On all but the wealthiest farms, the trees were seedling rather than grafted varieties, and they received little if any care. Most families couldn't spare the labor. But even if they could, they had no practical way to get their fruit to a market where it might fetch a worthwhile price. As the New York farmer and author Crèvecoeur said of his own harvests, "Perhaps you may want to know what it is we do with so many apples. It is not for cider. God knows, situated as we are it would not quit cost to transport it even twenty miles. Many barrel have I sold at the press for a half dollar. As soon as our hogs are done with the peaches we turn them into our apple orchards."[23]

North America's weather, extreme and volatile by European

standards, also discouraged commercial fruit growing. Even in the relatively mild Carolinas, a few unexpected freezes destroyed most of the colonies' prized orange and lemon trees. In the northern reaches of the frontier, only fruit varieties imported from Russia could survive the winters.[24] Then again, a lack of fruit was probably not these farmers' greatest concern, except when it deprived them of homebrew. Most came from countries where the common people had not traditionally eaten much fresh fruit anyway, except for a few brief weeks in summer and fall. Even the French communities in New York and the Great Lakes regions, known for their giant, prolific pear trees, no doubt consumed the bulk of their fruit in dried, baked, distilled, or otherwise unfresh forms.

In the early nineteenth century the value of freshness in fruit began to appreciate rapidly in the United States, at least in and around major cities. Partly this reflected dietary fashion. At a time of warm Franco-American relations, the country's early leaders and young urban sophisticates were enamored with French cuisine, including the habit of eating fresh fruit at mealtime.[25] As in Europe, exotics were prized and pricey; hothouse pineapples might cost two or three dollars apiece. Esteemed varieties of pears and peaches went for top dollar in the chi-chi shops of New York, Boston, and Philadelphia. Outside the cities, orchards provided gentlemen farmers with a fashionable pastime—horticultural societies, expositions, and journals were all thriving by mid-century—and commercial farmers with the income needed to offset rising land prices.[26]

For these suburban fruit growers, the advantages of proximity eroded as transport improved. In the 1830s the railroads brought Manhattan fruit from increasingly distant hinterlands: first New Jersey and the Hudson River Valley, then Virginia, the Carolinas, and Georgia. The peach and strawberry seasons stretched from weeks to months, while steamers brought lemons and oranges from Italy and Florida in the dark of winter. All the cargo traveled unrefrigerated, and shipments were often delayed. Nonetheless, the railroad's arrival set off a fruit-planting explosion, especially in the U.S. South. Only after farmers began losing fortunes on spoiled shipments did they start to appreciate that it took more than a boxcar to get peaches from Norfolk to New York in sellable condition.[27] Then came the summer and fall gluts, when prices dropped so low—even in the supposedly top-dollar big-city markets—that farmers let their fruit rot rather than pay to have it picked or shipped. Urban demand was more limited than they realized, and probably not helped by the widespread belief that fruit carried diseases such as cholera.[28] By the time such fears had faded and refrigeration had eased the most acute hazards of fresh produce marketing, many of the South's boom growers had already gone bust. By that time as well, the railroads had reached the West Coast, entirely transforming the country's commercial fruit landscape.

Gold Hunters

The Spanish introduced many fruits to the New World, but few places gave them a warmer welcome than Alta California. The winters were mild, the valleys fertile, and irrigation turned even the driest regions green. In the eighteenth and early nineteenth century, Indian workers in the Franciscan mission gardens tended just about all the fruits found in Spain, and then some: grapes, apples, pears, peaches, apricots, olives, plums, cherries, figs, pomegranates, oranges, and even bananas in San Diego.[29] Compared with settlers in the colder parts of North America, the *padres* ate well, enjoying an abundance of fresh produce year round. But colonial California was sparsely populated and impossibly far from major urban markets. Apart from passing sailors and trappers, the missions' fruit crops found few buyers.

Then came the Gold Rush, and hungry men arrived by the boatload. Food prices in general skyrocketed, but fruit's value as a safeguard against scurvy made it an especially precious commodity. A dozen cherries might cost a dollar, a peach three dollars. Scrambling to meet demand, merchants shipped grapes by boat from Los Angeles to San Francisco, and farmers in Santa Clara and San Jose grafted fresh branches onto the mission gardens' old pear trees. Nurseries from Oregon to Europe did a brisk business shipping seedlings to California; their customers preferred dwarf varieties because they bore fruit fastest. Back on the East Coast, news of California growers' earnings seemed only slightly less incredible than reports of their huge fruits and prodigious trees. Letters from the Golden State described peaches that measured twelve inches around, strawberries as big as plums, and melon vines that yielded $10,000 an acre.[30] Even if such claims were exaggerated, the New England–based journal *Horticulturist* noted in the late 1850s, it seemed clear that California was not just a pleasant place to grow fruit but also a profitable one.

Yet even as the gold grew scarce, California's growers planted as many trees as their orchards could hold. In 1857, one observer predicted a day "not far distant when fruit will be an important crop for raising and fattening swine." And he was right, for by 1861 peaches cost a penny a pound, and many rotted where they fell.[31] Prices eventually recovered, but the glut demonstrated once again fruit growers' vulnerability to boom-bust cycles. Growers had to invest thousands of dollars just to prepare their vineyards or orchards, and then wait three to five years before they yielded any returns. They couldn't easily change course if market prices soured, nor, once their fruit began to ripen, could they wait for prices to improve. Not surprisingly, California's early growers embraced fruit preservation as a means of self-preservation. Given the favorable climate for sun-drying raisins and prunes, some anticipated that the state would soon dominate the world market in dried fruit.[32] But its farmers weren't yet in any position, geographically or otherwise, to make a living off freshness.

This situation began to change once the railroads reached California in 1869. They soon became not just the state's vital link to East Coast markets but also, thanks to federal land grants, its biggest property owners and best-funded boosters. Southern Pacific (SP) alone held title to 11.5 million acres, and it hoped to sell many of them as orchards—for what better way to ensure future shipments of high-value cargo?[33] To do this the SP established fruit "colonies" complete with irrigated plots and free on-site technical assistance. As one developer explained, "few settlers would have the requisite special knowledge of fruit culture."[34] Few such settlers, in fact, knew anything about farming, period. They were lawyers, dentists, teachers, businesspeople, and retirees. They did not want a farmer's life so much as they wanted the lifestyle of the California fruit grower.

They learned about this lifestyle from popular magazines (especially *Sunset,* founded by the Southern Pacific in 1898), books and pamphlets published by local governments (with titles such as *Solano County: Land of Fruit, Grain and Money,* and *Campbell: The Orchard City*), and fruit-filled demonstration rail cars, such as "Placer County on Wheels," which toured the country in 1891. The San Francisco publisher William Nutting employed cutting-edge color lithography to illustrate *The Vacaville Early Fruit District* so that readers would appreciate the state's true beauty. "Its fresh, vigorous and truthful accounts," he wrote in the preface, "will be as fascinating as the *Arabian Nights* to the thousands who are looking with longing eyes to our great state from their land of blizzards, cyclones, thunderstorms and sunstrokes." More concretely, he expected the full-color depictions of prosperous orchard communities to "help fill all California with happy, healthy, intelligent fruit growers."[35]

Whatever the medium, the message was clear: California fruit growers enjoyed health, wealth, and the world's best climate—provided they had some wealth when they arrived. As Nutting warned, those with "brains and ambition but no money had better stay

East." But the returns looked unbeatable. Prospective buyers heard stories of land purchased for $100 per acre that yielded $1,000 per acre in fruit sales. Could such profits last? Given the hundreds of thousands of immigrants arriving in the United States each year— many from countries where people ate a lot of fruit, such as Italy— the state's Board of Trade forecast a sunny future. "Overproduction we do not think at all likely to occur," it said of the fruit industry in 1889. "There is much truth about the wonderful products of California that is stranger than fiction."[36]

Portrayals of dirt and toil were strangely absent from booster pamphlets like Nutting's. Instead, they emphasized that hired help was cheap and readily available. As Nutting put it, "The fact that Chinamen can be had for the disagreeable parts of the work makes fruit-growing more profitable and easier to carry on than if such were not the case."[37] In later years Filipino and Mexican laborers largely replaced the Chinese, but the work itself changed relatively little. Even as other kinds of agriculture became increasingly mechanized, fruit trees still had to be hand-tended and harvested. And growers' prosperity still depended, more than they wanted to admit, on access to a cheap and seasonal labor force.

In the boom years of the late nineteenth century, many of California's new arrivals were too busy buying land to worry about the cost of labor. Real estate prices soared, as did the acreage devoted to fruit.[38] Even speculators who had no plans to settle on their land planted orchards. So did wheat farmers, once they realized that it no longer paid to grow grain in the Golden State. Improvements in cross-country transport encouraged the switch to high-value perishable crops. By the century's end, 95 percent of the state's deciduous fresh fruit shipments traveled in ice-packed refrigerator cars and reached the East Coast in only ten days.[39] Between 1880 and 1895, fresh fruit shipments increased fifteen-fold; citrus shipments alone increased twenty-five–fold.[40] Within three decades California had gone from the eighth-ranked fruit producer in the United States to the first, and fruit had become the state's top industry, earning $28 million a year. The wealth flowed into growers' communities such as Riverside, for a while the richest town per capita in the United States. Tourists marveled at its grand mansions and manicured parks—evidence, the *California Citrograph* later said, that "a deep appreciation for the beautiful is not inconsistent with commercial success in orange growing."[41]

Beauty Contests

By putting so much fresh fruit on the market, California growers helped to transform the American diet. Their citrus crops especially made wintertime fruit a less exotic commodity; for many turn-of-the-century households, oranges weren't just for Christmas anymore. Greeted by increasing abundance and variety in the marketplace, American consumers ate more fresh fruit even before they learned of its nutritional benefits. One woman, testifying before the Massachusetts cost-of-living commission in 1910, described this new desire as almost mystical: "We are all grown luxurious; little by little it has come about, and we are almost unconscious that a change has been made. When I first began to keep house ten years ago, we ate cereal, eggs and coffee for breakfast, with fruit occasionally instead of cereal; but now we must have grape fruit every morning . . . We are no better off as far as being nourished goes, but we just somehow want it, and two grapefruit cost as much as did the whole breakfast ten years ago."[42]

It wasn't the sheer volume of fresh fruit that made consumers "just somehow want it," but rather how California growers managed and marketed its freshness. Early on it became clear that they needed to take drastic measures on both fronts, because refrigeration had not solved all their problems. For starters, it did not guarantee that their produce would actually reach the East Coast in sellable condition. Delayed and misrouted shipments were frequent and ruinous. Even without scheduling problems, crates that started the journey containing a few bruised or nicked fruits often arrived a fuzzy, rotten mess. Nor did refrigeration get rid of glut. On the contrary, it became a national rather than a local plague, as the spread of iced railcar service inspired orchard planting everywhere from Washington to Michigan to Georgia. Shipments were entirely uncoordinated, so the fruit wholesale docks in New York or Boston could be bare one week and buried in overripe peaches the next. Not least, growers' distance from their markets left them dependent on wholesale commission agents to find buyers. These merchants only kept the records that suited them. They could easily lie about the condition of the fruit on arrival, or how much it sold for, or whether it sold at all. When fruit was abundant, they often accepted kickbacks from some shippers and let others' deliveries perish.[43]

In short, the natural advantages of fruit production in California offered no protection against the hazards of long-distance distribution. The state's growers needed, as one trade official told them in 1897, to "organize and advertise."[44] Some had already organized to lobby for certain common interests, such as express freight trains and subsidized irrigation. But no group invested more in advertising the state's fruit than the California Fruit Growers Exchange (CFGE), founded in 1893 by a group of Riverside orange growers. Although technically a growers' cooperative, it operated more like a corporation. A number of its founders came from business backgrounds; others practiced law or medicine. Whatever their experience, they fit *Sunset* magazine's description of the orange grower as less a farmer than a manufacturer "whose raw material . . . is transformed into the finished product by living trees instead of machinery."[45]

One of their top priorities was to make sure their finished products, like other manufactured goods, could ship to distant markets without deteriorating en route. In 1904 they hired the USDA's top refrigeration expert, G. Harold Powell. His recommendations were simple—pick and pack only the best fruit, and pre-cool it before shipping—but apparently so effective that the CFGE made him its general manager. Under his leadership the group became a massive, vertically integrated international enterprise. It bought California forest land and manufactured orange crates from its own timber. It built semimechanized packhouses, developed markets for citrus by-products, and maintained sales offices nationwide as well as in Canada and Britain. And in 1907 it launched a brand: Sunkist.[46]

Consumers by that time were accustomed to brand-name tea, cereal, and other packaged foods. But branded fresh fruit was unheard of. So was fresh fruit advertising, which the CFGE first tested in Iowa. Bombarded with $6,000 worth of publicity in 1907 —mostly banners and train cars emblazoned with oranges and the refrain "Oranges for health—California for wealth"—Iowans ate 50 percent more oranges after the ads than before. Not surprisingly, the next season the CFGE expanded its advertising budget. A few years later it hired the twenty-four-year-old whiz-kid ad man Don Francisco. One of Francisco's first tasks was to advise the co-op's members on the design of their crate labels. Traditionally shippers used images that they personally liked—pictures of wild animals or pets, for example, or California scenery. But Francisco told them, "Our advertisements must sell fruit—not land or railroad tickets."[47] He recommended simpler designs and more suggestive brands, such as "Have One," "Handsum," and "Habit." Not surprisingly, many of the labels featured images of seductive women holding equally enticing fruit.[48] Over and over, Francisco emphasized that looks mattered: "Every illustration is calculated to arouse a desire to buy or eat oranges . . . [and] in advertising things to be eaten, beauty is not to be sacrificed."[49]

By the end of World War One, the CFGE spent a million dollars annually on Sunkist advertising. In 1919–1920, full-page color ads appeared in more than 119 million copies of nine major maga-

From an early date, California orange growers found that uniformity and flawless beauty helped sell fruit.

zines, most of them aimed at women, the main food buyers. That year's campaign also included newspaper ads, posters, billboards, recipe books, calendars, and a movie, *The Story of the Orange*, that screened in 2,000 cinemas. In addition, the CFGE's salesmen encouraged grocers to present Sunkist-branded fruit as its own best advertisement. "With very few products is the value of fresh, attractive displays so necessary and effective a sales factor as with fruits," Francisco wrote. "A display of fruit itself is better than a display of printed cards."[50]

Francisco recognized that some co-op members were skeptical

about paying to support such campaigns. So he made his case in terms of real estate, something that concerned them all. He argued that advertising protected growers' "enormous investments" in orchards and vineyards simply by driving up the value of their produce. Since the CFGE began pushing the Sunkist brand, he wrote in 1920, prices for fruit-growing land had increased in value by 500 percent in some areas. Advertising ultimately did more than sell fruit, he said; it "uplifted the grower's daily life."[51]

More basically, advertising served as "fertilizer to better the soil of consumption."[52] And in an era when this soil was still nearly untouched, the fertilizer worked. By the late 1920s, Francisco boasted, the orange had "dethroned the apple as 'King of Fruits.'" The average American ate fifty-five of them a year, up from thirty-two a year before Sunkist began its ad campaigns.[53] Increased nutritional awareness helped. The citrus industry contributed to this awareness at least as much as it benefited from it, billing oranges as protection against everything from colds to bad moods to a vague digestive syndrome known as "acidosis."[54]

Other California growers' groups imitated Sunkist's promotional tactics. The California Pear Growers Association, for example, rounded up scientific studies showing that Bartletts contained iron, calcium, phosphorus, and vitamins B, C, and G (the "growth vitamin"). Market research also found that consumers probably didn't need all this detail up front. Since they already preferred eating pears fresh and "out-of-hand" (rather than cooked), the research recommended ads featuring simply the Bartlett, "whole and life-size—in its most attractive form."[55] Though modest in its scope and health claims compared with Sunkist advertising, the pear growers' publicity got results. After blitzing Philadelphia and Boston with ads and storefront displays in 1921, the two cities' share of the East Coast fresh pear market doubled.[56] Back in California, meanwhile, fruit growers' demand for labels and ad copy helped turn San Francisco into the capital of color lithography. Firms such

The California Pear Growers Association sponsored contests among East Coast grocers, offering cash prizes for the most attractive displays of California Bartletts. (*California Pear Grower,* September 1922)

as Truang Lithograph and the Schmidt Lithograph Company ran their own ads in growers' journals, emphasizing the "pulling power" of their designs.

So advertising worked, but it couldn't work magic. An ad that created desire for fruit worked only if the *real* fruit did not disappoint. Otherwise, Francisco and other ad men warned, the brand became a "buyer's danger signal."[57] Growers obviously couldn't guarantee consumers the perfect health or domestic bliss portrayed in some of their publicity. But they could strive for fruit as pretty and tempting as the ads. Marketing experts suggested that they really had no choice. Especially in big urban markets, the USDA said in a report titled "Consumers' Fancies," "the appearance of an apple is everything and taste nothing, unless the purchaser was once a country boy and enjoyed the freedom of an orchard . . . Since the farmer supplies townspeople and city people . . . he must not govern himself in his business operations by standards based upon country life and country living. He must be prepared to raise pretty red apples stuffed with cotton if his customers want them."[58]

To be fair, growers didn't neglect taste altogether. They took measures to prevent decay, dehydration, and cold damage. They bred varieties for higher sugar content. But they came to accept that such efforts would go unappreciated unless their fruit conformed to what one USDA report called the "general principles" of successful food marketing: "Gloss, polish and luster are wanted. Things should be large and, when applicable, of plump appearance; they should be uniform in size, shapely, and with ornamental lines. A convenient and showy package is appreciated." "Remember," the report concluded, "one of the great buyers of the products of the farm is the human eye."[59] But the eye was a tough customer when it came to fresh fruit. It wanted perfect, predictable beauty, the kind that did not actually grow on trees. As one CFGE ad man pointed out, "Nature is a notoriously poor standardizer."[60] It's also full of creatures that can quickly render fruit ugly if not inedible.

The CFGE and other cooperatives pursued two strategies for making sure that Nature didn't interfere with marketing. First, they established strict quality grades based on size, shape, color, and degree of flawlessness, and permitted only the best fruit to bear the brand label. Because a brand promised not just high but also standardized quality, the CFGE went to great lengths to ensure that its members graded their fruit correctly. It inspected orchards, packhouses, and crates, and urged constant vigilance, lest a few bad oranges destroy the Sunkist brand. Another growers' co-op, the California Fruit Exchange, adopted the motto "Every grower an inspector."[61] California growers' organizations also welcomed the federal government's adoption of food grades after the First World War. National standards for a No. 1 orange or pear, they reasoned, would help highlight the superiority of California products.[62]

Second, California growers pursued aggressive pest control. As Steven Stoll shows in *The Fruits of Natural Advantage,* California looked less like a fruit farmers' paradise once the pests, many imported along with the seedlings, found their way into the monoculture orchards:

> The enemy was ubiquitous and formidable. Pear and quince trees were known to host fifteen different pests that attacked their leaves, buds, flowers, and fruit. Grapevines had fourteen known enemies including San Jose scale, grape scale, black rot, flea beetle, berry moth, and leafhopper, as well as phylloxera . . . The many other insects laying eggs in groves up and down the state had names that growers learned to fear: woolly apple aphid, codling moth, cottony cushion scale, red scale, red spider mite, and Hessian fly—all introduced into California since the Gold Rush and all economically threatening.[63]

In the early years, growers had few weapons to combat these foes. So when cottony cushion scale infested citrus groves across southern California in the 1870s, they pooled their funds and hired a USDA entomologist. Their peers in Monterey and Santa Cruz counties did the same when codling moth attacked.[64] Growers also lobbied for state-supported pest-control research at the University of California and defended their use of some of the more toxic innovations, such as lead arsenate (otherwise known as Ortho).[65] When East Coast health inspectors reported arsenic-dusted California Bartletts to the USDA's Bureau of Chemistry, the president of the Pear Growers' Association told members to keep spraying or risk devastating losses from codling moth.[66] The moth, he rightly figured, was a bigger adversary than the government, which agreed neither to ban lead arsenate nor to publicize the fact that contaminated fruit had already been sold and probably eaten.[67] In turn, growers found ways to minimize the likelihood of run-ins with city health inspectors. Fruit-washing machines, for example, wiped off the worst residues while adding an attractive shine.

California's fruit growers weren't the only ones dousing their trees with toxins in the early years of the twentieth century.[68] Nonetheless, they took pride in their organized and uncompromising pest-control campaigns. "Probably in all this far-flung world," boasted the *Pacific Rural Press* in 1926, "no business of any kind anywhere has been . . . more progressive, better planned, or more conscientiously carried out than the spraying of California's premium fruits and vegetables."[69] The *Press* frequently reminded growers that producing beautiful fruit was, in fact, a business:

The early-blooming trees are painting the valley and foothill expanses with their gorgeous colors and filling the country air with the deliciousness of their odors. These are the art and poetry of our most profitable and esthetic industry . . . the practical aspect is clearly this: unless the fruit grower does everything he can to bring the coming crop of fruits up to the perfection of which it is capable . . . the grower may as well get all the joy and comfort he can from the blossoms, for there may not be much but distress and disappointment in the fruits that will follow.[70]

Beauty as Brand

Eventually, grading, branding, and advertising produce became not just a California specialty but a global norm. Well before this happened, however, the fruit trade reached global proportions. By the 1890s, millions of stems of Central American bananas arrived in U.S. ports each year. So too did boatloads of Sicilian lemons, which for decades out-competed California's in American markets. And in London, Paris, and other northern European capitals, ocean steamers and express trains delivered all kinds of fresh fruit from as near as Provence and as far as South Africa, Argentina, and New Zealand.[71] Much of this fruit arrived during seasons when consumers hungered for its refreshing sweetness. Much of it, connoisseurs conceded, was undeniably good-looking. Altogether, it posed a formidable threat to local producers, especially those who'd worked hard to corner the lucrative market in counter-seasonal produce.

Around Paris, few had worked harder at this than Montreuil's fruit growers. Since the abbot Roger Schabol first wrote about their elegant espaliers in the seventeenth century, they'd turned three-quarters of the town's territory into plaster-walled, solar-heated gardens. They planted the most fashionable apple and pear varieties and beds full of the showiest flowers. Year after year, they harvested espalier peaches early, late, and often. Their best fall varieties cost five francs (about a dollar) apiece, or about fifty times more than ordinary summer peaches.[72]

In the meantime, factories and worker housing had gone up in the fields that once separated their hillside gardens from the eastern edge of Paris. The costs of land and hired labor had risen. The long-tilled soil needed heavy fertilizing, and hard-worked trees needed replacing.[73] Montreuil's growers had also changed. After generations of catering to the tastes of the Parisian haute bourgeoisie, some had become pretty bourgeois themselves. They were educated, entrepreneurial, and ambitious; many were immigrants from the Burgundy region. All were determined to parlay Montreuil's famed horticultural traditions into modern commercial cachet. "The sap that runs in our trees must remain liquid gold," wrote the Montreuil author Hippolyte Langlois in 1875.[74]

In 1878 the town's growers formed the Montreuil Regional Horticultural Society. One of their early priorities was to stop unscrupulous merchants from selling cheap Mediterranean peaches as *pêches de Montreuil*.[75] But it soon became clear that fruit shipped from the south would pose tough competition, since it enjoyed a naturally longer and cheaper growing season. So rather than try to survive on the mass market, Montreuil's growers went still further up-market. They aimed to convince elite consumers, in and beyond Paris, that *their* fruit was still the most beautiful of them all.[76]

To do this, they expanded a practice originally developed to fight off an infestation of snout moth *(pyralidae)* in the mid-1880s. Although the Montreuillois had nothing against sprays (tobacco juice from the city's cigarette factories worked on certain pests) they found that the best shield against the snout moth was a paper bag, wrapped around each immature fruit. Known as *ensachage*, this method also protected against weather damage and tended to produce larger, juicier, and more tender-skinned fruit. Not least, *ensachage* increased fruit's photosensitivity. Colorless while inside the bag, a fruit's skin took on extra-brilliant hues once exposed to sunlight, usually two or three weeks before harvest. Alternatively, covering the naturally light-skinned Calville de Blanc yielded apples as pale and luminescent as pearls.

Admittedly, it took a lot of labor to wrap paper bags around each fruit on each tree. But Léon Loiseau, the Montreuil Horticultural Society's longtime president, calculated that it was worth it, especially if women took on "this simple and non-taxing work."

MONTREUIL-SOUS-BOIS
Cueillette des Pêches

Producing some of the world's most beautiful and costly fruit was a full-time family business in turn-of-the-century Montreuil, France, as this postcard shows. (Copyright Collection du Musée des Murs à Pêches, SRHM)

Growers could also economize by having family members make the paper bags out of newspaper, book pages, or train schedules. Overall, he argued, the costs of *ensachage* amounted to "almost nothing" relative to the value it added. Plus, it freed growers from depending solely on the whims of well-heeled Parisians; fruit of such rare beauty could be exported all over Europe and beyond. "Thanks to the transportation and communication that is increasing every day," he wrote in 1903, "let us not fear overproduction. Let us produce as always . . . our agents will find foreign outlets."[77]

So in Montreuil, they produced. The women worked especially hard, since their responsibilities included not just putting bags on the fruit in springtime—the quickest could do 3,000 a day—but also selling it at Les Halles in the summer and fall. Along with other family members they also harvested the fruit and prepared it for sale. To lure the city's most demanding clientele, they brushed fuzz off the peaches, polished the apples and pears, and wrapped all the fruits in tissue or grape leaves, to highlight their shape and bright coloring. In winter, a relatively relaxed season, growers' families folded and glued tens of thousands of paper bags.[78]

The same competitive pressures that fueled all this hard work also shaped the Montreuillois's attitude toward new technologies. An 1895 newsletter framed the benefits of chemical fertilizer in terms of a global struggle: "If twenty years ago our region had a monopoly on the sale of flowers, fruits, etc, thanks to our proximity to Paris . . . these days we face competition not only with the rest of France but also foreign countries . . . The dispossessed of yesterday have become the favored of today. The most faraway countries now succeed in producing the most . . . How to recover our former superiority, or at least compete against our commercial rivals? We must put in practice scientific findings."[79]

Science suggested that Montreuil needed not just fertilizer but also refrigeration. Unlike many other French farmers, the town's growers thought this technology posed a threat as well as an opportunity. Well aware that refrigerated steamers were hauling in fruit from the southern hemisphere, Léon Loiseau urged his fellow growers to battle the "invasion" by getting their own *frigos,* so that they could store fruit into late fall or even winter. He also called for France to invest in more refrigerated transport. As an experiment, he tried shipping some of his own peaches to New York on a *frig-orifique*. Despite the hot weather and turbulent sea, he reported, they arrived "absolutely intact and were found delicious." Back in Montreuil, the grower Monsieur Weinling owned one of the first cold storage chambers in the entire French fruit industry. It kept peaches for six weeks, during which they tripled or even quadru-pled in value. One witness claimed (somewhat improbably, given how poorly peaches handle extended cold storage) that "neither appearance nor taste revealed that these fruits had sojourned in a refrigerator."[80]

As part of their strategy to build new markets abroad, Loiseau and other members of the Montreuil Horticultural Society partici-pated regularly in international expositions. Claiming to represent the best of French horticulture, the society sometimes managed to get its fruit shipped to the expositions free of charge (and refriger-ated), courtesy of the French railways.[81] But once they arrived, the society's members made sure that the Montreuil name hung promi-nently over the pyramids of apples and baskets of peaches. From London to Milan to St. Petersburg, their exhibits regularly won top prizes.[82]

Partly because of all this public exposure, the Montreuil methods caught on in other towns and regions. Although individual grow-ers still guarded certain trade secrets, some of the society's mem-bers spoke and published widely.[83] Articles about the profitability of *ensachage* appeared often in popular gardening magazines such as *Vie à la Campagne* ("Country Living"). Over time, its novelty wore off. In the meantime, though, some Montreuil growers carved out an even more precious niche: they began using sunlight and paper to stencil images onto the pale, sensitive skin of their bagged fruit. It's not clear where they got this idea. Loiseau claimed that Montreuil growers had practiced fruit *marquage* ("marking") for several decades, originally just as a hobby. But references to simi-lar practices date back centuries, suggesting that the more literate growers might have learned *marquage* from a book.[84] In any case, they made it a local specialty.

The marked fruit of the Montreuillois first won renown at the 1894 Saint Petersburg exposition, where they presented the czar of Russia with an apple stenciled with his own portrait. King Leopold of Belgium, Edward VII of England, and Teddy Roosevelt received similar fruits. Initially the stencils were homemade, applied to the fruit with egg white or *bave d'escargot* (snail slime). Later, local craftsmen manufactured stencils on a larger scale, and growers produced fruit emblazoned with the names of restaurants, resorts, and gourmet groceries, among them the Tour de l'Argent, the Casino de Deauville, and Fauchon. While some stencil designs appealed to tourists and holiday-goers, others were strictly limited editions. A handful of Montreuil's more ambitious growers even developed film negatives on the skin of their fruit. Loiseau himself admitted that *la photographie sur fruits* was not particularly profitable, given the work and materials it required.[85] But it made for an impressive and flattering gift, especially if the fruit featured a portrait of the recipient.

The French word *marque* means not just "mark" in a generic sense but also, more commonly, "brand." In a way, this is what the Montreuil growers tried to create, through *marquage* as well as through participation in international expositions. Obviously they didn't promote their "branded" fruit as a mass-market commodity, as Sunkist citrus did. They didn't pay for professional ad men or radio spots. Rather, they saw the artistry of *marquage* as advertising in itself. In a market of expanding choices, it encouraged elite consumers to seek out and pay top prices for all Montreuil's fruit. And they did. At the turn of the century, a tissue-wrapped 500-gram *pêche de Montreuil* cost half of a worker's daily wage.[86] In the years before World War One, Montreuil's fruits commanded similar prices in London, Berlin, and Moscow. Even in the Depression years, its prized Calville de Blanc apples sold in Paris for the

A few examples of the "fruit photography" practiced by Montreuil growers. ("Photographies sur fruits," from *De l'ensachage des fruits,* by Léon Loiseau, 1903)

equivalent of 15 euros each; the children of Montreuil growers delivered them regularly to the Russian embassy.[87]

Selling luxury fruit did not bring fabulous wealth so much as security and modest upward mobility. Successful growers built comfortable homes and educated their children; some bought or rented additional orchard land. But as the eastern edge of Paris filled up with factories and their workers in the 1930s, rising real estate prices increased the temptation to sell.[88] So too did the Communist Party's victory in 1935 municipal elections. The new leaders showed little interest in preserving either the fruit growers' profession or the plaster walls that made it possible. Expropriations and demolitions began in the late 1930s and continued after World War Two. By then, the town's premium-priced fruits were a tough sell anyway. Mass-market fruit had become more attractive—as more producers in France and neighboring countries adopted "American" grading and packing practices—and consumers' priorities had shifted. They spent more of their income on family vacations and

household durables—furniture, cars, and, ironically, refrigerators—
and less on luxury perishables.[89]

Although the 1951 census still counted more than 100 commercial fruit growers in Montreuil, their numbers dwindled over the
next few decades. Meanwhile the town attracted a new generation
of immigrants, many of them from Africa and East Europe. By the
late twentieth century, most of the peach walls had come down,
replaced by the drab but functional housing and sports complexes
typical of Paris's eastern suburbs. Today the Horticultural Society's museum and garden, still maintained by volunteers, lies in the
shadow of a high-rise apartment building.

Pomona Postmodern

In the history of fresh fruit, Montreuil played a very different and
more minor role than California. California's growers encouraged
consumers to turn what was traditionally a luxury into a daily staple; Montreuil's growers struggled to preserve the idea that *some*
fresh fruit still deserved luxury status. California growers standardized, advertised, and sprayed in order to break into distant markets; the Montreuillois innovated to defend against such incursions.
But for growers in both places, beauty was strategic. It both sold
fruit to urban consumers and staved off urban development; it justified staying on the land. In both places it worked only for a while.
In California, growers' communities quickly became the frontier's
suburbs. From Riverside to Vacaville, settlers poured in long after
California stopped trying to attract them with "health and wealth"
booster pamphlets. Sprawl drove up fruit growers' property taxes
and mortgage rates, increasing the pressure to make land pay. After
World War Two, California's orchards and vineyards were consoli-
dated in some places and paved over in many others. Anyone who's
visited Orange County recently will know that they don't produce
many oranges there anymore.[90]

Smaller farms on the urban edge survived by producing even higher-value crops, such as strawberries, exotic vegetables, or, increasingly, organics. And although the organic label once looked like a way *off* the beauty treadmill—a way to earn a premium from "natural" purity rather than sprayed and waxed perfection, this didn't hold true for long. At the end of the twentieth century, as organic production exploded and organic sales moved into mainstream outlets, the market got pickier. Unless growers went back to face-to-face commerce (which many did, at farmers markets and through "subscription" schemes) they had to produce fruit that was as pretty as it was pure.[91] This helps to explain the dazzling spectacle at the entrance of any Whole Foods.

Organic or otherwise, supermarkets now greet us with the gleaming harvest of what has come to be known as "permanent global summertime": the worldwide system of standardized production, high-speed transport, and climate-controlled storage that has largely destroyed the seasonality of fresh produce and with it, many would argue, its taste.[92] The focus of innovation has therefore shifted, as consumers demand both more and less than simple beauty in fresh fruit.

On one hand, consumers want fruit that not only looks good but also tastes good, or at least better than the stereotypical supermarket selection of rock-hard peaches, bland apples, and sour plums. In other words, the conventional wisdom of early twentieth-century market experts—consumers would be satisfied with "shiny red apples stuffed with cotton"—no longer holds in a world of endless snack options. So apple breeders have developed varieties that are pretty and sweet and keep well (like the Red and Golden Delicious) but also have crunch—an important measure, in consumers' mouths, of apple freshness.[93] One of the most successful early results of such research was the Pink Lady, a variety described as having the blushing skin and creamy flesh of a movie star.[94]

On the other hand, consumers increasingly want fruit that they do not even have to look at to eat. In the late twentieth century, as convenience became the top priority for all kinds of food products, the market for "fresh-cut" produce took off. Fruit proved less amenable to this sort of processing than vegetables, partly because many popular kinds, notably apples, quickly turn ugly when wounded. After years of experimentation, in 1999 a German company offered a solution in the form of NatureSeal, a calcium citrate formula that keeps a cut apple looking more or less fresh for up to twenty-eight days. Companies ranging from McDonald's to Whole Foods quickly began selling bags of NatureSeal-treated apple slices.[95] Supermarkets also upped their selection of lower-tech "fresh-cut" products, such as melon cups.

The fact that more and more fruit is sold in bite-sized pieces doesn't mean that its overall aesthetic appeal no longer matters. On the contrary: now the beauty of the fruit must literally go more than skin deep. In fact, this is one of the main reasons that not just ordinary consumers but also the world's biggest apple buyer, McDonald's, have fallen for the Pink Lady. It is shapely to the core, and its cut flesh, with or without NatureSeal, stays fresh-looking for an uncannily long time. Again, not unlike a movie star's face.

five

Vegetables

HIDDEN LABOR

The circumstances of gardeners, generally mean, and always moderate, may satisfy us that their great ingenuity is not commonly over-recompensed.
—Adam Smith, *The Wealth of Nations*, Book 1, Chapter 11 (1776)

"I'll just have a salad." How many times have you heard or said it? *Just* implies modest wants. It implies that you're not asking for much—not much food, not much effort—when you ask for a dish made mostly of raw vegetables. But it's a misleading modesty, and not only because lettuce isn't so light when tossed with cheese, croutons, and a ladle of Thousand Island dressing. I learned this once when I visited the home of an old friend in San Francisco at what turned out to be his children's dinner hour. His wife, tired and harried, asked what I wanted to eat. I said I was fine, not hungry. She insisted. I hesitated. She looked annoyed. How about just some salad? I suggested. Immediately I saw my mistake. No salad existed; it had to be made. Refusing all help, she took a bag of baby lettuce from the fridge and began to wash it, seemingly one leaf at a time. This was San Francisco, after all; salads were serious business. Drying and dressing proceeded at the same painstaking, passive-aggressive pace. Stuck watching her labor on my behalf, I wanted to crawl under the table with the toddler.

Ordinarily we don't see much of the labor that goes into a salad. This is partly because the ingredients are often produced far away, and partly because of how they are sold to us. Compared with the processed goods filling the middle aisles of the supermarket, lettuce and other vegetables in the fresh produce section seem the most natural and unmanufactured of foods. Yet even if a salad consists of nothing more than a chunk of iceberg or a plate of pre-packaged mesclun, a lot of work went into making it so naturally appealing— maybe not in the kitchen, but earlier in both the life of the lettuce and the history of an industry.

Since the early twentieth century, this industry has worked hard to sell the world on the virtues of salad and fresh vegetables more generally. Now dominated by multinational conglomerates, its historic capital is Salinas, California. In the 1920s the farmers of Salinas grew rich off iceberg lettuce, the same crop that ruined Adam in Steinbeck's *East of Eden*. Unlike him, they recognized the value of marketing. Taking advantage of middle-class Americans' growing interest in diets that would keep them young, slim, and energized, the growers promoted iceberg as a fat-melting ball of vitamins.

With vitamin science in its infancy, few questioned these claims. But California's lettuce growers faced other challenges. Many of the same consumers who wanted "light" foods also wanted to lighten their kitchen workload. As more and more processed goods promised to free the American woman from drudgery, growers had to do the same. How to turn a fragile garden crop into a commodity as convenient as canned soup? They approached this problem on three fronts, as did many other fresh produce growers. While iceberg lettuce is in certain ways an odd green, its history offers insights into modern consumers' conflicted desires for a kind of freshness that is carefree yet shows they care.

On the technical level, growers sought to make their crops as sturdy, standardized, and seasonless as any canned good. Lessons learned from iceberg they later applied to less obviously industrial greens. In their marketing, they tried to convince consumers that tossing a salad was as easy as opening a can, and much better for them. They played on widespread assumptions about not just the nutritional benefits of fresh vegetables but also the moral superiority of meals *prepared* fresh. And on the political front they fought to preserve their access to the cheap, migratory, and usually foreign labor force that had always worked on California's fruit and vegetable farms. Their reluctance to give up an exceptional status—as industrial-scale producers not bound by industrial labor laws— sometimes brought violence to the lettuce fields. But this traditional reliance on an extremely cheap and vulnerable workforce was never uniquely Californian, except perhaps in the degree that it attracted media attention beginning in the 1930s. Among the wealth of fresh vegetables now available in a typical supermarket, some of the most convenient, attractive, and priciest options come from the world's poorest countries. More than ever before, adding value to freshness depends on the least valued labor.

The Kitchen Garden

The foods we call vegetables are a diverse lot. At the broadest level, keeping them fresh after harvest requires keeping them just barely alive and breathing. The faster their cells metabolize, the faster they lose flavorful sugars and acids, as well as the water that makes them juicy or crisp. Certain vegetables, like sweet corn and peas, really are best straight from the kitchen garden, because at room temperature much of their sugar turns into mealy starch within a few hours. Others, such as broccoli and asparagus, gradually convert their sugars into tough, defensive skins and stalk ends. Among leafy vegetables, shape matters: head lettuce and cabbage last longer than their leafy varieties because their inner leaves respire more slowly. Roundness, however, doesn't protect against all spoilage. A bit of bacteria can quickly turn a crispy head of iceberg into a brown ball of slime.[1]

While refrigeration slows wilting and rot, it can also damage the flavor, texture, and appearance of some vegetables. The most cold-intolerant varieties are actually tropical or subtropical fruits, some more obvious than others: tomatoes and other members of the nightshade family (eggplants, peppers), cucurbits (squashes, cucumbers), avocados, olives, and even green beans and pea pods. Most of these faux-vegetables can tolerate or even benefit from a few days in the fruit bowl, with some important exceptions. Recall that fruits serve two purposes for the plant: first to protect its seeds, and then to spread them, usually via an animal's digestive tract. To ensure dispersal, most fruits have evolved to become irresistibly ripe just before they rot. Members of the legume family, however, do not disperse their seeds through seduction. On the contrary, they just dry out and eventually split open. Rapid breathers, they shrivel and lose sugar soon after harvest, yet suffer cell damage and discoloring if kept long at cold temperatures.[2]

Ironically, some of the most perishable and frigophobic vegetables are now among the most globalized, in that they are produced in remote rural areas of Africa, Latin America, and Asia, and shipped across continents and oceans. The science of vegetable shelf life is now as sophisticated as it is economically vital. But it's also a new science relative to other kinds of botanical knowledge. In many parts of the pre-industrial, pre-supermarket world, fresh vegetables traditionally came from the kitchen garden. Whether it took the form of French aristocrats' showy, symmetrical *potagers,* the space-saving raised beds of the Japanese, or African villagers' plots of okra and peppers, the kitchen garden served the needs of the cook. It didn't necessarily provide for much of the household sustenance, but it did add color, variety, and perhaps spice to the daily meal. It produced what the anthropologist Sidney Mintz has called the "fringe" foods, complementing and enriching the core starch, whether that was maize or rice or barley. The garden's proximity to the kitchen made it easy to tend and harvest on a daily basis. The cook could pick just enough herbs for the evening meal, or just the tomatoes that ripened in an afternoon. And with the garden so close, she hardly had to worry about whether the produce was fresh. At most she might worry about having enough hands to help dry, pickle, or can the seasonal surpluses.[3]

A cook with enough hands could also prepare veritable feasts of freshness. Consider the platter-sized "Grand Salad" described by seventeenth-century English cookbooks. Served as a first course in the country homes of the wealthy, it mimicked a formal garden. Small mounds of shredded meat, mushrooms, cucumbers, and boiled eggs were arranged on a bed of greenery, topped by fruit slices and flower petals, and crowned by a celery-stalk tree. Even less elaborate salads might contain a mix of young greens and herbs (common varieties included spinach, sorrel, purslane, lettuce, cress, mint, rocket, and sage) as well as nasturtium and marigold buds. None of these ingredients lasted long, and none traveled far.[4]

In regions with mild winters, salad eating didn't necessarily stop when autumn ended. In Europe, growing plants in greenhouses dates back to at least the sixteenth century. The seventeenth-century French cookbook *Délices de la campagne* recommended planting purslane "all the months, in order to always have tender greens for the salad."[5] The Englishman Thomas Tryon's *Pocket Companion* also offered recipes for locally grown, more or less fresh, winter vegetables:

A Sallad for Winter.
Parsley, Old Onions, Endive, Cellery, Lettice, Sorrel and Colewort plants, Seasoned with Salt, Oyl, and Vinegar, it is an excellent and warming salad.

Another
Take Cellery, Endive, Spinnage and Lettice, and half a head of Garlick in it, seasoned with Salt, Vinegar and Oyl, this is a brave Sallad. Sallads are good at all times, but most proper

from end January to the first of July, then again from September till December, and indeed all Winter, if the Weather be open.[6]

As a vegetarian, Tryon probably ate more salad than most English people in the seventeenth century. But his belief in its "wholesome" and health-preserving qualities was not unusual. His recommendation of winter salads also reflected a growing disregard for traditional humoral theories of diet. There was nothing strange or "hurtful" about eating "cold" greens in winter. On the contrary, "All such Herbs as do grow and continue fresh and green do retain their true natural virtues."[7]

Even after industrialization created vast populations of consumers who had no kitchen gardens, most fresh vegetables in urban markets came from nearby. The early nineteenth-century German economist Heinrich von Thünen (1783–1850) developed an entire theory of urban-area land use around two assumptions about fresh foodstuffs.[8] First, he assumed that vegetables, like milk, were too perishable to transport any distance on an oxcart. They didn't become deadly as quickly as milk, but fragile varieties lost market value if they traveled too far or too long. Second, he assumed that fresh vegetables were in fact valuable enough crops to justify cultivating them on prime urban-edge real estate. A farmer himself, Von Thünen probably realized that the poor and working classes in northern European cities ate relatively few fresh vegetables, except cheap root crops. The most lucrative market for local truck farmers (the term refers not to the vehicle but to the French word for truck or barter, *troquer*) would be the more affluent urban consumers—those who could send their servants to buy fresh supplies daily.

In the nineteenth-century United States, urban markets for fresh vegetables were also stratified, especially at colder latitudes. At Boston's Quincy Market in January 1835, *Hovey's American Gar-*

dener reported sales of not just onions and root crops but also broccoli, cauliflower, celery, three varieties of cabbage, four varieties of squash, spinach, and even "some fine heads of lettuce." But the lettuce, most likely greenhouse grown, was "quickly taken" at twelve cents a head, while cauliflower sold for fifty cents apiece. They probably ended up on the tables of affluent Bostonians, who at the time considered salads and "lighter" dishes to be fashionably French. For everyone else, turnips and carrots cost forty cents per bushel. They had probably spent at least a couple of months in a farmer's root cellar, but they worked fine in stew. As with the roots and tubers that have carried people through centuries of hard times, their age mattered less than their cheap, gut-filling substance. Freshness could wait until spring.[9]

Even as refrigerated storage and transportation spread across the United States in the late nineteenth century, the seasonality of the fresh vegetable supply faded slowly, at least compared with the dramatic changes sweeping through other perishable food trades. Although Thomas F. de Voe's *Market Assistant* mentions that the southern states sent "rare vegetables" to New York in early spring, the journal *Hovey's American Gardener* complained that the winter selection in Boston had changed little in three decades.[10] The reasons were not mysterious. For one, many vegetables just spoiled too fast to ship from a warmer place or to store between seasons. For another, they weren't worth enough to justify either expense. Unlike red meat, vegetables weren't coveted by workers and commandeered by armies; unlike eggs, they weren't used for winter baking, and they lacked the status, showiness, and gift value of off-season fruits. A third reason lay in American vegetable cookery, which had not yet thrown off the English tradition of extreme boiling. *Hovey's* suggested that professional horticulturists were simply uninspired by "plain kitchen gardening," since their carefully tended produce might well be unrecognizable once it reached the table.[11]

The Vegetable Age

In the early twentieth century, produce merchants, plant breeders, and enterprising farmers began to look at fresh vegetables with new interest. Although many vegetables were as tough to ship and store as ever, rising demand made the risks look more worthwhile. In fact consumers grew so keen on greens that *Western Grower and Shipper*, the California produce industry's main trade journal, dubbed the 1910s and 1920s "the Vegetable Age." This epoch was defined by new eating habits as well as by new notions of health, beauty, and fulfillment.[12]

One new development was a growing concern about the dangers of "overnutrition." In the past, critics of gluttony (Thomas Tyron among them) typically pitched their screeds at the elite few who could afford gluttonous eating. By the turn of the century, the United States had more than enough food, and manufacturers and chain retailers made it ever easier to buy and consume. Many farmers' children had taken office jobs. According to a new wave of nutritional advice, these "brain workers" needed regimens suited to their sedentary lives. Too much food or the wrong diet would hurt not just their health but also their on-the-job efficiency.

Nutritionists' preoccupation with efficiency was not in itself new. In the 1880s the chemist E. O. Atwater, founder of what became known as the "New Nutrition," convinced many American scientists, policymakers, and social reformers that the ideal diet was one based on the cheapest and most compact fuels. The human body ran fine on protein and energy, he argued, and foods such as wheat flour and stew beef kept it running much more economically than even the humblest greens. Take cabbage, for example: a housewife could get four pounds for ten cents, yet she was feeding her family only 460 calories and less than an ounce of protein. If she spent the dime on wheat flour instead, she'd get nearly six times as much protein and nearly twelve times as many calories.[13]

Although vegetables added "relish and variety" to the diet, Atwater recommended that poorer consumers save their money and instead learn "skillful cooking and tasteful serving." The government's own home economists agreed. A 1907 USDA study of American dietary patterns reported that working-class families in New Jersey spent almost 15 percent of their food budget on oranges and celery. These items "undoubtedly added to the attractiveness of the diet," the study conceded. "It is true, however, that such foods could have been omitted . . . without materially changing its nutritive value, while the cost of the daily food would have been considerably lowered."[14]

Within ten years, New Nutrition appeared thoroughly out-of-date. First, it looked like a surefire recipe for constipation and "dyspepsia" more generally. While these were hardly new complaints, a series of influential food fads led many middle-class Americans to seek relief in "natural" and less-processed diets. Kellogg's Battle Creek Sanitorium put patients on whole grains and fruit, while raw food restaurants in California served up "beauty salads" and sun-dried breads. And Horace Fletcher, founder of the famous 100-chews-before-swallowing fad, urged followers to cut back on meat, a "once-digested, half-decaying" food, in favor of "first-hand food elements as fresh from the heart and breast of Mother Nature as possible."[15]

Second, Atwater's meat and starch diet didn't look particularly slimming. Since his first writings on the New Nutrition, beauty ideals had shifted dramatically: instead of voluptuous women and stout men, American society in the late 1910s and 1920s admired slenderness in both sexes. Diet books like Lulu Peter's best-selling *Diet and Health: A Key to the Calories* offered recipes, menus, and advice on how to avoid unbearable hunger pangs on 1,500 calories a day. Dry toast, clear bouillon, cottage cheese, and "hot water flavored with coffee" counted among the staples. But so did green leafy vegetables, and lettuce especially. After all, two heads of iceberg contained only 100 calories, or 150 with a half spoonful of dressing. Their bulk also helped dieters "keep regular." Altogether, as one 1920s weight-control manual put it, leafy greens counted among the "friends of the fat . . . you can eat liberally of them, have that much desired full feeling, and still feel at ease with the scales."[16]

Third and most important, the discovery of vitamins in the early 1910s showed that the body needed more than simply sufficient fuel to run properly. Exactly why it needed vitamins was not clear; the earliest studies only determined that lab rats languished without them. These studies also found that vitamins existed in the foods known to prevent scurvy, pellagra, stunted growth, and even the vague seasonal malaise known as "spring fever." Citrus fruit and fresh milk clearly contained them but so did the leafy vegetables once dismissed as "watery" and frivolous. Indeed, vitamin researchers theorized that the "leaf-eating" habits of the Chinese explained their apparent good health, despite relatively little animal protein in their diets.[17]

In the United States, the most influential of these researchers was E. V. McCollum. Born in a sod hut in eastern Kansas, he went from studying dairy cow diets in Wisconsin to leading Johns Hopkins's newly founded Department of Chemical Hygiene, where he isolated the first four vitamins: A, B_1 and B_2, C, and D. He also authored several popular works, including *Newer Knowledge of Nutrition*. As the book's title suggests, McCollum advocated a diet radically different from Atwater's. He turned to his rural roots to explain why humans could not live off starch and meat alone:

We know that in the early settlement of many of the states the people suffered great hardships. With little capital and no food reserve their winter diet was generally very simple and monotonous. There is good reason to believe that it was chemically unsatisfactory for the maintenance of health. After a period of several months during each succeeding winter they felt "run down." . . . With the coming of warm weather sorrel pies, wild onions and dandelions grew in abundance . . . The cows, due to their starved condition, had been dry all winter, became fresh and milk was available. The few hens in the barnyard added worms, insects and tender grass to their diet and began to lay eggs. The garden furnished fresh vegetables . . . which were eaten with a relish which can be appreciated only by one who has for a period been semi-starved. The tired feeling disappeared about this time.[18]

"Vitamines" entered scientists' vocabulary in 1912; by the early 1920s Americans learned that their bodies depended on these invisible elements "as surely as the steam-engine depends on steam." Foods containing them were deemed "protective" because they were thought to compensate for the deficiencies of white flour and other processed staples. As the guidebook *Eating Vitamines* warned readers, "highly refined foods are safe taken only when plenty of milk and green foods are taken with them." Like McCollum, the guidebook's author deemed milk a more perfect food overall than vegetables, noting that "the human stomach can't accommodate as much green food as the cow's seven, and she passes her store on to you in milk." Still, given that green vegetables added to the diet fiber, minerals, and balance—a key principle of the Newer Nutrition—McCollum urged Americans to "eat of them liberally."[19]

And what better way to eat them than in salads? McCollum's message didn't spread through his many writings and lectures alone. Chefs adapted their menus to suit health-conscious diners. At Chicago's posh Edgewater Beach Hotel, ladies who lunched could choose from a wide variety of elegant greenery, including the best-selling Doctor's Salad (lettuce, tomato, watercress, cottage cheese, chives, and cream cheese) and an all-raw Health Salad that "if masticated properly will prove beneficial to all who eat it, no matter what their complaint, fancied or real." They could also buy their own copies of the *Edgewater Beach Hotel Salad Book,* which provided further information on the vitamin and mineral content of "Nature's most prolific crops." This was just one of many books and magazines that likened fresh salads to age-erasing tonics. "Lettuce means health," it said, "and health means youth."[20]

Green Marketing

In 1916, as McCollum began to publish his findings, Moses Hutchings began to plant lettuce. He started with just a couple of acres on his farm in Central California's Pajaros Valley, not far from Salinas. He'd never heard of vitamins, but he'd probably heard that truck farmers to the south were making money from a new kind of crisphead lettuce they called "Los Angeles." According to local lore, an L.A. dealer imported the seeds from Paris, where it was known as Batavia or "cabbage" lettuce. Later plant breeders would call it "New York" lettuce (though little grew there) and then just "iceberg." It didn't look anything like the soft, floppy salad greens found back east. It was round, solid, pearly pale, and shiny. It looked more mineral than vegetable. It didn't look like anything the valley had produced before.[21]

Most farmers in the Pajaros and Salinas Valleys grew commodity crops such as alfalfa, sugar beets, and beans, often with a few dairy cattle or fruit trees on the side. When Hutchings tried to buy crates and nails so that he could ship his lettuce by rail, local merchants wouldn't fill his orders until they saw that his crop really existed. Yet ultimately Hutchings got his lettuce harvested, packed, and even photographed before sending a wagonload to a place where he knew people ate salad—San Francisco, a little over a hundred miles away. Apparently it sold well enough, for he planted lettuce again the next year, and so did a few of his neighbors. In 1919 one of them, Louis DeLaney, tried sending a carload to New York. He'd probably heard that some southern California growers were already shipping "Los Angeles" head lettuce across the country; in fact one of this variety's greatest strengths was its durability. But the rail lines out of Salinas weren't fast or well-iced, and Delaney's carload met the same fate as the lettuce in *East of Eden:* when it arrived in New York after twenty days in transit, it had turned to slop.[22]

Unlike Steinbeck's unlucky Adam, though, the region's farmers stuck with lettuce. By the mid-1920s, they were profiting from not just the vitamin craze but also a new generation of refrigerated rail cars. Equipped with ice bunkers that cooled from the top down, they greatly increased the survival rate of cross-country lettuce shipments. In their rush to plant "green gold," the farmers of Salinas plowed under their beans and alfalfa, sold off dairy cattle, and uprooted fruit trees.[23]

The value of the valley's irrigated acreage more than doubled between 1920 and 1930, but that was nothing compared with the profits lettuce generated. In a single harvest a grower could earn more than the price of the farmland—and lettuce could be harvested twice a year. Packing sheds, box makers, and ice plants went up in and around Salinas; soon local shippers consumed more ice than New York City and sent more telegraphs than San Francisco.[24] Suddenly known as a town where jobs were plentiful and millionaires not so rare, Salinas saw its population of 3,500 triple within ten years. Meanwhile, even as lettuce acreage shrunk around sprawling Los Angeles, California became the nation's top producer, and iceberg the nation's top lettuce crop. By the end of the decade the state accounted for two-thirds of the total shipments.[25]

As lettuce got big, growers got organized. Following the example of the state's citrus farmers (founders of the Sunkist brand), they formed the Western Growers' Protective Association in 1926, and launched an advertising campaign two years later.[26] Their early ads made the most of easterners' perceptions of California as a land of cloudless skies:

> The Sun is the mother of us all . . . when you eat a portion of lettuce—say, half a head—you are also taking a sun-bath, an internal sun-bath. For this lettuce is grown under the smiling, sunny skies of the great Far West. Day after day the ardent sun irradiates it—shoots myriad of rays into it . . . in short, puts up a package of sunshine for you. This is Nature's way. And when your home skies are dull, and the sun never peeps out all day, or shows only a pale, wan face, you can still take your internal sun-bath. For Iceberg head lettuce is at your grocer's in winter as well as in summer. Every day of the year you can serve Nature's concentrated sunshine on your table.[27]

It's now well known that iceberg lettuce contains fewer nutrients than most leafy greens.[28] In the 1920s, though, even McCollum described all leafy greens as more or less equally protective. Given the large gaps in vitamin knowledge, the lettuce growers made bold claims. Iceberg, their ads said, was not just low-calorie; it actively toned the body and removed "superfluous flesh." It also contained a youth-prolonging "mystery vitamin," unnamed and "still under investigation." A photo of swimsuit-clad young women frolicking in a snowfield suggested that this was a mysterious vitamin indeed.[29]

In addition to their own paid advertising, the growers enjoyed plenty of free publicity. The *Washington Post* quoted doctors who described salad as a "health food par excellence," while the *Los Angeles Times* called lettuce a "Godsend" "to those of us who surreptitiously remove coats and bags before stepping on the penny scales." Writing about the stars' "quest for sylphlike figures," one of the paper's Hollywood correspondents noted that Jane Wyatt often treated herself to "half a head of plain lettuce, minus even the

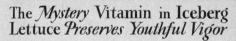

Early advertising for iceberg lettuce suggested that it was a vitamin-rich miracle food. In fact it contains fewer nutrients than most fresh greens. (*Ladies' Home Journal*, January 1930, p. 155)

dressing."[30] And the growers could always count on good press from Christine Frederick, the editor of *Good Housekeeping* and one of the era's most respected home economists. "On every side we hear that vitality, vim, vitamines and vegetables go hand in hand," she wrote in *The American Restaurant* in 1928. "Instead of paying $2 a bottle for Dr. Whoosis Bitters . . . we prefer to buy blood purifiers at the vegetable stand and the salad counter."[31]

No-Fuss Fresh

During the 1920s, per capita consumption of lettuce increased by more than 260 percent, from 2.3 to 8.3 pounds annually.[32] In an era when ideas about healthy diets and desirable bodies were changing, lettuce appealed both for what it contained (fiber and supposedly vitamins) and for what it lacked (calories). These qualities it shared with many fresh vegetables, some of which saw more moderate boosts in popularity. Yet lettuce also offered another, less common virtue: *convenience*.

The appeal of easy-to-fix freshness reflected at least two broader socioeconomic trends. First, as industrialization drew more and more working-class women into factory and then office jobs, hired help in the middle-class kitchen grew scarce. In *More Work for Mother*, Ruth Cowan estimates that up to one-half of households in pre-industrial America had live-in domestic servants. In New York City in 1888, the census recorded only 188 servants per 1,000 families; by 1920 this ratio had dropped to 66 per 1,000. Eventually the "cleaning lady" largely replaced live-in domestic service, and she typically didn't cook.[33] New household technologies eased certain tasks but also reinforced the idea that a woman could and should manage her home, and especially the planning and preparation of meals, by herself.[34] Not surprisingly, liberation from "drudgery" became a strong selling point, not just for appliances such as the gas stove, but also for foods that didn't even need the stove.

Lettuce growers knew this. From an early date the trade journal *Western Grower and Shipper* identified the "young modern housewife" as a promising customer. As the main food shopper in her family, the journal reported in 1932, "she is going strong for salads and vegetables that are prepared without much cooking."[35]

Second, with growing affluence came more possibilities and pressures for middle-class women to participate in social activities outside the home. In other words, as leisure time became more central to maintaining a marriage, family, and social standing, it also became busier. Between bridge parties, weekend drives, club meetings, *and* keeping house, middle-class married women—whether or not they held paid work—formed the vanguard of what Harvey Levenstein dubbed the "revolution of declining expectations" in domestic cookery. They took out their white tablecloths less often and began serving guests "family-style" meals, prepared as much as possible in advance.[36]

Canned and other processed foods of course benefited from this "revolution," ultimately more than did most kinds of fresh produce. But surveys in the 1920s suggested that consumers bought them despite rather than because of their taste and perceived nutritional value.[37] By contrast, one home economist noted, fresh vegetables offered domestic cooks as much opportunity for "showing their ingenuity" as did fancy baked goods, yet they required potentially much less time.[38] Written in 1911, this comment didn't necessarily apply to lettuce. As Laura Shapiro observes in *Perfection Salad*, many pre-war salad recipes required an architect's precision. They called for lettuce cut into "ribbons of uniform width," symmetrically layered and prettily dressed.[39]

Yet simpler options abounded in cookbooks such as *The Calendar of Salads; A Thousand Salads* (which offered thirty-three variations on the stuffed tomato); and the many pamphlets published by makers of salad dressing and other convenient accoutrements (such as Jell-O).[40] By the mid-1920s, salads as reckless as the Caesar, with

The Western Growers' Protective Association promoted tossed salad as
easy and fun, especially when served in one of its extra-large salad bowls.
(*Printer's Ink,* February 10, 1938, p. 19)

its raw eggs and torn romaine, had become fashionable restaurant
fare. At home, iceberg was easy to wash, chop, and store. And once
dinnertime arrived, the salad toss became a family affair. "Don't
hide the secrets of your salad bowl behind the four walls of the
kitchen," *Good Housekeeping* told readers in 1927. "Carry all the

fixins' to the table and there share the mysteries of its making with your family and guests as well." For even more fun, the magazine suggested turning over the tossing to the "master of ceremonies"— the husband.[41]

Within a few years, many Americans had abandoned the European tradition of eating salad after the main course in favor of the "old California custom" of eating it before or even as the main course. "Visitors often comment on this apparently screwy idea, but . . . soon become addicts themselves," noted *Western Grower and Shipper;* "it isn't just another Hollywood fad."[42] Altogether, the reinvention of the salad, from a dish associated with French haute cuisine to one that epitomized "California casual," reflected a much broader change in American eating habits, one that extended beyond the home kitchen. A 1932 nationwide survey of hotels and restaurants reported that meat orders had decreased by 45 percent since 1920, while fresh vegetable orders had increased by 35 percent, and salad orders by 110 percent. The only faster-growing food category was sandwiches, sales of which increased by 215 percent in the same period.[43]

The western vegetable industry did not watch this change from the sidelines. As the authors of the restaurant survey noted, the industry excelled at "supersalesmanship."[44] It took advantage of Americans' weakness for foods that promised natural solutions to modern problems ranging from too little exercise to too much work. But the effectiveness of the "sun food" ads ultimately depended on the substance behind them. The real labor took place in the fields and the packing sheds, ensuring that lettuce lived up to its shimmering image.

Power Farming

In the early days of the Salinas lettuce boom, the crop appealed partly because it didn't seem to demand much work. It flourished in the valley's mild climate and rich soil. Compared with beans or alfalfa, it grew quickly and paid extraordinarily well, even on just five or ten acres.[45] In between harvests, newly prosperous growers had time for neighborhood barbecues, "lettuce balls," and backroom dice games, where bets reportedly reached $10,000 a roll.[46] They golfed with shippers and ice plant owners. The boom, in other words, built not just fortunes in Salinas but also a tight-knit community. One lettuce shipper described the town's seasonal dinner dances as "a frolic among the 'big family,' which is what the fruit and vegetable crowd feels itself to be at such times."[47]

This "big family," however, didn't include the people who actually labored in the lettuce fields. Even before the Depression, many of these workers lived in wood-and-tarpaper shacks on the edge of town.[48] And even among the growers, competition and rising costs quickly created a clear hierarchy. By the end of the 1920s, most of the smaller farmers produced under contract for a new breed of large, vertically integrated grower-shipper companies. These companies had the resources needed to negotiate East Coast markets, and this gave them considerable authority over local lettuce industry affairs ranging from wage rates to seed varieties to packaging standards. As market conditions deteriorated in the early 1930s, *Western Grower and Shipper* warned small growers to adapt or go under. "The 'gravy train' of the early and middle '20s has been unloaded, and the American public is not billing out anymore," the journal's editor wrote in 1931. "This means that we will say goodbye to many who have ridden along on the wave of increasing popularity of vegetables in the diet—that is, unless they are willing to get into line."[49]

Lettuce had become a serious business, and in fact the largest grower-shippers came from the business end of the fresh produce trade. They were not born farmers but led the nation in "power farming"—the use of tractors and other machines that, according to some, gave lettuce production "a speed and sureness paralleling industry."[50] In reality, these machines only gave them power to cultivate more land more quickly. This enabled them to expand south into Arizona, and thereby ship lettuce fifty-two weeks a year. Sureness, however, was another matter. Tractors did not increase the grower-shippers' control over volatile markets. Nor did they increase their clout vis-à-vis competing food industries, retail buyers, or regulatory agencies.

Machines didn't even master the iceberg lettuce itself. The crop was bred to travel well, and it does. But its durability depends entirely on its condition when packed. Ideally each head should be harvested only when hard and dense. Since heads mature at different rates, they have to be handpicked over a period of days. Mowing is not an option. Nor can iceberg be harvested too soon after rain, irrigation, or even morning dew, because its water-filled leaves are too crisp and fragile. If they break, they slime. Lettuce left too long in the field, by contrast, loses market value just as quickly. So does lettuce that is not thoroughly chilled soon after harvest. The iceberg's firm head, ironically, holds the field's heat all too well. Even when packed in ice, it takes several hours to lower its temperature enough to slow decay.[51]

Power farming amounted to cultivating the kinds of power needed to access what *Western Grower and Shipper* called the industry's key resource: "an ample, fluid and unfailing supply of labor." Growers called this labor *unskilled,* which implies replaceable and therefore cheap. Yet they also believed that only certain kinds of people made suitable "stoops," a name describing the posture imposed by the short-handled hoe. "The vegetable industry requires a class of stoop labor that is impossible to get without using either Mexican, Filipino or Japanese," wrote one grower, noting that immigration laws had already dried up the supply of Japanese workers. Growers argued that not even much higher wages would attract native-born Americans to field jobs; some suggested that certain foreign races were physically and temperamentally better suited for them anyway.[52]

Jobs in the packing sheds, by contrast, went primarily to white

Americans. Both women and men trimmed and graded the lettuce, while only men packed and iced the heavy wooden crates. Many shed workers enjoyed year-round employment, and all earned more than field workers. By the end of the 1920s they were unionized and considered the "aristocrats" of the West's farmworkers. Their high wages were due partly to strong demand during the boom years and partly to a tactic previously employed by California's migratory "fruit tramps": if wages were threatened, work stopped at peak harvest.[53] The perishability of the crops, in other words, gave them a measure of power over individual employers.

The Depression threw this power into question. With thousands of Dust Bowl refugees camped in shanties outside Salinas, growers assumed that plenty of people would accept lower pay. Up to a point, they were right. In 1933 they cut the hourly wages of Filipino field workers (who were not unionized) from forty cents to fifteen cents. In the sheds, they cut women's wages from forty to twenty cents an hour, and men's from sixty to thirty-five cents. The Grower Shipper Vegetable Association (GSVA), formed in 1930, claimed that the industry's survival depended on these measures; already many of its members had gone out of business.[54] Demand for lettuce was slumping and competition from other states was rising. A the same time, they faced high and non-negotiable costs for rail transport, icing, and land, most of which they leased from corporate landowners. In this squeezed position, labor appeared to be the only cost they *could* control.

In September 1936, a still-unknown John Steinbeck reported in *The Nation* on the grim consequences of the grower-shippers' logic. California agriculture had already seen dozens of strikes during the past three years, many of them violent. In Salinas, Filipino strikers had their camp raided and burned.[55] In the fall of 1936, wrote Steinbeck, there was "tension in the valley and fear for the future."[56] The lettuce harvest was about to begin, and 3,200 shed workers were set to strike. Although as packhouse workers they had the same legal protections as factory workers (the same was not true of fieldworkers), months of negotiations over a union contract had failed. The grower-shippers, who'd previously boasted of their industrial character, now insisted that industrial labor policies did not apply to producers of perishable foodstuffs. As the GSVA announced in a local newspaper, "This Association is favorable towards betterment of working conditions when warranted . . . However it must call attention to the fact that the vegetable business is solely agricultural and *cannot by its nature* be regulated as a manufacturing business. Consequently we cannot entertain or discuss questions pertaining to closed shop, guaranteed hours, seniority or like matters."[57]

Elsewhere the GSVA listed the reasons the union had nothing to complain about: employers provided radios in the sheds, sponsored baseball tournaments, and even let workers smoke on the job.[58] Not surprisingly, Steinbeck cared not so much about the details of union contracts as about the historical significance of an impending conflict. Things had changed since the days when only foreign male migrants worked on California's fruit and vegetable farms. The Dust Bowl refugees came with families, and many planned to stay. Most important, they were "undeniably American and cannot be deported . . . and they cannot easily be intimidated." For their part, the grower-shippers had twelve million dollars' worth of lettuce—more than three-quarters of the nation's fall supply—that could not wait. So the GSVA was determined to break the strike, first with backup workers and then, if necessary, with force. "The large growers," wrote Steinbeck, "are devoting their money to tear gas and rifle ammunition."[59]

Days after his article appeared, the growers put these weapons to use. The "Battle of Salinas" started with lettuce—five flatbed trucks full, headed for the packing sheds. Strikers threw rocks and bottles at the trucks; the police responded with gas grenades; and two days of street fighting ensued. To take on the strikers, the sheriff created a "citizens army" of 2,500 deputized local men and teenagers; high school students were reportedly recruited to make hickory wood clubs.[60] The growers also brought in a noted anti-communist army reserve officer, Colonel Henry Sanborn, to coordinate arrests and raids and ensure, above all, that the lettuce shipped. And ultimately it did. Once many of the strikers were jailed and it became clear that they could not stop the harvest, they gave up. East Coast markets got their iceberg.[61]

Although the lettuce industry averted a financial disaster, its image fared less well. A three-year investigation by the National Labor Relations Board probably had less effect than the immediate press coverage. Headlines described Salinas in "a state of war"; photos showed head-crackings, broken store windows, and bloodied lettuce. Even if the 1936 Salinas lettuce strike was nowhere near the era's biggest, longest, or most deadly, the violence in "the nation's salad bowl" clashed sharply with the virtuous image of salad itself.[62] Press coverage of the incident also primed the public for a spate of critical magazine articles and books, among them Carey McWilliams's work *Factories in the Fields* and Steinbeck's *Grapes of Wrath*. As Steinbeck's novel climbed the bestseller charts in 1939, *Western Grower and Shipper* noted bitterly, "California's agricultural labor problems have been given a thorough airing before the court of public opinion in recent months."[63]

The negative media coverage didn't necessarily hurt sales of California lettuce. But it did shine a harsh light on the industrial nature of a crop otherwise portrayed as a spontaneous product of soil and sun. And it revealed how an industry that harvested such a wealth of fresh produce for American consumers offered little but poverty to its American workers. At a time when organized labor enjoyed greater popular legitimacy and legal protections than ever before, this raised a troubling question. Did the perishability, seasonality, and high value of the crops—in short, their *freshness*—justify denying workers their rights as citizens?

A Revolution in Cool

Labor unrest in the lettuce fields made headlines again in the early 1960s. Under the leadership of Cesar Chavez, the United Farm Workers' strikes, marches, and boycotts all challenged California growers' contention that their industry was inherently unsuited to industrial labor laws. Eventually the state decided in the farmworkers' favor. With passage of the 1975 Agricultural Labor Relations Act, farm workers won rights to many of the protections long enjoyed by the rest of the country's workforce.[64]

Or at least some of them did. By the mid-1970s, only about 20 percent of California's farmworkers held U.S. citizenship, a significant drop from the immediate postwar era. The shift was not gradual; nor was it merely a by-product of Americans' growing prosperity. Rather, it came quickly after the invention of vacuum packing transformed the labor of lettuce—and that of many other kinds of fresh produce—all the way from the field to the supermarket. By all accounts this wasn't the aim of the inventor, Rex Brunsing; he was a hired hand himself. He worked as a vegetable traffic manager for one of Salinas's biggest shippers and simply wanted to speed up traffic.[65] Initially, he slowed it down. As one magazine later described it, "Back in 1946, motorists traveling the Castroville-Salinas highway reported some weird goings-on. A strange maze of boilers and pipe fittings were being rigged up. The project . . . looked like a poor man's attempt to build a rocket ship capable of reaching the moon . . . [they] were indeed looking upon a pioneer step in man's attempt to shoot for the stars. In this case the goal was to cool vegetables without using ice."[66]

Brunsing's "project" was a giant vacuum tube. With ice, it took between twenty-four and to seventy-two hours to chill a carload of sun-warmed lettuce down to the near-freezing temperatures needed for reliable cross-country shipping. But Brunsing found that if he rolled the same lettuce into his vacuum tube, evaporation sucked the heat out in a flash (later the technique came to be known as "flash-cooling"). It took less than thirty minutes to chill iceberg, half that time for spinach.[67] And unlike ice, the vacuum cooled the heads from the inside out. This reduced the risk of internal "sliming" and kept the heads cooler longer—so long, in fact, that vacuum-chilled lettuce packed "dry" in cardboard cartons reached the East Coast in fresher shape than ice-cooled produce.

These reasons alone made supermarkets among the earliest converts to vacuum packing. Besides the obvious appeal of longer shelf life, they wanted goods delivered in lighter containers. As they expanded into the nation's suburbs, they were hiring more and more women as shop clerks. As employees they were generally cheaper, but they could not easily lift the cumbersome, ice-packed wooden crates, each of which held four-dozen lettuce heads and weighed up to a hundred pounds.[68] From the supermarkets' perspective the cardboard box, which didn't require a man with a forklift to move, was itself a labor-saving technology.

In Salinas, shippers were also looking to cut labor costs. Since 1943, Public Law 45 (and later P.L. 78) had delivered them a regular supply of temporary Mexican field laborers, known as *braceros*.[69] As agricultural guestworkers the *braceros* had few legal rights, but they were legally bound to individual employers, ensuring that their labor stayed cheap even as they grew more skilled. By contrast the shed workers, most of whom were U.S. citizens, were considered factory employees under federal labor law. Their unions took advantage of local labor shortages to negotiate large and regular pay increases.[70]

In 1950 *Western Grower and Shipper* warned the unions that they were pushing the industry toward labor-saving mechanization: "Representatives of labor, in their eagerness to increase the take home pay of their own members, should give very serious consideration to placing the last straw on the camel's back before the break comes."[71] The "break" came with vacuum cooling, though not be-

Salinas packhouse workers circa 1940, shortly before vacuum cooling made their jobs obsolete. (Courtesy Monterey County Historical Society, Inc.)

cause it reduced packing man-hours. Rather, it got rid of the most costly labor. With no need for ice or crates, the tasks of trimming and packing the lettuce could be moved out of the sheds, where unionized workers earned $1.32 to $1.62 an hour, and into the fields, where *braceros* earned $.87 an hour. Altogether, vacuum cooling cut costs by nearly 30 percent.[72]

Despite the dramatic savings, Salinas's grower-shippers hardly rushed to invest. They didn't need to understand the physics of vacuum evaporation to appreciate how quickly it could ruin large quantities of valuable produce. Brunsing's jerry-rigged pilot plant provided all the proof they needed. Lacking temperature controls, it turned out more than a few crates of frozen iceberg. Shipping pre-cooled but un-iced produce was also risky, albeit for the opposite reason: unless the cartons traveled in rail cars with fan circulation—rare before the early 1950s—they stood a fair chance of getting cooked en route.[73]

In 1949, Salinas's one commercial vacuum cooler had so little business that its owners kept it running even with no lettuce inside, just to keep up appearances. Three years and a few union contracts later, local shippers overcame their doubts about Brunsing's invention. Then the changes came fast. By 1956, eight new vacuum cooling plants chilled 92 percent of the region's lettuce (up from 20 percent in 1952). *Braceros* were trained to pack lettuce in the fields. Meanwhile, most of the sheds closed down, costing thousands of jobs and creating "a localized depression" in Salinas.[74] By 1957, foreign guest workers accounted for 70 percent of Monterey County's agricultural labor force, up from 4 percent in 1950. Across the Southwest, the switch to vacuum cooling had similar effects.[75] With the passing of the ice age, the produce of the western lettuce industry got fresher and its labor much cheaper.

Super-Marketing and Shelf Life

Supermarkets' preference for cardboard boxes was not a trivial detail in growers' decision to abandon ice. Self-service grocery chains such as Piggly-Wiggly and A & P had been around for decades, but in the postwar era not even the largest fresh produce shippers could ignore their demands, much less their advice. Only through supermarkets could they hope to know and influence the all-important Mrs. Consumer. The American woman, *Western Grower and Shipper* told readers in 1946, "now spends nine-tenths of the American consumer dollar—and in food, she spends all of it. It behooves the vegetable grower to get acquainted with this lady, for all changes that now loom up in the vegetable industry . . . eventually come down to what has happened to her."[76]

Growers could see a few things that had happened: she'd gotten a refrigerator, a car, and she'd discovered self-service shopping. One result was a sharp decline in the number of food stores; 81,000 disappeared between 1939 and 1946, even as the number of medium and large chain stores quadrupled. Another was a marked change in the shopping experience. *Western Grower and Shipper* understatedly called it "a far cry from the general store of 1875. The merchandise is all accessible, nearly all packaged for self-service, the customers shop in wide aisles, brightly illuminated—with network music, and discreet commercials over a store broadcasting system!" But a third trend was more disquieting: purchases of canned and frozen foods had shot up so dramatically since the 1930s that Helen Goodrich, the California growers' expert on consumer trends, asked, "Are we building a race of processed food eaters?"[77]

Growers had worried about this trend for a while, especially once canned food makers began claiming—with the backing of the American Medical Association—that their products were as nutritious as fresh produce. They also worried that many American salads did not actually contain much lettuce, relative to ingredients like Jell-O and canned corn. And they'd fought back with the usual kinds of promotions—print ads, recipe books, promotional sales of giant wooden salad bowls, and plugs on popular radio shows. In the late 1930s, *Western Grower and Shipper* reported that the cookbook author and radio show host Ida May Allen mentioned lettuce "an average of ten times per broadcast," while other shows made and served salads to live studio audiences.[78] But after the war, as consumers increasingly both learned about and bought their food at supermarkets, such promotions weren't enough. Marketing experts warned growers that their produce would be "displaced" by more durable goods unless they adapted them to the new retail environment.[79] Yet what did this mean? How could they make their fresh produce as appealing to the housewife as frozen or canned alternatives—while still keeping it fresh? What kind of freshness did she really want?

Supermarkets had one word for the growers: pre-packaging. Again, this wasn't a totally new idea; chain stores in southern California had sold "Sunny Sally" bagged spinach and salad greens since the mid-1930s. But in the early postwar era, the vegetable industry's modernizers saw plastic wrap as an innovation as valuable and necessary as the refrigerator car. As one put it, "Fresh produce is the last great food packaging job that remains to be done. It will be difficult, but you can bet it will be done! For it is progress—and American."[80]

Pre-packaged produce certainly suited American supermarkets. Compared with "naked" produce, it was easier to buy, display, and track; it looked neat, sold faster, generally kept longer, and suffered less "consumer abuse." Shoppers who squeezed, peeled, and picked their way through the produce bins were a pesky problem for the self-service stores; plastic wrap offered a polite solution.

The main argument for pre-packaging, however, was that the shoppers themselves wanted it. A 1947 nationwide survey of 2,367 housewives (sponsored by Union Bag and Paper Corporation) found that 64 percent answered "yes" to the question of whether they "preferred to buy their products prepackaged and pre-priced"; 21 percent answered "no," and 15 percent were indifferent. Nearly three-quarters of those who responded "yes" also said they'd pay a penny extra for a packaged lettuce or bunch of carrots. Perhaps most important, pre-packaging was most popular in cities whose supermarkets had carried it the longest, such as Newark and Columbus (where approximately three-quarters of housewives approved of the packaging). In San Francisco, already a city known for its pristine produce, less than half the survey respondents liked

Miss Prepackage of 1952, pert, pint-sized, red-headed Virginia Gibson, singing and dancing Warner Bros. star, and Queen of the Second Annual Conference and Exposition of the Produce Prepackaging Association in Columbus, Ohio, is seen with a jumbo collection of prepackaged fruit and vegetables.

Plastic-wrapped fresh vegetables looked glamorous but posed practical challenges. ("Miss Prepackage," *Western Grower and Shipper,* July 1952, p. 19) the idea. Retailers took this as evidence that consumers just needed time to appreciate the value of plastic wrap.[81]

For fresh vegetable growers, the message from supermarkets was clear: package or perish. Just as with vacuum cooling, many balked at the cost and complexity. Because they dealt with living, breathing organisms, the wrong wrapping would quicken death, not prolong it. But alongside this technical challenge, the fresh vegetable industry faced a marketing imperative. They had to make sure that their own customers, the supermarkets, appreciated that the very rawness of fresh produce—the fact that it took at least a little work to make a meal—mattered to the housewife. "The salad is a 'convenience food' which can be prepared in a matter of minutes," *Western Grower and Shipper* explained, "[yet] convenience food must not be too easy since the homemaker needs, they say, 'to feel she has done a little something on her own.'"[82]

In a speech about the industry's packaging research, C. B. Moore, the director of the Western Growers' Protective Association, emphasized this point to a national meeting of food retailers in 1947:

> Today, American homemakers want to feed their families as wholesomely and as economically as possible; but they also want and intend to have the time and energy to create a fuller cultural and spiritual life for their husbands, their children and themselves. In short, the American woman is primarily interested in making a better home. More than streamlined, step-saving kitchens, Mrs. America wants food products which do not require long hours of kitchen work to prepare. These food products must be nutritious, tasteful and economical. She has no wish to do away with the home-prepared meal with her husband and children gathered about the dinner table. Quite the contrary.[83]

In 1961 the Salinas grower Bud Antle rolled out a product that he was sure both supermarkets and homemakers would appreciate. Shrink-wrapped iceberg, he claimed, lasted longer and looked better than "naked" or merely bagged iceberg. Unlike other kinds of plastic wrap, the polystyrene film didn't suffocate the lettuce. It also benefited from the latest in labor-saving technology: a giant motorized field-packing rig, ridden by workers who wrapped and sealed the lettuce heads while other workers picked them.

Among lettuce growers, shrink-wrapping set off what *Western Grower and Shipper* called "the hottest controversy since the introduction of the carton and vacuum cooling." Part of the controversy centered once again on labor, because the *braceros'* contracts prohibited them from riding machines. Antle's field rig clearly reduced man-hours, but it might also require more expensive workers (Antle was hiring women) and more amenities, such as field toilets. Nor were growers sure whether to invest immense amounts (the rig alone cost more than $20,000 in 1961 dollars) on a packaging process that might backfire. Retailers might abuse the promise of longer shelf life, destroying the "fresh factor" that attracted consumers to lettuce in the first place. Even if that didn't happen, it was clear that shrink-wrap was, well, very clear. "The close fitting, transparent wraps glaringly display any defects or minor damages to the head," noted *Western Grower and Shipper,* so more lettuce had to be culled.[84]

Ultimately, of course, Antle was right: supermarkets did like shrink-wrapped lettuce. Besides its looks and longevity, they liked that it was shelf-ready. It saved them the work of trimming the tough outer leaves from each head of naked lettuce. Consumers also took to the new packaging, and soon other California growers rushed to offer it. As for Bud Antle, his company (now Tanimura and Antle) went on to become one of the world's biggest producers of lettuce and other vegetables. Few in the food industry gave more credence to the famous line in the 1967 film *The Graduate:* in the world of fresh produce, there *was* a great future in plastics.

Who Shrunk the Veg?

The pre-packaging revolution proved a long and incomplete one. Contrary to some predictions, naked vegetables still had a place in most twenty-first century supermarkets. As Whole Foods and other upscale chains demonstrated, their color and machine-sprinkled dewiness did more for shopper loyalty and brand image than even the dazzling selection of frozen dinners. Yet if unadorned freshness still *attracted* customers, sales figures showed that many of them ended up buying packaged "fresh-cut" alternatives. Annual sales of these products went from near zero in the mid-1980s to $5 billion twenty years later. In the early 1990s, the fresh-cut market grew at a rate of 95 percent a year. Industry observers were impressed but hardly mystified. Once plastic-wrapped fresh produce came to seem as natural to consumers as boxed cereal, it was easy to introduce new and improved forms. And no improvement was easier to sell than convenience.[85]

Figuring out exactly how to package fresh-cut produce, however, was not so easy. Chopped into pieces, vegetables respire and lose moisture much faster than when whole. They need to be protected but not suffocated. Mixed salad greens are especially complicated, because different lettuce varieties respire at different rates. The first breakthrough came in 1988, when Steve Taylor, the grandson of the longtime Salinas grower Bruce Church, developed the breathable Keep Crisp bag. Pumped full of nitrogen, it kept oxygen levels low—which prevented browning—while flushing out carbon dioxide. Effectively it put the same technology used in controlled-atmosphere warehouses into a plastic bag. Once retailers saw that the bag really did keep cut greens crisp for up to two weeks, Taylor and his new company, Fresh Express, became the next Salinas super-success story. Sales went from $80 million in 1989 to $200 million in 1994 to $1 billion in 2003. By 2006 salad mixes accounted for more than 8 percent of all supermarket fresh produce sales, more than any other single category of fruits or vegetables.[86]

Later innovations have taken packaging to new levels of sensitivity, economy, and endurance. A "Fresh Hold" label, for example, both seals a bag and permits oxygen and carbon dioxide transfer through internal micropores. "Intellipac" packaging, made of patented "smart" polymers, becomes more or less permeable in response to temperature changes. Fresh-R-Pax microwave-safe pouches and trays absorb moisture on one side, breathe on the other, and promise to keep even the soggiest produce, such as fresh-cut tomatoes, looking good for at least two weeks.[87]

Most consumers probably wouldn't notice these high-tech features, and for good reason.[88] The produce, after all, is supposed to look fresh-cut; the packaging should not suggest that one's salad has actually spent a fortnight sitting in a highly engineered environment. Looking through the plastic, consumers might instead notice two things about the greenery inside. First, it has become much greener and more colorful all around. Right around the time the Keep Crisp bag hit the market, sales of iceberg went into decline while those of romaine and other leaf lettuces took off. Iceberg was still the country's most popular salad green in 2006, but per capita consumption had fallen by one-third since 1989, even as overall lettuce consumption increased.[89]

The news that iceberg actually contained very few vitamins might have accounted for part of the shift, but far more important was the new status attached to salad choice, and to food choice more generally. For certain segments of U.S. society in the 1980s and 1990s—post-counterculture baby boomers and young urban professionals—rejecting the postwar generation's mass-produced meat-and-potatoes diet in favor of presumably healthier, fresher, and more natural foods showed taste and distinction. To these foodies, iceberg looked like the Wonder Bread of the lettuce world. Instead they nibbled on mesclun, the mix of baby greens introduced to American food culture by Alice Waters, chef and owner of the famous Chez Panisse restaurant in Berkeley.[90]

Originally cultivated and delivered to the restaurant by local organic farmers, mesclun quickly caught the attention of a new generation of Californian agro-capitalists. Besides its elite appeal (and corresponding price tag), mesclun possessed qualities suited for the mass market. Once washed and bagged, it was as convenient as chopped-up salad mix. Foodies were busy people, after all. It was also easier to keep fresh, because the leaves were harvested whole. Not least, the harvest itself could potentially be made much easier. Paradoxically, the very lightness of baby lettuce makes it better suited to machine handling than bulbous, easily bruised iceberg heads. Mesclun can be mowed. It can also be blown, shaken, and spin-washed.[91]

California's most successful mesclun moguls, Drew and Myra Goodman of Earthbound Farms, took several years to develop suitable machinery. But by the time they did, Earthbound had expanded from 2.5 acres in 1984 to 26,000 by 2006, becoming the world's largest organic fresh vegetable producer. In an apparently un-ironic allusion to the California lettuce industry's history of Darwinian corporate consolidation, Earthbound partnered with Tanimura and Antle to create Natural Selection Foods. Although its pristine image suffered when some of its organic bagged spinach was found contaminated with *E. coli* bacteria in 2006, the company otherwise seemed to have found a mechanical solution to the lettuce industry's longstanding labor problem. It seemed to prove that attractive, fresh, and (usually) healthful vegetables no longer demanded much drudgery in either field or kitchen. They could spring from nature to plate, untainted by human toil.[92]

Or could they? The second thing consumers might notice about the pre-packaged produce was the prevalence of "baby" vegetables more generally. These included not just the ubiquitous baby carrots (which are really just sanded pieces of adult carrots) but also extra-fine *haricots verts* ("French beans") and snow-peas, and micro-varieties of zucchini, corn, and pattypan squash. Their ancestry, once again, is vaguely French; first appearing in *nouvelle cuisine* restaurants in the 1970s and 1980s, they drew on a much older French appreciation for *primeurs,* the first vegetables of spring. Once they reached supermarkets in the United States and Britain, they were not associated with either formal dining or springtime. On the contrary; trimmed and petite, baby vegetables took less time to cook than frozen peas, and not much more expertise. And like frozen peas, they were available even in the dark of winter. Retailing for upwards of seven dollars a pound in the United States (and even more in Britain), baby vegetables offered convenience and elegance in an irresistibly cute if pricey package. As one Manhattan shopper put it in a 1985 *New York Times* article on the new fad, "in a quiet way, baby squash and baby carrots are taking over the earth." Supermarkets profited from the conquest, and in some cases helped arrange it. They sought out suppliers and encouraged the development of new varieties. The British chain Marks and Spencer takes credit for the world's smallest runner bean: the fifteen-centimeter-long Weanie Beanie.[93]

As with salad mix, improved packaging helped keep baby vegetables looking youthful. But expanded production and trade depended even more on advances in air transport, and especially the 1970s advent of big-bellied passenger jets like the 747, which had ample room for commercial cargo. Airport cold storage came somewhat later.[94] These developments mattered because most baby vegetables are too perishable to spend long in transit, yet too expensive to produce in the countries that consume them. The United States imports most of its supplies from Central and South America (Guatemala and Peru count among the major producers), while Europe counts on its former African colonies. A few Southeast Asian countries export baby vegetables to both the East and the West.

While most baby vegetables come from countries with either mild or "counterseasonal" winters (as defined by the North), weather alone doesn't determine where they're grown. Because their market value hinges on their flawless appearance, they demand constant attention and gentle handling. The *haricot vert,* for example, must be protected against wind and hail, watched and pruned so that it does not grow crooked, and harvested at precisely the right time. Even a twenty-four-hour delay and the bean grows too big and fat. In California or Spain this kind of care would cost much more than it does in Africa or Guatemala, and then it still might not result in adequate attentiveness. Among merchants who trade in these semiprecious crops, it's a rule of thumb that a measure of desperation makes for good discipline. People who have no other way to make money will take better care of their vegetables than people with choices.[95]

As a result, some of the highest-value crops are produced in some of the most unlikely places—places that would not seem the logical choice if delivering freshness were the sole priority. Burkina Faso, for example. At first glance, it seems doubtful that this country could produce anything perishable for export: it's stuck in the middle of West Africa's drought-prone Sahel and is one of the poorest nations on earth. Refrigeration is scarce, as are paved roads. As a former colony of France, though, Burkina Faso has both direct flights to Paris and nearly a hundred-year history of growing food to French tastes. Growing *haricot vert* for French colonials used to be a form of forced labor. Since the early 1970s it has been the country's most important "non-traditional" export crop produced by small farmers around a scattering of donor-funded irrigation projects. When all goes well, it's a much more profitable crop than cotton, the country's biggest foreign-exchange earner.[96]

Yet things often don't go well. Geography is partly to blame; some of the major production zones are several hours from the airport in Ouagadougou, the capital city. The country's green bean merchants targeted these regions not just because they had irrigation but also, paradoxically, because they were remote. Closer to the city, farmers can grow cabbage and tomatoes for the urban market. "It's difficult to find people who'll work as hard as the *haricot vert* requires," said one trader. "So I go farther out to find quality." And the green beans harvested in these regions are often of amazing quality: uniformly slender, straight, and smooth-skinned, they'd meet with any Parisian chef's approval. Whether the beans reach Paris, however, is another matter. If a truck breaks down or a plane arrives a few hours late, the beans wither. At that point, they're worth less than the cardboard cartons they travel in. It's not uncommon for several tons of produce to perish on the runway. Farmers usually bear the brunt of the losses, but everyone in the business has at some point watched freshness die.

Not all baby vegetables come from such remote places, but most come from areas where a lack of alternatives ensures cheap and diligent labor, as well as a willingness to abide by the market's demands. Most of Guatemala's producers of fresh vegetable export crops are highland Mayan peasants. In a country where a tiny elite owns most of the land, they're lucky to have a half-acre. They're also lucky to have any schooling, yet they must make sense of complicated U.S. expectations governing on-farm hygiene and pesticide use.[97] If a border inspector finds excess pesticide residues in a single box of Guatemalan micro-zucchini, the United States could suspend imports of all the country's fresh produce. It has done so before.

In Zambia, foreign investment in export vegetable plantations took off in the mid-1990s, after the near total collapse of the country's copper industry left it as poor as Burkina Faso. Selling almost exclusively to British supermarkets, Zambia's baby vegetable industry had to comply with strict standards not only for food safety and "traceability" but also for worker welfare. The farms won't employ anyone who looks remotely like a child laborer (defined as someone under sixteen). They have clinics and childcare onsite, and housing for some of their workers. They offer more amenities than what you'd traditionally find on a California vegetable farm. But a woman picking baby corn in Zambia in 2003 earned only about a dollar a day. And her job was in demand.[98]

That countries as poor as Zambia produce such pricey and perishable vegetables could be seen as either an opportunity—for dollar-a-day jobs are better than none—or an outrage. It has been portrayed as both. Still, it is important to realize that the transnational trade in baby vegetables represents simply an extension of vegetable growers' longstanding dependence on a foreign migrant workforce. This dependence is so established that we take it for granted. Yet it's worth considering how our very ideal of freshness in vegetables—as a natural, even evanescent quality—has contributed to the historic undervaluing of the human labor that produces them. It's an ideal encouraged by supermarkets and other dealers in fresh produce, because it permits them healthy mark-ups for qualities that cost them little or nothing. The real cost has always been borne by the people whose work we don't see.

six

Milk

BORDER POLITICS

Nature intended . . . milk to be partaken while fresh at
the fountain of its production.

—M. J. Rosenau, *The Milk Question* (1912)

The drive to my milk supply is short but telling. I live in Hanover,
New Hampshire, a two-minute jog from the Connecticut River. Fifteen miles downstream, Pat and Mary McNamara keep two hundred Holsteins on a farm that Pat's parents bought for four thousand dollars in 1950. Their property rolls down to the river on one
side and stretches up into pine forest on the other. On a warm evening in early summer it's hard to imagine a more bucolic setting,
especially compared with the car-clogged tangle of big box stores,
fast food chains, and freeway ramps you have to drive through to
reach it. Over the past several years, the strip has expanded along
with the region's population. At the farthest, newest end of the
strip, just past Wal-Mart and Home Depot, a concrete company
does a brisk business in the raw materials of suburban sprawl.

The McNamaras are among a growing number of dairy farmers
who bottle their own milk. While their returnable glass bottles look
old-fashioned, the McNamaras see them as a tool for facing modern challenges. Since the days when their grandparents farmed,
milk prices and milk consumption have generally dropped. Yet costs

have risen, especially the costs of keeping cows on prime suburban real estate. Simply selling bulk milk would not pay the bills, Pat says: "We process in order to be able to farm." Like many farmers who depend on local markets, they also spend a lot of time cultivating local loyalties. Each year the family hosts a stream of school groups, and thousands come to their "open farm" days. Visitors take home pints of chocolate milk and memories of a setting they would hate to see swallowed up by a strip mall.

In northern New England, this landscape of barns, pasture, and rolling fields of corn and hay doesn't just look nice. It draws tourists and second-home buyers who together generate far more revenue and jobs for the region than cows do.[1] That's one reason state governments in this region want to keep dairy farms around, even if they don't produce milk nearly as cheaply as the behemoth, ten-thousand-cow dairies in the Far West. Another reason is that if the farms disappeared, milk would have to be trucked in from much farther away.

Technically this is easy enough. In many countries, most of the milk supply comes from afar, arriving as powder or heat-treated cartons. But in the United States milk remains the most local of supermarket staples. Historically its own perishable nature kept it from traveling far, at least in its fresh liquid form. Yet even after the spread of refrigerated transport, government policies treated fresh milk like clean water: as a beverage so important that it had to be kept within easy reach of city markets. This view reflected new ideas about the value of fresh milk, and for public officials it posed a new problem: namely, how to preserve urban "milk-sheds" against the pressures of urban growth. It also raised questions about the boundaries of food markets more generally. By the early twenty-first century, these questions mattered more than ever. Anxieties about globalization fueled nostalgia for the days when most foods, not just milk, came from nearby. Calls to "buy local" increased even as many farmers found it impossible to *stay* local.

Beneath the easy appeal of local food—freshness counts among its many supposed virtues—remained the tougher issue of who could produce and consume it. Who ultimately gets to be a local?

Milk and Microbes

Humans have relied on other animals' milk for food since at least 5,000 B.C. In regions too cold or arid for agriculture, such as the Sahara and parts of Central Asia, milk was *the* food of nomadic peoples, providing far more regular sustenance than meat. Across Eurasia, milk-bearing animals ranked among peasants' treasured possessions, turning otherwise useless forage into much-needed fat and protein. In medieval England, dairy products were the "white meats" of the rural poor. Across the pre-industrial world, dairying peoples showed their appreciation for milk in their creation myths, in their literature and art, and in their protection of the animals themselves.[2]

By the turn of the twentieth century, milk had become what some experts called the single greatest public health threat facing Western industrial societies. Physicians warned of the perils hidden behind "the veil of opaque whiteness"; newspapers and social reformers demanded public action to address the "milk problem." What exactly was the problem? With its rich liquid mix of protein, fat, and sugar, milk's greatest appeal was also its greatest menace. Milton Rosenau, a Harvard Medical School professor and public health official, summed it up in his 1912 book, *The Milk Question:* "bacteria loves milk," he wrote, "almost as much as the baby does."[3]

This wasn't news. Centuries before Pasteur discovered bacteria, dairying peoples recognized that milk was vital in more than one sense. It wasn't just nourishing; it was alive and volatile. Its qualities changed quickly and dramatically when heated or shaken or even if just left to sit. People also knew that some changes made milk distasteful or even deadly, while others made it even more delicious and nourishing than when fresh from the udder. Milk's value to dairying societies depended on how well they avoided the former and mastered the latter. For most, this meant not consuming much milk in its fresh, raw form.[4]

Instead, they learned to control the elements living in milk—what we now know as members of the *Lactococcus* and *Lactobacillus* genus of bacteria. Many of them also live in the human gut, among other places. In milk they serve as fermenting agents. By converting lactose (milk sugar) into lactic acid, these bacteria not only produce the tart taste and thick texture associated with yogurt, kefir, and other cultured dairy products; they also slow the attack of microbes that cause rot and disease.

Fermenting (or souring) milk is the simplest and probably oldest method of extending its edible life. In warm climates, nomadic peoples soured their milk just by keeping it in animal-skin bags, where the lactic acid bacteria flourished. West Europeans preferred buttermilk, the tart, low-fat liquid left over after churning. And from Egypt to the Balkans to India, people ate *leben, kisselo mleko, dadhi*—all basically yogurt, made by adding a "starter" of fermented milk to fresh milk. The hundreds of recipes for fermented dairy products worldwide testify to humans' longstanding appreciation for the power of well-managed bacteria to turn fresh milk into safer, more durable, and perhaps more healthful foods.[5] Yet in all its forms, fermented milk lasts only several days, unless well chilled. It is also relatively bulky, which historically limited how far people attempted to carry or trade it.

Butter and cheese, by contrast, can be made to travel far and keep well. For many dairying peoples, these foods figured more importantly in the daily diet—and certainly more importantly in commerce—than did fresh raw milk. This was especially true during the seasons when their animals had no nursing offspring and relatively little to eat, and therefore produced little if any milk. In northern climes, cows normally went "dry" for much of the fall and winter unless kept on a rich diet of grains and root crops, which was an option only for the wealthy or for farmers supplying urban markets. Most households, however, aimed to turn as much fresh milk as possible, whenever possible, into products that could be eaten or sold throughout the winter.

The great diversity of such products (France alone famously boasts several hundred cheeses) shows that milk is as fertile a medium for human inventiveness as it is for microbial growth. But who made what cheese depended as much on the resources available to the cheesemakers—including, most basically, how many animals they had—as on where they lived or what kind of milk and cultures they used. In general, households with only one cow or a few goats made cheese in small batches (typically after one night's and one morning's milkings) and consumed or sold it quickly. In France, cheeses such as Camembert and Brie fell into this category, as did fresh *chèvre*. In England, "green" cheese referred to any soft, young variety. The simplest were drained curds, typically made with skimmed milk.[6]

Only larger operations—monasteries and feudal estates, for example, or later village cooperatives—had enough milk, labor, and alternative food sources to make the kind of aged, low-moisture cheeses suited for long-distance trade or months of storage. By the seventeenth century in England, farmers in some counties specialized in cheeses for the London market. Varieties such as Cheddar, though associated with particular localities, were the very opposite of local foods. Farmers made them for bulk sale, with individual cheeses often weighing in at well over a hundred pounds.[7]

Both butter and cheesemaking traditions crossed the Atlantic to colonial North America, along with Old World dairy cows. As in Europe, on-farm dairy processing tended to be women's work, and for some it became a major business. Both before and after the Revolution, New England's cheese and butter shipped the length of the eastern seaboard and beyond, from Quebec to the West Indies. By 1796 the United States exported 2.5 million pounds of butter and 1.8 million pounds of cheese annually, some of it traveling as far as Africa and China.[8] Given that onboard refrigeration was still several decades off, the butter had to be preserved in a mix of sugar, salt, and saltpeter (potassium nitrate) and the cheese carefully aged. Ideally, according to an early nineteenth-century cheesemakers' guidebook, it would end up "sweet, full of spirit, and free from every nauseous flavor."[9] Not all cheese met such standards; often it ended up crawling with maggots. But Europeans praised the overall quality of New England's dairy products. Export markets for farmstead butter and cheese were especially important to the newly settled, sparsely populated northern states, since they had no local markets of any size.[10]

The Making of a Milkshed

Liquid milk also traveled during the era before the railroad, but not very far. As in England and continental Europe, dairy farms clustered in and around cities.[11] Colonial Boston had milk ferried in from Charlestown, just across the river. In Manhattan, milk came from farms in what is now Midtown, including the blocks occupied by Saint Patrick's Cathedral and the Empire State Building. Landless city-dwellers also kept livestock. In Boston, goats were outlawed by the mid-seventeenth century but cows remained, and their owners sold extra milk to neighbors. Even in 1830s New York, one visitor remarked that "there are more ways than one of keeping a cow . . . These animals are fed morning and evening at the door of the house, with a good mess of Indian corn boiled with water; while they eat they are milked, and when the operation is completed the milk-pail and the meal-tub retreat into the house, leaving the republican cow to walk away, to take her pleasure in the hills, or in the gutters, as may suit her fancy best."[12]

While urban Americans seem to have liked fresh milk when they could get it, this was not necessarily very often or very much. Exactly how much is difficult to say before 1850. Most historians agree that people consumed little fresh milk relative to butter and other dairy products. Certainly both cows' low yields and slow transport limited how much milk was even available. The average New Yorker in the 1840s probably used no more than a third of a pint per day, including whatever went into coffee or baked goods.[13]

In the second part of the nineteenth century, demand for fresh milk skyrocketed, partly due to a sharp decline in breastfeeding. Women of all classes and backgrounds turned to cow's milk to supplement or replace breast milk, despite the stern warnings of pediatricians and social reformers. Not all had the same reasons. Working-class women had to return to work; middle- and upper-class women had to resume their busy social schedules. In the past it was not uncommon for neighbors and relatives to nurse one another's babies. But this practice faded as American society became more urban and family life more centered on the nuclear household.[14]

At the same time, cow's milk became more readily available, even in the poorest neighborhoods. Corner shops ladled it into customers' bowls and pitchers, while peddlers' wagons made pre-breakfast deliveries. Until the 1870s, much of this milk came from "swill dairies," where cows lived off the slurry waste of adjacent distilleries and breweries. Distillery grains are actually nutritious, and in dried form farmers still use them as a feed supplement. Swill dairies, though, fed cows little besides hot fermented slop. Never leaving their three-foot-wide stalls, many grew too weak and sick to stand and had to be hoisted aloft with special harnesses for milking. Often their tails rotted and fell off.[15]

One of the earliest critics of swill dairies was the New York temperance campaigner Robert Hartley, who in the 1830s encouraged farmers just outside the city to send more of their presumably purer "country milk." Lack of transport, though, limited how much they could deliver. Swill dairies still produced most of the city's milk in 1858, when the muckraking newspaper *Frank Leslie's Illustrated* launched a much more graphic exposé. Full-page ink drawings of fearsome milkmen and filthy stump-tailed cows disgusted consumers well beyond Manhattan. Boston banned its own swill dairies the following year; remarkably, New York City's remained legal until 1873, and Chicago's until 1892.[16]

If nothing else, milk from the swill dairies was often quite fresh, simply because it came from only a few miles or blocks away. Consumers knew as well as health experts did that nearness mattered when it came to milk. Many of them had emigrated from countries where goatherds drove their animals door-to-door and milked them in customers' homes. So when the railroads first began hauling in milk from the countryside in the 1840s and 1850s, not everyone saw it as a great improvement over in-town supplies. Who knew how old it was, or how many people had handled it? The second concern was at least as worrisome as the first, because more handling meant greater likelihood that the milk would be dirty, diluted, skimmed, and then adulterated to hide all its other flaws. Milk dealers' deceptions were legendary; even milk sold as the freshest possible—"warm from the cow"—might actually have been boiled (to stave off souring) and delivered warm from the fire.[17]

Despite such concerns, railroad milk accounted for the bulk of supplies to major cities by the last quarter of the nineteenth century. The outlawing of swill dairies was far from the only reason for the shift. For one, raising cows even in the most cramped quarters was hardly an efficient use of downtown real estate. For another, the demand of booming populations quickly outstripped whatever urban or suburban dairies could produce. Not least, as railroads extended service across the rural Northeast, enterprising middlemen went out seeking supplies.

One of the most successful was Harvey P. Hood, a Vermont farm boy who headed to Boston in 1844 and found a job delivering fresh-baked bread. The pre-dawn hours must have suited him, for after a year with the baker and a short stint as a shopkeeper, he bought a door-to-door milk delivery route. He soon discovered one reason that many milk peddlers diluted their stocks: the supply from nearby farms was unreliable at best, and often scarce when demand ran highest. Hood nonetheless managed to build up his trade, and by 1856 he had saved enough to buy his own farm in Derry, New Hampshire. He bought neighbors' milk in addition to producing his own, and he took it all to Charlestown on the 7 a.m. daily train.[18] Initially he loaded the forty-quart cans himself, then rode with them to and from the city in the same day. He did his accounts at a desk in the freight car on the way back. Unlike many dairy farmers, he delivered even on Sundays—babies, he said, didn't know what day of the week it was.[19] This punishing work schedule helped Hood become one of a handful of railroad milk dealers who dominated the Boston milk market by the century's end. Each monopolized the use of ice rail cars along particular "milk runs" stretching into northern New England. Since farmers had no other way to get their milk to Boston, they had to accept whatever prices the dealers offered.[20]

Many farmers did accept, for the railroads otherwise brought them mostly hard times. Carloads of cheap bounty from the West undercut New Englanders' markets for everything from grains to wool to pork. The rails also carried away the young and ambitious to city jobs and frontier farms. By 1880, Vermont, New Hampshire, and Maine had together lost more of their native-born sons and daughters to out-migration than any other three states in the nation. Farmland prices collapsed, selling for under $10 an acre in the 1890s, down from $100–$200 an acre a quarter-century earlier.[21] Dark forests reclaimed abandoned pastures. Visitors described fallen-down barns, rotting churches, inbred villages, and a

Harvey P. Hood (seated), founder of New England's largest dairy company. Hood's business expanded along with the region's rail network. (Photograph from the H. P. Hood Collection. Courtesy of Historic New England)

population left sullen, haggard, and prematurely old by an unrelenting diet of biscuits and salt pork. In 1897 an *Atlantic Monthly* article titled "The Future of Rural New England" saw reason for the exodus: "The very deadness and dullness within exert[s] a strong expulsive force . . . as soon as a boy has become able to walk, he has walked away."[22]

But the Holstein population, at least, was booming. Even before the Boston milk dealers began seeking supplies in the hill country, cows earned more than most crops. In the second half of the nineteenth century, the opening of small-town creameries and cheese factories offered farmers an alternative to processing milk themselves. The creameries' butter was often considered superior to farmstead, because their centrifugal separators could churn butter from fresh rather than soured cream. Early factory cheeses proved less commercially successful, partly because many were made with lard rather than butterfat.[23] Still, local butter and cheese manufacture encouraged New England farmers to focus their energies on raising cows. Vermont soon became the most dairy-dependent state in the Union.[24] Its farmers, like those in New Hampshire and Maine, welcomed the arrival of the milk runs. When big-city dealers like Hood began offering better prices than the creameries and cheese factories could pay, most farmers signed on to sell them their milk. It looked to them like one of the very few commodities that they could produce without fear of western competition. Milk's perishability would protect them, at least for a while.

To produce more milk, New England farmers planted more feed crops. Rather than let their animals go "dry" during the winter, they fed them a diet of hay and grain to keep them fat and lactating year-round. The new schedule kept farmers busy. In 1910 the Vermont Department of Agriculture reported that farmers in the state were "neither building fences nor houses, nor painting their buildings, or doing anything except buying a few cows and shipping their milk to Boston."[25] This was not entirely accurate; many rural

Vermonters had turned their homes into inns for tourists. "Farm holidays" appealed to city folk seeking relaxation, pleasant scenery, and fresh food (though many were disappointed by the pork-and-biscuit fare). The pastoral landscape of northern New England looked idyllic and old-fashioned to them, but in fact it was new and entirely commercial.[26] It was a landscape defined less by the needs of the home and local market than by the unceasing thirst of the cities.

Poison from the Countryside

In some ways, American cities' milk supplies had greatly improved by the early twentieth century. The swill dairies were gone, and in the rural areas larger herds and higher-yielding cows ensured year-round abundance. Iced rail cars brought loads in daily from two, three, even four hundred miles away. Service was also more convenient. Even fifth-floor apartment dwellers could count on fresh supplies at breakfast, delivered by a milkman who climbed the stairs while his horse pulled the wagon to the next building.[27] Schools and factory cafeterias served milk, and quick-lunch shops sold it by the glass. And when prices of other foods climbed, milk remained economical, especially relative to its nutritional value.[28]

But milk remained as deadly as it ever was, especially for babies and young children. Urban infant mortality rates averaged more than 15 percent nationwide, and in some mill towns and cities— including the nation's capital—they topped 25 percent. Despite ongoing improvements in sanitation, water supplies, and health care, babies stood much less chance of surviving in cities than in rural areas, where infant mortality rates averaged around 11 percent.[29] Diarrhea and related intestinal ills accounted for the most deaths, especially in hot weather. They went by many names: cholera infantum, marasmus, intestinal inflammation, dysentery, even just "the summer complaints of children."

Although it was often impossible to pinpoint the cause of these ailments, cow's milk ranked among the primary culprits. Evidence from across Europe and the United States showed that bottle-fed babies died at much higher rates than their breast-fed peers, especially if the milk came from dubious sources. A 1903 New York City study confirmed the popular view that diarrhea-related deaths peaked in summer, when milk was often not properly chilled either in transit or at home. Indeed, in *The Milk Question,* Rosenau called the average household icebox a "a snare and a delusion . . . [and] more often a household incubator. Instead of being cold, it is only cool; sometimes actually warm."[30]

People had long known that milk spoiled more quickly at warm temperatures. But now scientists' microscopes revealed exactly how fast it deteriorated over time and place. One intrepid New Jersey dairy inspector, Samuel Sharwell, proved this point by tracking four cans of milk "from cow to consumer." Starting at four a.m. in Baldwinsville, New York, he took the cans from four "ordinary" farms (none had refrigeration) and accompanied them to Newark, 340 miles away. At the station platform, the first samples revealed that the milk already contained up to 50 times more bacteria than the Newark standard of 100,000 per cubic centimeter. Then he boarded the iced freight car. "There are several points of difference between a Twentieth Century Limited and a milk train," he remarked. "But if one is following milk over the country he must ride with it. So I found myself in semi-darkness among the milk cans bumping southward over the Lackawanna tracks as speedily as such trains usually travel." Along the way the temperature rose from 50 to 65 degrees Fahrenheit, despite the ice, and the milk's bacteria count increased exponentially. Arriving in Newark at midnight, the cans went to four shops, where Sharwell took a final set of samples the next morning. By then the milk contained on average almost 500 times more bacteria than the legal limit.[31]

Sharwell didn't test for the five other "plagues" scientists had linked to milk: typhoid, diphtheria, scarlet fever, septic sore throat, and tuberculosis. While diseased cows were sometimes the source of infection (especially in the case of TB), public health officials worried at least as much about milk's long haul to market. Over greater distances milk not only grew less fresh but also passed through more environments and human hands, encountering more opportunities for infection. In addition, wrote Rosenau in *The Milk Question,* the long haul created a dangerous moral buffer:

> When the producer and consumer were near neighbors and closely acquainted with each other, the one had a personal interest in the product he furnished the other . . . the separation between the two has [now] lulled the conscience of the producer. He no longer sees the results of his acts, and is skeptical concerning the dire consequences . . . The intervening distance, time and circumstances help to strengthen the immunity of the producer in his own opinion. He is very apt to claim that the infection was introduced or the poison developed after the product left his hands. . . . It is human nature to concern ourselves more about things we make for our friends and neighbors, whom we know, and see frequently, than it is for some far-off foreigner. This phase of the situation nourished the taproot of many of the difficulties found, not only in the milk question, but in the entire food supply of the world.[32]

Rosenau saw plenty of local evidence of this global problem. Although the passage of the 1906 Pure Food and Drug Act helped to reduce fraud and adulteration in the rest of the food supply, milk was routinely implicated in Boston-area epidemics. Even Harvard students fell victim to the "white poison." After strep throat swept through the campus in 1908, tests of the milk found "pus and abundant streptococci."[33]

Despite milk's dangers, most public health experts agreed with Rosenau that "we cannot do without it." Even doctors and social

The Long
vs.
The Short Haul

70 percent of city babies get their food through a tube 60 miles long.

It takes about 36 hours — often 42 hours — for the milk to run from the cow end of the tube to the baby end of the tube.

This tube is open in many places and baby's food is frequently polluted. It is often wrongly kept in overheated places.

Then there may be a diseased cow at the country end of the tube.

And Yet Some People Wonder Why So Many Babies Die!

On the other hand the mother-fed baby gets its milk fresh, pure and healthful — no germs can get into it.

To Lessen Baby Deaths Let Us Have More Mother-Fed Babies.

You can't improve on God's plan.

For Your Baby's Sake — Nurse It!

Warnings about the dangers of "long-haul" milk had little effect on sky-high urban infant-mortality rates. (*Weekly Bulletin of the Chicago Health Department*, reprinted in M. J. Rosenau, *The Milk Question*, 1912) reformers working to lower infant mortality rates felt that it was no longer enough just to promote breastfeeding. Something had to be done about the commercial milk supply. But what? Here the consensus ended. On the surface, debates about the "milk problem" centered on science and legislation. The root issue, though, was the growing distance between producers and consumers.

On this issue, milk reformers fell into two main camps. Neither wanted to bring cows back to the cities or even necessarily the suburbs. As the author of *The Health of the City* observed, "the increasing value of real estate . . . only too often places the dairy farm in some damp, undrained spot."[34] Instead, one camp saw long-haul milk as an unavoidable hazard that technology—namely pasteurization—could render harmless. The other favored legal measures to shorten the haul and therefore lessen the risks associated with transporting milk. Beneath these different approaches to reform lay different ideas about what made milk fresh.

Pasteurization kills bacteria with sub-boiling heat, typically between 145 and 162 degrees Fahrenheit. In milk it also stops the enzymatic activities that contribute to spoilage. Louis Pasteur first demonstrated the process on wine in the 1860s, and it was applied to milk shortly afterward. For years pasteurization didn't get far out of the laboratory, relative to the growing scale of milk commerce. In the United States, some dairymen in the late nineteenth century used it to delay souring and save money on ice. Many scientists debated its efficacy. Pasteurization's most ardent supporter, however, was neither a scientist nor a dairyman, but the owner of Macy's, Nathan Straus.

Although Straus and his wife, Lina, supported many charitable causes, the death of two of their own children—including a daughter whose death they blamed on bad milk—gave the campaign against infant mortality a particular urgency.[35] Beginning in the early 1890s, Straus consulted a handful of pediatricians and concluded that pasteurization was the best way to stop "the slaughter of babies" in New York City. He was so convinced of this approach that he bankrolled an entire private-distribution system. He bought milk from vet-inspected farms, pasteurized it in his own state-of-the-art laboratory, distributed it in refrigerated containers to depots around Manhattan, and sold it below cost. Besides cutting infant mortality, Straus hoped that his system would set an example and ultimately convince the city to open its own pasteurized milk depots.[36]

Selling the milk to New Yorkers was easy. After all, the retail baron knew that location was key—almost as key as low price. On the East 3rd Street Pier, a popular spot for family outings, he built a pavilion where women and children could enjoy cold milk in hot weather. It stayed open until midnight—later than Macy's. At storefront depots in poor neighborhoods, take-home bottles came with ice. In parks, penny-a-glass milk attracted all ages and both sexes, including (Straus boasted) men who would have otherwise quenched their thirst in pubs. On one day in July 1898 at the City Hall Park depot, a reporter observed milk dispensed at a rate of eight glasses a minute.[37]

Selling pasteurization to public officials and the broader clean-milk movement initially proved much tougher. Scientific opinion remained split through the 1910s, and even supporters often described pasteurization as a stopgap measure—a way to purify milk until it could be delivered truly pure. Opponents raised several concerns. First, they doubted its effectiveness. Scientists did not yet know the precise "thermal death point" of different milk-borne pathogens. They did know that many of the dealers and farmers who pasteurized their milk cared little about precision; they just wanted to delay souring. Second, opponents of pasteurization feared that it was all *too* effective at killing healthy bacteria and nutrients, leaving milk "devitalized." They believed that invalids couldn't digest pasteurized milk, and that it caused scurvy and rickets in babies. Third, many in the clean-milk movement believed that pasteurization would encourage "dirty habits." Farmers could continue shipping germ-ridden milk if they knew it would eventually go through the cooker. Some of these concerns never went away, as the contemporary raw-milk movement shows.[38]

Many opponents of pasteurization promoted "certified" milk, which the pediatrician Henry Coit introduced to New York around the same time that Straus opened his first depots. Certified milk was raw but came from farms inspected by medical commissions to ensure that they met exacting standards of hygiene. The bacterial count of the milk itself could not exceed 10,000 per cubic centimeter. Public health officials agreed that certified milk was clean enough for clinical purposes—Coit's original intent—and within several years, hundreds of medical commissions were inspecting dairy farms. But certified milk cost about twice as much as "market milk," so poor families could rarely afford it. At its peak it accounted for less than 1 percent of the total supply.[39]

Mary Putnam, the founder of the Massachusetts Milk Consumers Association (MMCA) and the sister of Harvard President Abbott Lawrence Lowell, did not think that either pasteurization or certification went far enough. She urged mothers to buy certified milk if they could afford it, but her ultimate goal was to make fresh, clean, raw milk available to all Massachusetts consumers. A second goal, which she believed would advance the first, was to get as much milk as possible from the state's own producers. As head of the MMCA she fought for a bill that would authorize the state's Board of Health to inspect all dairy farms supplying its milk, including those outside its borders. At the time inspecting dairies fell to cities and towns, though only about a quarter actually bothered. Lax local control was not a major problem in the state's rural areas, where many people milked their own cows. Boston, though, got only about 20 percent of its milk from local sources. The rest came from more than fifty miles away, and mostly from out of state.[40]

Born into one of Massachusetts's "first families," the wife of a prominent attorney, and a well-known child health advocate in her own right, Putnam had little trouble finding high-profile backers. The Harvard physician Rosenau became an especially important ally. So too did George Ellis, an agricultural college administrator and also an owner of one of the few dairy farms left in Newton, an affluent Boston suburb. Ellis lent his name to the clean-milk bill to show that it was not merely a consumer's cause.[41]

Most of the groups that endorsed the "Ellis Bill"—civic leagues, the Springfield Baby Feeding Association, the Boston Fathers and Mothers Club, and a long list of labor unions—did in fact represent urban consumer interests. Among rural voters, Putnam's campaign raised hackles. "I am not in sympathy with all the noise about milk," a resident of Lawrence wrote to her. "I was raised on a farm; I am acquainted with at least 5 dairies near Lawrence. Every one of the five is clean . . . Keep up your noise and in a few years Massachusetts will not have enough milk produced to feed even the little babies."[42]

Other letters to Putnam accused the MMCA of scaring consumers away from milk, making it even harder for the state's dairy farmers to stay afloat. As one critic wrote to Putnam in 1911:

Why have so many producers sold off their cows and gone out of the business? . . . What can you expect of the farmer if you don't give him a square deal? All we ask is to make a living honestly which you will not let us do when you cause so much talk on the unhealthiness of milk. Why has the consumption of milk fallen off. . . ? Simply because people like you and especially newspaper reporters have had so much to say about the danger from the use of dirty milk . . . Instead of that why don't you cooperate with the producers and tell people the value of milk as a human food and educate them to using more of it and to be willing to pay a living price for a good product.

When you are ready to dip into your own pockets and pay for a sanitary, healthful, nourishing quality of milk you will get it.[43]

In 1915 the Boston Chamber of Commerce confirmed that "agitation has lessened public confidence" in milk, and the public was drinking less of it. Putnam alone probably agitated many mothers who heard her lecture about the "horrid little germs" that grew in less-than-fresh milk.[44] At the same time, she struggled to win over the state's dairy farmers. As she said in a speech before the Farmer's Institute, "There is no opposition between our interests! . . . I have come here today in the hope . . . of convincing producers that we consumers feel that we want and need the same thing that you want and need. . . . We want milk produced as near our homes as possible. We want to bring back all the cows that have left the State, and put our dairies on a sound basis, so that—more and more—the milk consumed in Massachusetts will be produced within her borders."[45]

The MMCA did lobby for state-of-origin milk bottle labels, as well as for railroad rate reforms that would favor in-state supplies. The group's testimony in favor of the rate reforms, though, stressed the need to protect the consumer's food, not the producer's livelihood. It suggested that long-distance trade was fine except when it brought "stale" milk: "Where the free interchange of stoves, shoes, coats, millenary, etc., between various parts of the country is to be promoted, it may be well to allow railroad tariffs to encourage long hauls and much railroad business, but in the case of milk, the service of the public, to which the railroads are bound, leaves no room for doubt that we should have short hauls and fresh milk."

Putnam reassured farmers that the Ellis Bill would not override local inspectors, nor would it require costly new equipment. Massachusetts consumers just wanted and deserved "good, plain, clean farming," she often said, and once the Ellis Bill restored their confidence, "the demand for milk will increase by leaps and bounds."[46]

Above all, she emphasized, the bill would protect the state's dairy farmers against competition from producers in northern New England, where land, labor, and feed all cost less. This argument assumed that such producers were also less hygienic and therefore less likely to pass inspection. Putnam occasionally said as much, but at least one dealer in long-haul milk did not agree. Testifying against railroad rate reforms, the H.P. Hood Company called the Commonwealth's supply "the meanest milk we receive . . . as a rule Massachusetts farmers are not clean. Their barns are dirty, their milk is not up to the requirements."[47]

The Ellis Bill never did win the support of Massachusetts's rural districts. Perhaps farmers saw through the rhetoric about some states having clean farms and others dirty ones. They probably realized that such stereotypes were political conveniences, created to win votes. Massachusetts farmers knew that their counterparts in Vermont or New Hampshire could produce milk just as clean as theirs—or even cleaner, if they had more land to pasture their animals. They also must have known (even if Putnam did not) that milk's freshness depended much more on proper refrigeration and prompt delivery than on what side of the state border the milk was produced.

By the time clean-milk legislation made it through the Massachusetts assembly in 1914, it had been so weakened to satisfy agricultural interests that some of its original backers, including Rosenau, revoked their endorsements.[48] Nonetheless, infant mortality rates were already declining rapidly in Boston and other cities. Although historians still debate why babies stopped dying in such large numbers, pasteurized milk no doubt helped, just as it helped reduce adult death rates from diseases such as typhoid. Outbreaks dropped markedly in New York City after it passed a mandatory pasteurization law in 1914. Over the next several years other cities followed suit. By the mid-1920s, about 75 percent of New England's milk supply was pasteurized.[49]

Large companies such as H.P. Hood pasteurized even before the law required it, because it helped them expand both their supply and their distribution networks. Many smaller dealers could not afford the new technology and soon left the business. Ultimately pasteurization didn't just protect consumers against certain diseases. As Melanie Dupuis shows in *Nature's Perfect Food,* pasteurization protected milk itself against the dangers inherent in an industrial food system, turning it into a substance that no longer needed to be handled through personal connections. It could ship longer distances and pass through more hands—or increasingly, machines— and still arrive "fresh" (albeit cooked) at the consumer's doorstep.[50]

Legislating the Local

The most remarkable aspect of technological change in the milk industry is how little it affected the basic geography of trade, at least in the United States. New England still gets 95 percent of its supplies from either its own farms or New York State.[51] Admittedly, the country's urban milksheds have expanded since the 1920s, and tanker trucks long ago replaced the railroad's milk runs. But given how much fresh food routinely crosses continents and oceans, why do milksheds even exist? Economic theory says that it is simply more efficient to produce a bulky perishable commodity near its main market. But that doesn't explain why farmers can continue to produce this commodity on land where it would also be efficient to build, say, tract housing. Milksheds persist because of state intervention that dates back to the Depression, a time when dairy farmers protested plunging prices by waging strikes. Violence and hardship highlighted the need for more "orderly" marketing of the country's most unstable staple.[52]

The mind-numbing details of federal milk-market policies matter less than their basic geographical logic. Here a bit of background helps. In the 1920s dairy farmers enjoyed good prices for their milk, despite the growing market power of big processors. Advertising, nutritionists' promotion of "protective" foods, and school lunch programs all drove up demand for dairy products.[53] Membership in dairy cooperatives also increased. Organized regionally, co-ops negotiated prices with the milk processors, who in turn received stable year-round supplies. Prices depended on end use and season: processors paid more for milk they sold *as* milk ("Class I") than they did for milk destined to become butter or cheese ("Class II") because the former had to meet higher hygiene standards. They also paid more in fall and winter, the traditional milk "dry" seasons, to account for higher feed costs.

For their part, the co-ops managed milk supply over both time and space. Milk produced nearest to urban markets went to fluid milk dealers; milk from farther away, where costs were lower and production still more seasonal, went to cheese and butter factories. Individual members received a "blended" price, regardless of where their own milk went. They also enjoyed a secure market. Given how little time milk can wait around for a buyer, this security mattered. But it was also fragile, for a couple of reasons. First, plenty of farmers did not belong to co-ops. While membership in a few urban milksheds (Baltimore and Washington) topped 90 percent by the early 1930s, only about half of Boston's milk came from co-op members, and in New York the number was even less.[54] In addition, the spread of the automobile was rapidly eroding the monopoly of milk-run dealers. Even farmers far from a rail line could now truck their cans to the nearest town or city.

When the Depression hit, all semblance of milk-market stability collapsed. Demand dropped yet supply surged, as anyone with a few cows tried to make a little money from them. Small dealers took advantage of this desperation to buy and sell milk cheap; big dealers responded by slashing their own payments to co-op farmers. On average such payments fell by almost a third between 1929 and 1932; in some markets, such as New York, they fell by more than half.[55] Of course, many farmers suffered during the Depression, but dairy farmers had less flexibility than most. Cows needed to be fed and milked twice daily regardless of market demand.

The fact that consumers had come to expect milk to arrive daily, however, gave farmers a degree of leverage. If fresh milk had become as taken-for-granted as tap water, what better way to get the public's attention than to turn off the tap? Dairy farmers' strikes—also called "holidays"—were not new. New York State farmers struck as early as the 1880s, and in 1910 New England farmers, demanding better prices, cut off more than 80 percent of Boston's milk for over a month. The city's dealers imported supplies from wherever they could find them, leading one local newspaper to describe milk shipped from New York City as "old enough to vote."[56] In the 1930s, dairy farmer strikes swept through the country. Aimed at cities rather than employers, they caused much more popular alarm than labor unrest elsewhere in the food system. A temporary shortage of peas or iceberg lettuce might go almost unnoticed, but a city faced with an imminent "milk famine" had to take emergency measures.

Sometimes consumers themselves mobilized; in 1938 the *Los Angeles Times* reported that "a large housewives' organization . . . may send militant women into home milk delivery wagons and even to the milking sheds on farms to keep the vital food supply from drying up." In Chicago, the *Tribune* published doctors' advice for surviving a milk strike "without danger to the health" (they recommended canned and dried milk, as well as cheese).[57] At the same time, processing companies scrambled for alternative supplies, and public health officials debated whether to allow the sale of milk from outside the traditional milkshed. Such milk might have to travel suspiciously long distances—say, from Wisconsin to New York—and come from farms unseen by a city's own inspectors. It also represented a betrayal of local producers. As New York's Health Commissioner Shirley Wynne said on the eve of a strike in May 1933, "While my sympathies are and always have been with the producers, my first concern is with the residents of the city of New York. If the threatened strike takes effect, I shall be compelled to enlarge the milkshed . . . This, I realize will be disastrous to the dairy farmers of our present milkshed . . . [but] we cannot permit any interference with the steady supply of so important a food as milk."[58]

Attempts to "interfere" with supply often led to violence, and usually huge amounts of spilled and spoiled milk. Picketers typically tried to force delivery trucks to dump their loads. In New York, a sheriff's deputy escorting a truck was crushed beneath its wheels. In the Chicago milkshed, strikers battled police, hurled explosives, and invaded non-striking farms, where they doused milk cans with kerosene. Outside of Milwaukee, police tear gas failed to stop strikers from seizing and dumping tens of thousands of pounds of milk a day; only Red Cross trucks carrying supplies for babies and invalids were allowed into the city. Even in sleepy Portland, Oregon, dairy farmers came to the pickets armed with fists and dynamite.[59]

Most of the time striking farmers suffered more than consumers, simply because cities found other sources of milk. Eventually, though, their protests convinced Congress to pass a series of acts aimed at stabilizing the nation's milk markets.[60] Over the short term these measures raised farmers' income. Over the longer term, they turned traditional urban milksheds into official marketing "orders," with boundaries defining who got paid what for their milk. These boundaries assumed not that nearby milk was fresher or safer but that fresh milk, period, always had to be available and affordable. To achieve this goal, both boundaries and prices had to take account of seasonal variations in supply and demand. Ideally, farmers closest to the city (also called "inner ring") would find it profitable to deliver milk year round, and more remote farmers ("outer ring") only when the market needed it. The rest of the year,

Striking New York State dairy farmers patrolled their towns to make sure no milk went to market. (Dairy Farmers Union Collection, Special Collections, Owen D. Young Library, St. Lawrence University)

the "outer ring" would supply nearby towns or manufacturing plants. Only during times of extraordinary demand would it make economic sense to turn to supplies from outside the marketing order. During the Second World War, for example, Boston's dealers trucked in millions of pounds of milk from the Midwest.[61]

From the beginning, marketing order boundaries were not just complicated but also intensely political, because they affected both retail milk prices and farmers' earnings. In general, farmers within a city's traditional milkshed wanted boundaries that excluded outsiders, in order to limit supply and keep prices high. This was the stance of the Chicago Pure Milk Association, an Illinois dairy farmer co-op that supplied the city's milk dealers up through World

War Two. As its name suggests, Pure Milk played to consumers' concerns about safety and freshness in order to keep dealers buying only its members' milk, despite the vast quantities produced just over the Wisconsin border. But such concerns faded as transportation and hygiene improved, and the incidence of milk-borne disease plummeted. During the 1930s, Chicagoans snapped up cheap long-haul milk sold in "cash and carry" stores on the city's outskirts. This evidence that consumers cared more about cost than about proximity helped Wisconsin dairy farmers convince the Department of Agriculture to enlarge the boundaries of the Chicago marketing order so that they, too, could receive Class I prices.[62]

Eventually most marketing orders spanned at least a few states; Boston, for example, became part of the New England order. The spread of the federal highway system made it possible to ship fresh milk farther and more cheaply, as did the shift from hand-carried milk cans to bulk tanks.[63] But these technical advances did not always overcome farmers' determination to keep local markets protected. In the Los Angeles area, for example, they defended themselves against both urban sprawl and rural competitors by incorporating and creating three "dairy cities" in the late 1950s. Located in the L.A. suburbs but zoned for agriculture, the 18 square miles of Cypress, Dairy Valley, and Dairyland became home to a herd of 75,000 dairy cows—at that time, the highest-density cow population in the world. Although feed had to be imported (some materials came from as far away as the Philippines), the milk itself traveled only a few miles to market.[64]

Clearly "local" didn't mean the same thing to everybody, and it still doesn't. Urban milk reformers such as Rosenau waxed nostalgic about the days "when the producer and consumer were near neighbors," ensuring freshness and presumably neighborly care. For mid-twentieth-century dairy farmers, by contrast, battles over who belonged in what ring or order—who counted as a local and where—all boiled down to pricing. This remains the case, though

The dairy industry's switch to bulk milk collection required farmers to install tanks in their own milking parlors. Many left the business instead. (Farm Security Administration, Office of War Information Photograph Collection, 1941. Library of Congress)

now a farmer's monthly milk check reflects an even more complicated calculus than it did in the 1950s. Marketing orders across the country, for example, factor distance from Eau Claire, Wisconsin, into their prices for Class I milk.[65]

Out of the muddle of milk pricing emerges one clear fact: fresh milk remains the only food regulated to keep it *near* its market, relatively speaking. The federal government subsidizes other agricultural commodities to encourage exports or keep out imports; some states, such as Alaska and Vermont, promote their own state "brands" nationally and internationally. But milk's legislated localness is unique and reflects both its own natural qualities and its historic importance in the American diet. It has also proven politically self-perpetuating, in that dairy farmers live and vote all over the country, creating a congressional "dairy bloc" much broader than other farm lobbies.

Cows Versus Malls

Ongoing government support has created a contradictory dairy landscape. On one hand, it has brought massive and ultimately ruinous excess. The federal marketing order system ensured sufficient year-round supplies by encouraging greater year-round production. This inevitably resulted in seasonal oversupply, which processors turned into less perishable dairy products. Markets managed surplus, in other words, by making cheese and butter. Unfortunately, this orderly arrangement did not plan for changes in American eating habits. Per capita milk consumption peaked in 1945. From then on, farmers continued to buy new machinery and feed and breed their cows to produce ever more milk even as cholesterol-conscious, cola-drinking consumers drank less and less of it.[66]

The federal government bought and warehoused vast quantities of surplus dairy products, unloading whatever it could on food aid and school lunch programs. By the early 1980s, though, taxpayer-supported "butter mountains" had become politically indefensible. The Reagan administration tried to stanch the flood of excess milk by cutting price supports and buying up entire herds, most of which ended up at the slaughterhouse. Relatively few New England farmers took advantage of this "Dairy Termination Program," but many went out of business. Vermont lost nearly 70 percent of its dairy farms between 1980 and 2003, leaving it with fewer than 1,500. In Massachusetts their numbers fell even more sharply, from 829 in 1980 to 167 in 2007.[67]

On the other hand, public support has sustained dairy farming in parts of New England where suburbanization and rising land values would otherwise have made it unviable.[68] Even in the once-depressed hill country of Vermont and New Hampshire, an acre worth five dollars in the 1890s might sell a century later for ten, fifteen, or even twenty thousand dollars. To encourage farmers not to sell, state "current use" programs protect them against escalating property taxes. Conservation agencies such as the Vermont Land Trust also pay dairy farmers not to develop their properties.[69] In both cases, the aim is to preserve the "working landscape" that has historically provided milk for the region's cities. Because of that history, it is easy to fold freshness into arguments for saving local dairy farmers—even though their milk could now travel from coast to coast and arrive with shelf life to spare.

Freshness and landscape preservation were thoroughly entangled in the debate over the federal Northeast Interstate Dairy Compact, a 1997 law that raised minimum milk prices for the region's farmers. Congressional opposition from Midwest dairy states doomed the program after only four years, but in the meantime it fired up urban New Englanders much more than did most farm bills. Critics claimed that higher milk prices hurt the poor; backers pointed to the compact's broader quality-of-life benefits, among them fresher supplies. Quoted in the *Boston Globe,* one Massachusetts congressman suggested that the feared alternative was not so much spoiled milk as a spoiled view:

Tiny, family-owned dairy farms have defined the New England landscape for generations: Black-and-white cows munching on green grass are the embodiment of country living for many. But a controversial effort to save the region's dwindling number of dairy farms faces an uncertain future today as Congress considers whether to continue a subsidy program for farmers . . . "Without the dairy compact, a lot of local dairy farmers will close up shop," said Representative James P. McGovern, Democrat of Massachusetts, a cosponsor of the bill which would reauthorize the compact and include five other states from New York to Maryland. "It also means fresher milk for New England consumers," McGovern said. "You shut down one of these dairy farmers, you could see a strip mall the next day. No one wants that."[70]

To many Americans, the strip mall had become both the source and the antithesis of freshness. Strip mall supermarkets counted fresh milk among their top-selling, highest-turnover items.[71] Together with fast food chains, they sold most of the fresh food consumed in the United States. But the media were full of stories about the processed nature of perishables ranging from bagged lettuce and gassed fish to McDonald's Apple Dippers. So were they really fresh? Purists didn't even consider them "real food." Instead they saw strip malls as displacing "real" food producers, including dairy farmers and their cows.

The typical Holstein is now an extraordinary producer, yielding up to eighty gallons of milk a day. But the regional economy arguably benefits more from her role as a consumer of hay, grain, and grass, the crops that give rural New England its pleasantly pastoral appearance. Without the cow, the land would revert first to overgrown fields and eventually to deep forest, the region's indigenous ecosystem. Already during the twentieth century, forests reclaimed much of the land cultivated in the nineteenth.[72] While this would be good news in many parts of the world, New England now has more forest than it wants. Too much forest bores tourists. It scares off second-home buyers. Cows munching on green grass, however, have the reverse effect. They add value to the land just by mowing it.

Ferment on the Farm

In the century since New England's inn hostesses struggled to meet their guests' expectations of "farm fresh" meals, the region has become a tourist destination as renowned for its cuisine as for its lakes, mountains, and outlet malls. It has also become home to many urban émigrés who take food—and especially local food—very seriously. Some have made it their livelihood, whether as chefs or farmers or craftspeople. Others have simply made eating local a way of life. Among the latter, many have joined or started locavore groups, linking the region into a larger movement that encourages eating within one's "foodshed." Like other locavores (a group in the San Francisco Bay Area takes credit for inventing the term), those in New England show their support for local food through various means, ranging from foraging classes and potluck dinners to legislative campaigns. Their signature events are periodic "challenges": a week or month when members eat only food grown within a hundred miles.

Robin McDermott founded a locavore group in Vermont's Mad River Valley in 2006, two years after she and her husband moved their software firm north from Connecticut.[73] She described herself as a "poster child" for the eat-local movement, because in her past life she didn't give a second thought to where her supermarket pork chops came from, or why they cost so little. Living in Vermont opened her eyes. "I think a lot of people who have recently moved here," she said, "moved because they loved the beautiful farmland. And they realize that if we don't support the local farmers, the farmers aren't going to be here. There'll be a tract of houses where there used to be farms." It helps that her town of Waitsfield already had at least one restaurant that specialized in fare made from locally grown ingredients. "We would go there every Friday night," she recalled. "We would leave there and I would say, gosh I feel so good. I feel so good after eating fresh local food."

While New England's locavores list many reasons for eating local, the region's restaurants, co-ops, and farmers markets tend to push taste and freshness over politics. "That pulls in a much bigger group of people," one New Hampshire locavore said. The challenge, of course, is to hold on to those people during the several months of the year when there's nothing fresh on the land but snow. The locavores each allow themselves a few "wildcards"—imported foods such as coffee, olive oil, or perhaps oranges. McDermott said she has gotten used to winter salads made from less-than-fresh vegetables, such as root cellar carrots. The area's locavores exchange information about food canning and drying, and about which farms sell cellar-stored vegetables during the winter. For the most committed, freshness is a seasonal pleasure.

Ironically, one of the few foods produced both widely and year round in New England is also one of the hardest to buy locally: milk. The Mad River Valley has five good-sized dairy farms, McDermott said, yet all ship most if not all of their production out of state. "It's the same thing that happens in every other town in Vermont. The Agri-Mark [dairy co-op] truck comes through, or the organic truck comes through, and they take the milk and it goes off, never to be seen again." McDermott and other townspeople don't see much of the farmers, either. Since they don't sell at the local market and work the typical dairy farmers' long hours, McDermott said, they are "totally disengaged from the community. There's no interaction with them at all."

Like other locavores, McDermott seeks out raw milk from the few local farmers who sell it. Raw milk is scarce partly because Vermont, like many other states, permits only on-farm sales, and then only in limited quantities. But such laws haven't stopped the estimated half-million American raw-milk drinkers from finding supplies. In Boston and Manhattan, farmers deliver bottles to locations known only to members of black-market "milk clubs." Some farmers sell their stocks as pet food (wink, nod). In states where people can only legally drink raw milk from their own cows, dairy farmers sell "herd shares" (purchasing just part of a cow usually suffices). And for many devotees, the only option is to drive a couple of hours each week into the countryside—and hope that their suppliers don't get raided and shut down by state authorities.[74]

Many members of the modern raw-milk movement, like the pure-milk campaigners of a century ago, care most about the health benefits. They point to studies that suggest raw milk can strengthen the immune system, improve digestion, and protect children against allergies and asthma. Others just love its rich, straight-from-the-udder taste—a taste perhaps enhanced by its semilicit status. For the preservationist group Rural Vermont, the top priority is to bring raw (or what it calls "farm fresh") milk above ground, so that dairy farmers can also reap its benefits.

Rural Vermont scored a victory in April 2008, when the state legislature passed a bill that increased the amount of raw milk a farmer could sell each day from twenty-five to fifty quarts.[75] This still amounts to only a fraction of a cow's daily yield. But farmers can earn at least three times as much from those quarts as they can from bulk milk sales. If farmers could sell more of their milk raw, McDermott pointed out, they could get by with smaller herds and fewer hours in the barn. They could get to know their customers. They could have "a more normal life." But she doubted whether her own farmer neighbors would ever cut their ties to the Agri-Mark truck. After all, it takes all their milk all year round. Not even the most loyal locals could necessarily promise that.

Milk Beyond Borders

In late January 2008, a herd of dairy cows took over Mexico City's historic Plaza de la República. The cows' owners, dairy farmers from the state of Hidalgo, had come to town to protest the latest stage of the North American Free Trade Agreement (NAFTA). They had company. At the first of the year, Mexico dropped the last of its tariffs on U.S. farm products, opening the door to increased imports of "especially sensitive" products such as corn, dried beans, sugar, and powdered milk. While the cows chewed their cud, the dairy farmers gave away 25,000 liters of fresh milk and called on the government to do something about prices that had dropped to 4 pesos a liter (about 38 cents).[76]

For many passersby, drinking milk from Mexican cows was as unusual as the cows' own march on the capital. Long before the dairy farmers' NAFTA protest—long before NAFTA even existed— Mexico had become a major market for nonfat dry milk (NDM), getting the bulk of its supplies from the United States. By the end of the twentieth century it was the world's largest importer. In a big, poor, semitropical country, NDM is admittedly practical. It doesn't need pasteurization or refrigeration (though it does need to be mixed with clean water). It can be easily transported and stored. It can be used to make a wide variety of foodstuffs, though much of it is simply reconstituted. And it's cheap, especially when processed with vegetable oils rather than butterfat. This has made it attractive to the Mexican government ever since it first began reconstituting large quantities of imported NDM in the 1940s and selling it to urban consumers at a subsidized rate.[77]

In rural areas, most government support for milk production dried up a while ago. Past projects to "modernize" the dairy industry brought high-yielding Holsteins and enriched feeds to some regions, where a handful of mega-dairies now operate much like their counterparts in Wisconsin and California. For the rest of Mexico's dairy farmers, reconstituted milk was a source of hardship—and street protest—even before the tariffs came down. It didn't help that the government allowed all kinds of NDM and whey-based formulas to be labeled simply "milk."[78]

The American farmers in the underground raw-milk trade and the *campesinos* in Mexico City might seem to have little in common. But both see their livelihoods threatened by the U.S. dairy policies that promote chronic overproduction.[79] These policies are riddled with ironies. For example, the milk marketing order system was originally supposed to help the kind of small dairy farmers who now feel most threatened. By keeping them afloat, the system aimed to keep Americans' milk supplies local, or at least regional. In this narrow sense, it worked. Over time, though, mounting surpluses increased the pressure to find global markets.

For the dairy industry, nonfat dry milk powder was the ideal export commodity. Lightweight and long-lasting, it disposed of a liquid once considered a cream by-product rather than a desirable drink in itself.[80] During World War Two the United States opened milk powder plants in all the major dairying regions, becoming the world's largest producer. Afterward, it faced an oversupply problem: although a variety of food and feed manufacturers used powdered milk, American consumers preferred theirs fresh. Conveniently, a thirsty market lay just across the border. Mexicans didn't actually drink much milk, compared with Americans. But the high prices and poor hygiene of city milk supplies were a source of worry for the Mexican government. With the opening of a state-of-the-art reconstitution plant in 1946 (financed partly by American capital), Mexico became a large and reliable buyer of U.S. milk powder.

Mexico wasn't the only market targeted for NDM exports. In 1947 the American Dairy Association predicted that the United States would become "the milk shed of the world."[81] Briefly, it was. Then Europe rebuilt its own herds and boosted its exports with generous subsidy programs. By the late twentieth century, tins of European powdered and condensed milk filled grocery shelves across the developing world. In most countries, milk from local cows couldn't begin to compete, especially since scarce refrigeration made it hard even to get to market. Instead, syrupy sweet *lait concentré* found its way into postcolonial coffee cultures everywhere from Hanoi to Dakar. Donor-funded dairies in West Africa turned imported powder into "fresh" milk and yogurt.

The U.S. response to the European dairy export subsidies was to create its own. The Dairy Export Incentive Program (DEIP), founded in 1985, pays "bonuses" to firms that find foreign buyers for U.S. goods. Early on, a handful of big dairy processors captured the bulk of the DEIP bonus money. In 1994, the Pennsylvania-based company M. E. Franks exported some $92 million worth of dairy products (most in the form of milk powder sold to Mexico and Algeria) and collected nearly $52 million in DEIP subsidies, more than one-third of the total given out that year. The irony of this policy is that the millions spent to compete against European dairy exports ended up in the pockets of a European firm—Ecoval, a Belgian multinational that bought M. E. Franks in 1992. This buyout also allowed Ecoval to ship European milk powder to the United States and then out again under the DEIP program.[82]

Companies like Ecoval benefit from the fact that relatively few of the world's people expect their milk to be either fresh or local. Even in dairy-rich countries such as France, Germany, and Switzerland, most consumers buy cartons of ultra–heat-treated (UHT) milk, which boasts a six-month shelf life.[83] In China, where milk has only recently become a staple in urban households, more than 40 percent of the national supply comes from Inner Mongolia's 800,000-cow Mengniu Dairy, founded in 1999 and already one of the world's biggest dairy farms. Besides UHT milk, the company sells a variety of "milk beverages" in flavors such as wheat, egg, and green bean. New Zealand, meanwhile, is the world's biggest exporter of milk products, despite being local to almost nowhere.[84]

In a fiercely competitive global milk market, the United States dairy industry depends heavily on Mexico—and not just to buy its goods. Even in northern New England, farmers rely on Mexican workers to milk their cows and muck their stables. In Vermont, Mexican workers first began to appear in the late 1990s and now account for roughly one-third of the dairy workforce. Perhaps not surprisingly, most come from the southern states where NAFTA reforms hit small farmers hardest, such as Jalisco and Chiapas. Almost all are undocumented, since the seasonal guest-worker programs used by New England's fruit and Christmas tree farms don't cover year-round dairy jobs.[85]

This is the final irony of U.S. dairy policies: Americans' access to fresh, local milk is ensured not just by a complicated set of federal and state laws but also by an unlawful, transnational flow of labor. In Vermont this labor force has become what one study calls "superlocal," though not by choice. Conspicuous in a state where 93 percent of the population is white and few speak Spanish, the Mexican dairy workers risk deportation just by taking a trip to Wal-Mart. So most live on the farms where they work and leave only rarely, usually relying on their bosses for rides. Life can get lonely and dull, especially in winter. The main appeal of the superlocal existence is that workers have few expenses or distractions, so they save more money.[86]

Tourists aren't likely to see much evidence of Vermont's Mexican dairy workforce. Driving down some of the state's two-lane highways, they'll see little evidence of the twenty-first century, period. The landscape of pastures and hayfields hasn't changed all that much since the railroads first pulled northern New England into the Boston milkshed. But the farmers whose herds maintain the landscape make no secret about their dependence on undocumented Mexican workers. Even a farm run by the governor's in-laws openly employs them. In an interview with a Vermont newspaper, the farm's owner praised their reliability and said that no one else around wanted to do dairy chores. These also haven't changed much, at the most basic level. "It's odd hours," he said. "You get dirty and can get kicked by a cow."[87]

American consumers have long liked the idea of keeping dairy farms nearby, so as to keep milk fresher and the view greener. But freshness is not so pretty when it's 4:30 in the morning and the only view in sight is a cow's soiled backside. Vermont's Mexican dairy workers would probably agree that localness has its limits. Given a choice, most of us don't actually want to get that close to the source of our food.

seven

Fish

What has pleasure come to? . . . The belly of gourmets
has reached such daintiness that they cannot taste a fish
unless they see it swimming and palpitating in the very
dining room.

—Seneca

It was lunchtime at the Boston International Seafood Show, and
every booth had bait. Attendees from all over the world and every
branch of the industry teemed around the sample trays, snapping
up crab legs, sashimi, smoked trout, clam chowder, and deep-fried
shrimp. The entire subterranean expo hall smelled like a fish and
chips shop at closing time.

Despite the stale air, freshness figured big at the seafood show.
Hundreds of booths promised it, whether they dealt in seafood itself or in one of the myriad technologies used to keep it fresh. Well-
rehearsed sales managers talked up the latest advances in ice slur-
ries, controlled-atmosphere packaging, high-speed filleting, and
super-frozen tuna. But few displays could compete with the barn-
size, bright blue banner of the Chilean fish-farming company Pesquera Los Fiordos. It reeled in attendees with just two words: *Think
Fresh.*

How fresh? Less than four hours from harvest to packing, said the young man at the booth. Passing out glossy flyers and free pens, he explained how the company harvests its salmon live, de-stresses the fish in holding ponds, then processes them "pre-rigor"—that is, before rigor mortis softens their straight-from-the-water muscles. These are the gold standards of the modern salmon farming industry, scientifically proven to produce fresh taste and texture. It's to be expected from Agrosuper, the Chilean conglomerate that owns Pesquera Los Fiordos. The country's biggest producer of pork, beef, poultry, and fruit, Agrosuper boasts control over every part of the food chain, from breeding through retail distribution. That explains Pesquera Los Fiordos's giant banner. It also explains its mission "to design, grow, process and commercialize SuperSalmon products."

Such statements raise the hackles of people who catch salmon for a living. Some of them have come to the seafood show from as far away as Alaska, hoping to find big buyers for their smoked sockeye and air-shipped fresh Chinook. Their booths are much smaller than Pesquera Los Fiordos's, their flyers less slick. They have photos, though. Pictures of their boats, of the crystalline bays where they fish, of salmon arched against blue skies and pine forests. Passers-by can get details about their fishing gear or seasons or the salmon's Omega-3 content, if they want them. But the bottom line is in the pictures. They boast of a quality that companies like Pesquera Los Fiordos cannot offer: wildness.

In the very long history of humans' catching, farming, and eating seafood, only recently has "wild" become such a lucrative and politically loaded label. This doesn't mean that wildness has not figured prominently in people's ideas about what constitutes fresh fish. On the contrary: fish-loving societies have long recognized freshness as not just a fleeting and precious quality in seafood but also, quite literally, a vital one. From Tokyo to Seattle, Hong Kong to Marseille, people want to taste the fish's untamed life. They'll pay a premium to partake of the energy and strength that helped it survive, as both a species and an individual. People in these places don't necessarily agree about the best ways to keep fish fresh, or how best to cook it (if at all), but they do share a historical relationship to this last wild food. As industrialization and affluence have made wildness of all kinds seem distressingly remote, the perceived vitality of wild fish has increased in both economic and symbolic value. It has become so valuable, in fact, that we humans will go to the most extreme measures to get it—to the ends of the earth, if necessary.

The Most Fragile Freshness

It doesn't take a sushi chef to appreciate why freshness matters in fish, or how quickly it disappears. Even in the refrigerator, where some perishables can last for weeks if not months, most seafood very soon shows and smells its age. Exactly how soon depends partly, of course, on when and how it came out of the water. A quick and nonstressful death helps, because the lactic acid produced by a struggling fish hastens deterioration. Shelf life depends also on physiology and ecology; fatty species spoil faster than lean ones, temperate species faster than tropical ones, and marine species faster than fresh-water fish.

In all cases, the basic problem lies in refrigeration's limited effects on the enzymes and bacteria that make good fish go bad. These culprits have evolved to function in cold aquatic environments, so the temperature of an ordinary fridge does not faze them. It just slightly slows their attack on the molecules that give fresh seafood its bright color, firm texture, and sweet taste. Together with oxygen, the enzymes do their business first, breaking up fatty acids and leaving fish softer and less shiny. Not all varieties suffer from the earliest stages of aging; some fishermen say that salmon meat actually needs a day or two to "relax." Once the bacteria get started, however, a fish quickly goes foul. Living naturally in its gills and slime (one good reason to rinse fish), these bacteria convert tasty amino acids and protein into a variety of unappetizing substances, such as ammonia, putrescine, and cadaverine.[1]

Compared with mechanical refrigeration, ice keeps fish colder, moister, and thus fresher. Submerged in ice chips or slurry, salmon can last for a week, cod for about two weeks, and warm-water species such as carp and tilapia for up to three weeks. So for many modern fishers and fishmongers, ice remains as indispensable as it was a century ago. Its enduring usefulness stands out in an industry always in search of newfangled techniques to keep fish fresh, or at least looking that way. But in much of the world ice is itself a relatively new tool for fishermen. Although the ancient Chinese and Greeks both knew that it helped preserve fish, it's not clear that anyone but elites had access to it. Outside of Scandinavia, European fish merchants only began using ice in the late eighteenth century, after a Scottish official for the British East India Company observed that Chinese fishing boats carried it onboard. Scottish salmon fishermen didn't begin carrying ice on their own boats until the second part of the nineteenth century. In tropical regions, ice was rare onboard smaller fishing vessels even in the mid-twentieth century.[2]

The intense perishability of fish has shaped its economic and culinary history in a few important ways. Most important, of course, it long limited how far fishermen or merchants could travel with dead catch, especially in warmer climates and seasons. This in turn limited the geographical reach of popular culinary traditions featuring fresh fish. A staple food for populations living directly on or near water (or ice, in Arctic regions), fresh fish was elsewhere either entirely unavailable, a seasonal treat, or a dish reserved for the tables of the rich and powerful. This was the case even in societies where it's now a central part of the diet.

In the West, the status of fresh fish as luxury food dates back at least as far as ancient Athens, where anyone shopping at the Agora's fish stalls was assumed to come from a wealthy household.[3] In

Rome, the conspicuous consumption of fresh seafood went well beyond mere marketplace purchases. Senators and patricians served it in lavish quantities to their banquet guests, giving pride of place to the rare and exotic. To ensure an abundance of fish even during the stormy season, they built huge basins and had them stocked with species from as far away as Syria. Lucullus apparently even had a canal dug through a mountain in order to fill his ponds with seawater. Rome's elites also arranged to have certain coveted varieties shipped live on barges from the Ionians; runners delivered them, still breathing and flipping, to expectant diners.[4]

While such excesses diminished with the fall of the empire, the eating of fresh fish remained a well-guarded privilege in medieval Europe. Royals, nobles, and religious communities raised carp and trout in ponds, which allowed them to enjoy fresh catches on Fridays and during Lent.[5] In Britain, some of the richest salmon streams ran through private land. From the tenth century onward, royal edicts further limited commoners' access to fresh fish, both by establishing closed seasons on spawning rivers and by banning sales of undersized fish.[6]

In parts of Asia, warm-water ecology and aquaculture together made fresh fish more widely available. Tropical reefs were bountiful year-round, and in some regions rice paddies doubled as fishponds. Still, hot weather kept most commerce strictly local, while high prices kept large ocean species out of reach of all but the wealthy. The same applied to raw fish. A luxury coveted as far back as the eighth century in both Japan and China, it demanded expert preparation and absolute freshness.[7] Eventually South China became home to the world's most extensive trade in live fish—perhaps not surprising in a land where elites were accustomed to having their dinner pulled straight from a pond. But in the days before motorized transport, even this trade didn't stretch far.

So fresh fish foodways, for all but the elite few, have stuck close to water.[8] Many port cities are known for dishes that highlight the freshness of what was traditionally local seafood: Tokyo has its *edomae* sushi, Hong Kong its steamed grouper, Marseilles its *Bouillabaisse,* Boston its boiled lobster, New York its raw oysters. Tourists seek out these dishes even if the main ingredients now often come from farther away than they do. Historically, though, it wasn't fresh fish that brought wealth to cities and regions, but rather the industries that turned it into a cheap, durable, and portable protein.

A history of fish-preservation techniques could fill a book in itself. Since ancient times, peoples across the world have sun-dried their catches on beaches, racks, and rooftops, smoked them over open fires and in specially built houses, and pickled them in vinegar and brine. Even the rice in sushi traditionally served as a fish preservative before it was an edible wrapper. Of all the age-old methods, though, salting most fueled the rise of the western's world industrial cities and nations. Long practiced all around the Mediterranean, salt-drying proved especially suited to the mass-scale preservation of the North Atlantic cod. Salt cod fed growing urban populations in both Europe and North America, as well as slave colonies in the Caribbean. The fish's seemingly inexhaustible abundance enriched the coastal New England towns where much of the catch was landed and preserved, and inspired the building of bigger and faster vessels. North Atlantic stocks became raw material for some of the world's first factory ships.[9]

As the use of ice to preserve all kinds of perishable foods increased in the late nineteenth century, markets for salt fish declined. But the scale of the catches only increased, as the British and other Europeans converted their floating "salteries" to giant freezer boats and equipped them with deeper, bottom-sweeping nets. The fleets' hunger for North Atlantic cod, salted or otherwise, reflected the decline of many fisheries closer to home. By the mid-nineteenth century Scotland's once plentiful salmon had grown rare and costly, in part because faster transportation led to new markets. "Steamboat and railway transit . . . converted salmon into a valuable commodity," wrote one account in the 1860s, "and such is now the demand . . . that this particular fish, from its great individual value, has been lately in some danger of being exterminated."[10] By the century's end, salmon had in fact disappeared from many of Britain's rivers.[11]

Human societies had depleted local waters before. At the height of the Roman Empire, the first-century poet Juvenal observed that fish were pulled from the Tyrrhenian Sea "before even being allowed to grow to full size." Centuries of salmon fishing regulations proved no match for the traps and nets used in rivers across northern Europe. In the *Wealth and Poverty of Nations,* Adam Smith described the exhaustion of nearby fisheries as though it were an inevitable result of economic growth:

> As population increases, as the annual produce of the land and labour of the country grows greater and greater, there come to be more buyers of fish . . . A market which, from requiring only one thousand, comes to require annually ten thousand ton of fish, can seldom be supplied without employing more than ten times the quantity of labour . . . The fish must generally be sought for at a greater distance, larger vessels must be employed, and more expensive machinery of every kind made use of. The real price of this commodity, therefore, naturally rises in the progress of improvement. It has accordingly done so, I believe, more or less in every country.[12]

What Smith did not anticipate was that "progress" would increase not only market demand for fish but also pressures on its aquatic environment. As the first industrial nation, Britain discovered just how quickly humans could make their rivers and harbors uninhabitable. The same port cities that turned peasants into wageworkers, that created wealth as well as "more buyers of fish," dumped their factory wastes and raw sewage into nearby waterways.[13] In *Harvest of the Sea* James Bertram wrote that "at one time there were famous salmon in the Thames," but "it is certain that much deleterious matter has been allowed to get into that stream, and also into that famous salmon river the Severn; and in the rivers of Cornwall I believe the hope of breeding salmon is faint in consequence of the poisonous matters which flow from the mines. Many rivers which were known to contain salmon in abundance in the golden age of the fisheries are now less prolific, from matter by which they are polluted, such as the refuse of gasworks, paper-mills, etc."

Intensified fishing took its toll on other species besides salmon. Bertram suggested that Britain's herring survived only because the country was no longer Catholic, and therefore observed fewer meat-free holidays. Otherwise, "the demand . . . would have been greater than the sea could have borne."[14] Britain of course was not alone in its growing capacity to both consume and kill off its fisheries. In continental Europe, North America, and Japan, later industrial revolutions had the same paradoxical effect. Economic growth generated demand but threatened the most proximate supplies. With the development of new preservation industries, though, no fishery was too remote.

Canning the Wild

What salt did for New England, the tin can did for the Pacific Northwest. It turned the region's salmon, long central to native diets stretching from Oregon to Alaska, into a global trade good. Like salt, it became the basis for an industry that nearly depleted certain fisheries and then tapped others in more wild, distant places. This industry also invested in technologies that preserved fish in seemingly fresher forms than canned. Not least, the canning of the Northwest's salmon put the area on the map, in more ways than one. It helped to create not only a regional economy where many livelihoods depended directly or indirectly on the catch but also a regional food culture that appreciated its freshness.

Canning also helped to destroy the indigenous foodways that preceded its arrival. Salmon's importance to the peoples of the North Pacific is hard to overestimate. Annual salmon runs fed densely settled communities, as well as a host of cultural traditions that both revered the fish and regulated its catch. "First salmon" ceremonies all around the northern Pacific Rim—including parts of Japan—celebrated the fish's seasonal arrival. These ceremonies thanked the spirit that the salmon was assumed to represent, and prayed for its continued generosity. They also marked the annual replenishment of the food supply. While people enjoyed fresh salmon during the run, the rest of the year they ate it smoked or sun-dried. Pounded into jerky, it lasted years and was traded far inland. Ground into meal, it made a rich porridge or pemmican.[15]

European fur trappers bought the Indians' dried salmon and found it almost indigestibly rich. Lewis and Clark, arriving in the Northwest in 1804, observed the fish-drying methods practiced along the Columbia River, noting that the pemmican was "very sweet tasting." What really struck them, though, was the abundance of salmon, which had already grown scarce in New England: "The multitudes of this fish are almost inconceivable." Historical estimates put the pre-contact annual run in the Columbia River alone at somewhere between eleven and sixteen million salmon.[16] How to exploit this bounty? The first commercial attempts to preserve the Northwest's salmon for shipment to the East Coast—an ocean journey of several weeks—found that the challenge was to keep it not just edible but also palatable.

Pickling didn't produce very satisfactory results, as Captain John Dominis found when he brought fifty-three rum barrels of briny salmon back to Boston in 1829. They earned him $742 on the wholesale market but then "retailed rather poorly." Nathaniel Wyeth, the inventor of the horse-drawn ice cutter discussed in Chapter

1, fared little better in his efforts to sell Bostonians on salt-cured Columbia River salmon. Salt was not readily available in the Northwest (later it would be imported from Hawaii), which was one reason the Indians didn't use it. Plus, they'd long known that the fish's fatty flesh, even dried, turned rancid if not kept tightly packed and protected from oxygen.[17] No wonder merchants reported that most barrels of salted salmon were "not in good merchantable order" by the time they reached the Eastern seaboard.[18]

Canning suited salmon much better than did the drying methods long used on cod and other lean fish. In the early nineteenth century, however, the preservation of food in vacuum-sealed containers was itself a new and imperfect technology. Invented by the Frenchman Nicolas Appert in 1809, the technique won a contest sponsored by Napoleon, who wanted the army supplied with more portable provisions. But the process was labor-intensive, the containers hardly convenient (Appert's glass jars broke too easily; early cans had to be stabbed open), and the end results often neither appetizing nor safe. Insufficiently heated or poorly sealed, early canned goods led to more than a few outbreaks of botulism.[19]

In the United States, the 1860s was a decisive decade for canned foods. The Civil War increased demand, and machine-cut cans reduced production time. William Hume, a Maine fisherman who had moved west to catch salmon during the California gold rush, saw how the technology could help him develop a market far larger than local mining towns. In 1864 he recruited his brothers and Andrew Hapsgood, a tinsmith and former schoolmate, to help him start up a salmon cannery like those already operating in Scotland and Maine. After two hard years on the Sacramento River—where the salmon had already been devastated by hydraulic mining—Hapsgood, Hume and Co. moved north to the Columbia.

There success came quickly. In 1866 they sold 4,000 cases for $32,000, sixteen times more than they had made the previous season. The brothers each built their own cannery, and other companies soon joined them. Canneries mushroomed alongside rivers and streams throughout the Northwest. By 1883, thirty-nine plants operated alongside the Columbia alone, and the region churned out nearly a million cases of canned salmon a year.[20] Pacific salmon quickly became one of the most global canned goods. Britain and its colonies were the biggest export markets, but the tins also found their way to Latin America and the Far East. In the United States, they reached parts of the country where people rarely ate fish, salmon or otherwise.[21]

Like the Far West's fresh fruit and vegetable farmers, the salmon canneries depended on immigrant workers. Greek and Scandinavian fishermen brought in the catch and Chinese laborers cleaned and packed it. The region's Native Americans, meanwhile, discovered that the treaties protecting their right to fish did not actually give them access to the fisheries. By the late 1880s, the Yakama tribe's Columbia River fishing grounds had become the fenced private property of white pioneers. To the north, Puget Sound canneries initially bought salmon from Lummi fishermen, then found that they could just as easily catch fish by setting traps directly in front of the Lummi's reef-nets on the Fraser River.[22]

Also like the West's fresh produce growers, the canneries contended with great uncertainty. The short and unpredictable annual run propelled fishermen and canners alike into a "race for fish," often resulting in colossal oversupply. Fish that couldn't be processed quickly clogged the traps and rotted on the docks. Some canners packed only the choicest parts of the fish—the sockeye's belly, for example—and threw away the rest.[23] In this sense they were very different from the Chicago meatpackers, who found commercial uses for nearly every part of the pig and cow. But such efficiencies appeared pointless to the early salmon canners. Their raw material cost them nothing to produce, and even if millions of fish went to waste one year, millions more returned the next.

Until they didn't return, or at least not in such numbers. On one river after another, the canners' annual pack began to decline. The Sacramento River pack peaked in 1882, the Columbia's in 1895, the Klamath's in 1912, the Puget Sound's rivers by 1915. Habitat destruction was partly to blame. Although can labels and advertising portrayed the waters of the Northwest as wild and pristine, in fact mining, logging, ranching, and dams had rendered many salmon spawning grounds either unlivable or unreachable. Even engineering accidents had devastating effects, as when a railroad crew's dynamite blast set off a giant rockslide on the sockeye-rich Fraser River in 1913.[24] But overfishing was also at fault, despite state regulations as old as the canning industry itself. Mainly consisting of poorly enforced restrictions on seasons and gear, such laws did little to slow the industry's feeding frenzy.

As the salmon's numbers diminished in the rivers of California, Oregon, and Washington, many fishermen invested in vessels suited for open-sea troll fishing. By 1915, 500 gas-powered trollers, some equipped with over 100 lines, crowded the mouth of the Columbia River; by 1919 there were 1,500. When states prohibited troll fishing within three miles of shore, the boats went farther out. It was a predictable response with a predictable effect; as in Europe and the northeastern United States, the salmon simply disappeared from many of the Northwest's rivers.[25]

The business of salmon, however, did not dry up. The canning boom drew immigrants and investment. Fishing and boat-building peoples from several European countries as well as the eastern United States settled in towns around Puget Sound and along the coast. Their boats filled the harbors, and their fresh catches supplied the markets of Seattle and Portland. So even though canning hastened the destruction of the native fish-based foodways and livelihoods, it helped make the Northwest into a place where fish, and salmon especially, mattered to a lot of people and companies. They didn't go away when local stocks declined; they just came to depend on fish coming from farther away, namely, Alaska.

The territory's first cannery opened in 1877, and for several years

Salmon canning was big business in early twentieth-century Seattle, but many locals preferred their fish fresh. ("Fish Market at Seattle Pier," circa 1917. *Seattle Post-Intelligencer* Collection, Museum of History and Industry, Seattle)

it had little company. Compared with the Pacific Northwest, Alaska was an expensive and risky place to pack salmon. The season was shorter, the weather stormier, and everything but the fish had to be imported. But once the Columbia River industry went into decline, Alaska became the new capital of canning. Despite the distance, the short annual runs drew the Northwest's fishermen by the thousands. By the late 1890s its annual pack dwarfed the entire production of the Pacific Northwest; by 1916 it accounted for more than three-quarters of world supply. Soon canned salmon generated more wealth for Alaska than all its other resources combined, including copper, gold, and silver.[26]

Labor-saving technologies such as the "Iron Chink" fish cleaner (notoriously named after the Chinese workers it replaced) helped expand production, as did investments by Del Monte, Libby's, and A & P, each of which took over large chunks of the industry. In 1903 the *New York Times* described canned salmon as "one of the most important food products of the world"—an admittedly exaggerated claim that the corporate canners, with their marketing clout, helped make truer.[27] Even as production exploded, though, the canners' own publicity was oddly modest. In the 1908 pamphlet *Interesting Facts about Canned Salmon*, the Alaska Packers' Association noted that "while there need be no present fear for the diminution of the salmon supply, the canned product will not increase in quantity, the fish being so much sought after in their fresh state . . . Fresh fish handlers with refrigerating appliances have invaded every salmon district . . . and they are paying more for the raw fish than the canner can afford."[28]

Sins of the Freezer

The packers' predictions were premature. Alaska's output of canned salmon increased dramatically over the next few decades, while exports of the fresh fish barely registered. In 1939, for example, canned salmon exports earned $34.4 million, and fresh exports less than $285 thousand. This isn't surprising given the remoteness of most of the territory's fisheries. Even with twenty-first-century technologies, shipping fresh salmon out of Alaska by boat is feasible only from the southeast, and only to nearby markets such as Seattle or Vancouver BC.[29]

Given the shipping options available in the first half of the twentieth century, freezing salmon was much more practical than canning it. Yet Alaska's 1939 exports of frozen salmon amounted to only $303 thousand. Although the territory's exports of frozen halibut were more substantial, the paltry figures for its most famous fish are remarkable, especially since the salmon selected for freezing were the most prized varieties—coho, sockeye, and king. Compared with canning, freezing required less labor and fewer materials, both of which had to be imported at considerable expense. It also kept fish in a form that, in theory, resembled and could be substituted for fresh, arguably giving it more center-of-the-plate appeal than any kind of canned fish (especially one widely used as army rations). Nonetheless, frozen fish suffered from a stubborn image problem. Even as techniques improved and the industry insisted that frozen fish was as good as, better than, or the exact same thing as fresh, consumers didn't trust it. How did a preservation technique so suited to fish's perishable nature earn such a bad name? As always, the answer lies less in the technology itself than in the way it has been used.

Admittedly, in the early days the technology was crude. In fact it was considerably less effective than Northern peoples' traditional practice of freezing fish in the winter air, which reached temperatures of minus 40 degrees Fahrenheit. Neither the ice-filled boat smacks and storage rooms used in the mid-nineteenth century nor the ammonia-chilled "sharp freezers" developed later were nearly as cold, and so they could not freeze their contents nearly as fast. When fish freezes slowly, the moisture in its flesh forms large ice crystals that eventually rupture cell membranes. Upon thawing, water leaks out, the cells collapse, and the fish tastes like mush. Its eyes and gills also lose their brightness, which turned at least some fishmongers against freezing from an early date. At London's 1883 International Fisheries Exhibition, merchants claimed that frozen fish, while not spoiled, had a "wizened and disagreeable appearance . . . as if it were stinking."[30]

Although most foods suffer from slow freezing, many of the efforts to develop faster methods focused on fish. In 1862 Enoch Piper of Camden, Maine, received the first U.S. patent for a fish freezing method. He froze salmon on racks beneath pans of ice and salt, which both chilled faster than ice or cold air and prevented salt absorption. After they froze, he dipped them in water to form a protective glaze. Patents for similar brine-freezing methods proliferated in the 1880s and 1890s. They were laborious and still not lightning fast (Piper's salmon apparently froze in twenty-four hours) but worked well enough to encourage the spread of fish freezing enterprises in New England and the Great Lakes Region, and later the Pacific Coast. By the mid-1890s, British Columbia was sending Europe a million pounds a year of frozen salmon, halibut, and sturgeon.[31] In 1902, the first freezing plant opened in Southeast Alaska.[32]

The expanding frozen fish trade did not necessarily reflect consumer enthusiasm. In 1908 James Critchell, an authority on the international trade in frozen and chilled foodstuffs, reported to an audience of refrigeration experts that London's imports of frozen U.S. salmon were negligible compared with those of beef, and for good reason. "Fish does not bear the freezing process without some loss of flavor," he said, "and from enquiries made, I cannot gather that frozen salmon is, or is likely to be, a great popular success." In the United States, home economists advised caution. "Never expect frozen fish to be a good or wholesome dish for your table," said a 1903 *Chicago Tribune* advice column. The columnist recommended only educated and seasonal purchases of fresh seafood: "Try to learn by reading books on fish and their habits when to expect each particular fish to appear in the market, and then you will not be misled by the market man."[33]

This was the crux of the problem. Although some of the prejudice against frozen fish was due to abuses it suffered en route from plant to plate—variations in temperature, improper thawing, inept cooking—even the industry's biggest proponents admitted that freezing often served as a "secondary" preservation method. This could mean one of two things. Sometimes it meant that fish was frozen as an intermediary step until it could be canned, pickled, or smoked. Tuna boats, for example, froze their catches in brine until they reached shore.[34] Often, though, freezing was used to salvage whatever failed to sell when fresh. This happened often enough that consumers came to believe that freezing itself spoiled their fish.

In 1927 Harden Taylor, a senior biologist at the U.S. Bureau of Fisheries, published "Refrigeration of Fish," a report that was really about freezing.[35] Calling it "the only method of preserving that keeps fish in essentially its original condition over long periods," he described recent scientific advances that he believed would soon win over the public to frozen fish. Among them was "quick freezing," an approach developed by a fellow fisheries employee, Clarence Birdseye. While Taylor's report focused on the technical details of quick-freezing, Birdseye himself recognized that marketing was half the battle. Consumers had to be convinced that proper freezing did not destroy freshness but rather captured it perfectly. He illustrated this claim with a classic fisherman's tale. Years before, while working as a fur trader in Labrador, he had observed Eskimos ice-fishing. He noticed that their catch "froze stiff" almost as soon as it was pulled out of the water into the −50 degree air. When thawed months later, he claimed, the fish were still alive.

This might or might not have been a true story; freezing fish alive is technically possible, though keeping them alive for months is unlikely.[36] Either way, it hardly described Birdseye's own method. He froze pre-packaged fish fillets, not whole fish, and he pressed them between two hollow ammonia-chilled metal plates. Held in such close contact with a surface of −25 degrees Fahrenheit, each package froze in about an hour and a half. Birdseye demonstrated the effectiveness of his method on haddock fillets, frozen at a small plant in Gloucester, Massachusetts. Later he showed that it also worked on boxed meat, fruits, and vegetables. Although he didn't have enough money to turn his experiments into a full-scale business, Birdseye's trial products attracted the attention of a firm that did. In 1929 the Postum Company, later known as General Foods, paid a reported 22 million dollars for the Gloucester plant as well as for rights to the patented "Birdseye process."[37]

Except for the pre-packaging, quick freezing drew on principles

"Gloucester is blazing a new trail in the producing and marketing of fresh fish products," said *Fishing Gazette* magazine of the Birdseye quick-freezing process. (*Fishing Gazette,* November 1927, p. 37)

developed decades earlier. But General Foods and Birdseye, who joined the company as director of a laboratory of his own, Birdseye Laboratory, pitched it as "a process wholly unlike anything the world as ever seen!"[38] Test marketing at shops in New England was encouraging—three-quarters of the consumers who tried quick-frozen fish fillets once reported repeat purchases—and attracted favorable media coverage. Indeed, some magazine articles on quick-frozen foods implied that they were, as a *Popular Science Monthly* headline claimed, "exactly the same" as fresh foods—just more convenient. In *World's Work,* an article titled "Time Stands Still for Food" marveled at out-of-season frozen raspberries, "fresh as fresh," and frozen spinach, "fresh in all respects save that there was no sand in it." *Popular Science Monthly* further blurred the borders between fresh and frozen in an article titled "Fish Kept Fresh 1500 Miles from the Sea": "'And a package of haddock,' a Kansas housewife tells her grocer. He hands her a sealed box, faintly cool to the touch. In the kitchen, she drops its contents in a frying pan. It is a fish fresh from Atlantic waters, now for the first time available to her through a new process of quick freezing. Frost still covers it after its 1500 mile journey by refrigerator car and fast motor truck."[39]

The article also noted that quick-frozen foods reached consumers' kitchens in more wholesome condition than fresh foods. Birdseye himself emphasized this point, noting that "quick frozen perishables are preserved at the height of their goodness, are sanitarily packaged, and are protected against deterioration throughout the storage and distribution processes."[40] At least in theory. General Foods soon discovered that the nationwide cold chain, even if it was the best in the world, was nowhere near ready for a deluge of quick-frozen products. Many shops only had freezer cabinets suited for limited stocks of ice cream; many consumers only had iceboxes with cold and less-cold shelves. Improperly chilled, quick-frozen foods quickly lost all claims to superior taste or nutrition.

General Foods recognized this problem, and so convinced thousands of retailers to buy or rent its specially designed Birdseye frozen food display cases (12,000 retailers had them by 1939—admittedly not many relative to the 600,000 shops nationwide). Greater visibility boosted sales only slowly; by 1945 Americans ate on average only two pounds of frozen foods a year.[41] And they still harbored doubts about frozen fish in particular. The 1947 edition of *Freezing Preservation of Foods* (written by two former Birdseye scientists) acknowledged widespread "public prejudice," despite ongoing improvements in seafood freezing, transportation, and storage. Like earlier writers on the topic, the authors blamed the

Fresh salmon on display in Seattle's Pike Place market in late summer, one time of year when wild stocks are abundant. (Author's photograph)

prejudice on bygone days when "often fish were frozen which were in such poor shape that they would not stand transportation and sale as fresh fish."[42] But were such abuses of the freezer really so far gone? After all, fish themselves were still seasonal and unpredictable. They still deteriorated quickly, especially if crushed in the giant nets used by many industrial-scale vessels. And market supplies still sometimes exceeded demand. If anything, the spread of freezer plants in the mid-twentieth century simply gave more fishermen and merchants a new way to solve an old problem.

Not surprisingly, consumers never really bought into the claim that frozen fish was just as good as fresh.[43] The industry did find growing markets in "institutional" meals (schools, fast food restaurants, airlines) and in the form of convenience foods like breaded fish sticks. But it failed to make seafood a spaceless commodity, as the industry once hoped. Kansans never ate as much fish as Seattleites, except when they visited Seattle. And even though that city has become home to some of the country's biggest freezing facilities, fishmongers at Seattle's famous Pike Place market reassure tourists that their twenty-pound Chinooks or sockeyes are indeed fresh. In the past, they stretched the truth. The "fresh"salmon sold during the long off-season was often actually freshly thawed. But then such deceptions became unnecessary. In the late twentieth century, salmon suddenly became abundant year-round, in Topeka as well as in Seattle. It was fresh, cheap—and farmed.

Wild and Branded

In the early 1970s, Norway's calm fjords became the birthplace of an aquacultural revolution. Drawing on the same selective breeding methods used for generations on cattle and poultry, Norwegian scientists bred an Atlantic salmon that converted feed to flesh faster and more efficiently than did any of its ancestors. Raised in giant netted cages, the fish multiplied on a Biblical scale. Norway soon became the uncontested world leader in farmed salmon production. By the end of the century the country's annual harvests had increased from 100 to 500,000 tons, and Norwegian fish farmers had become multinational corporations, raising salmon off both coasts of North America.[44] In the meantime, Scotland, Ireland, and even Chile, a country with no tradition of catching or eating the fish, had also become major producers.

The exploding population of farmed salmon reverberated through world markets, where by 2004 they accounted for 65 percent of salmon sales (up from 2 percent in 1980). In Europe, almost all the salmon came from farms.[45] Rarely in human history had a domesticated foodstuff so quickly triumphed over its wild cousin. For salmon breeders, it marked a great advance for both the fish and its consumers. After all, aquaculture had vastly increased the global supply of a species that humans had over-exploited for centuries. Prices had fallen while year-round availability had greatly improved. The consistent size, color, and mild taste of farmed product suited supermarkets, restaurants, and processors. Not least, the domesticated fish could be raised close to shore, harvested on demand, and kept alive until just before processing and shipment. In a word, it was fresh. The same could not always be said of wild salmon that spent several days in a gillnetter's hold before reaching dry ground.[46]

Many environmentalists and fishermen could not have disagreed more. They saw the plump, docile farmed salmon as a form of industrial pollution, as toxic as an oil spill and much more insidious. Because it tolerates captivity well and can be harvested year-round, the Atlantic salmon is the species of choice for the vast majority of the world's fish farms, including those in the Pacific Ocean.[47] But it doesn't always stay on the farm. Hundreds of thousands have escaped from pens off the coasts of British Columbia and Washington State. Contrary to predictions that this alien species wouldn't survive long in the wild, Atlantic salmon have been caught in several Pacific Northwest rivers. According to some fishermen, they have also made it up to Alaska, where salmon farms are prohibited.

The possibility that farm escapees could interbreed with wild populations and weaken their gene pool (especially if the farmed fish are genetically modified "supersalmon") was not the only reason environmental groups opposed salmon aquaculture. While initially dismissed as alarmist, their claims found increasing support in top scientific journals. Some studies found that salmon farms endangered wild stocks by polluting nearby waters and spreading pathogens such as sea lice. Others suggested that fish from the farms' pesticide-laden waters posed a consumer health risk. Still others warned that salmon farms' disproportionate use of other fish as feed could lead to the starvation and collapse of many wild species, including salmon.[48]

In the Pacific Northwest and Alaska, fishermen didn't need hard science to see that they, too, were an endangered species. As farmed salmon flooded the world market in the late twentieth century, their earnings plunged. The overall value of Alaska's catch—which accounts for most of the U.S. as well as the global supply of wild salmon—fell by more than 60 percent by 2002. Hardest hit were prices for sockeye, once known as Alaska's "money fish." The second biggest catch after pink salmon and the second priciest after Chinook, sockeye was mainly frozen and exported, much of it to Japan. There it ran up against competition from farmed Chilean salmon, and lost.[49]

Falling prices affected communities everywhere from Kodiak to Central California. Natives living in some of Alaska's most remote fishing grounds depended on wild salmon not just for food but also for their sole income. In Kuskokwim, earnings fell by 95 percent between 1988 and 2002; in Kotzebue they fell by 100 percent. In the Bristol Bay and Southeast Alaska fisheries, where most of the fishermen came from out of state, earnings fell by 84 and 62 percent respectively. Many got out of the business altogether, creating a glut on the region's fishing boat market.[50] In Seattle and the surrounding Puget Sound area, thousands of fishermen scrambled for extra work. They got jobs as carpenters, truck drivers, and schoolteachers.

A few fishermen, like Pete Knutson, joined forces with the environmentalists. The grandson of Norwegian boat-builders, he grew up near Seattle at a time when the job options were "either timber or fish, basically." After one summer working in a pulp mill, he opted for fish. He started migrating up to Alaska each summer in the early 1970s, when four months on a boat earned a decent yearly income. In the meantime he also got a Ph.D. in anthropology and took a job at Seattle Central Community College during the academic year. In the 1990s, Knutson began to take on the farmed salmon industry at every opportunity. Articulate and not afraid to ruffle feathers, he debated industry representatives at public forums at the University of Washington. He sat on panels with Jacques Cousteau. He wrote his own articles and was quoted in many others. He referred to farmed salmon escapees as "smart bombs."[51]

While the degree of Knutson's activism was unusual, his efforts to rebuild markets for wild Alaskan salmon were not. Indeed, the activism helped create the markets. Ironically, so did the farmed salmon industry itself. By aggressively promoting fresh salmon, packaging it in convenient new forms and making it a product that retailers everywhere could stock year-round, the industry drove up demand on a global scale. In Europe, the main market for Norway's farmed salmon, consumption increased by more than 450 percent between 1989 and 2004; in the United States, it quadrupled. By 1999 salmon had become the country's favorite fresh fish and its third favorite seafood overall, after shrimp and canned tuna. And this was before reports on the health benefits of Omega-3 fatty acids (found in salmon and other high-fat fish) boosted sales even further. Given that the average American still ate only two pounds of salmon a year by 2006 (compared with seventy-three pounds of chicken and sixty-three pounds of beef), the industry saw plenty of room for growth.[52]

Most of this increased demand was for fresh rather than canned, cured, or frozen salmon. This posed certain challenges for purveyors of the wild Alaskan variety. It also posed opportunities for Alaska Airlines, the only company with regular flights to the state's remote fishing communities. For decades the airline had flown perishable goods both to and from towns like Yakutat and Nome, and its cargo staff was accustomed to the seasonal excitement of the salmon runs: the extra flights and last-minute schedule changes, the frantic customers, the ravens that attacked cartons of chilled kings while they sat on the Sitka runway. But by 2005, the airline was carrying so much fresh seafood that it added new "combi" carriers to its fleet (planes with moveable walls and seats, so that extra cargo could replace passengers on the main deck) and upgraded its booking system. It was high time for a change, said Shannon Stevens, the cargo manager otherwise known as the company's "fish lady":

> In the past . . . the freight forwarders and fishermen would bring product to Alaska Airlines, drop it off and say "get it there." Our claims were very high and people weren't happy. So . . . we incorporated a fish desk. You call our 1-800 number, you have different options. If you're going to book a live animal, you push this button, if it's going to be for seafood, you push 5. You'll get connected to an actual person who'll book the product and tell you "yes, I can get it on flight X to Seattle out of Anchorage, and then on flight such and such to L.A. or San Francisco" . . . So we now know when the product's coming through, meet the airplanes and do our best to get it into a cooler or onto its connection.[53]

One of the airline's best-known fresh cargoes comes from the Copper River, off the Prince William Sound. Fillets of the river's fresh king salmon retailed for thirty-seven dollars a pound in 2006, and even at those prices consumers lined up to buy it. Some may have heard that the salmon that migrated up the long, cold, rugged river were both extra-strong and extra-rich in healthy, tasty fat. Some may have wanted to be among the privileged few who could buy this coveted fish when it arrived fresh on the market each year. They almost certainly knew *when* it arrived, because timing was key to the success of the Copper River brand. A Seattle fish dealer,

Jim Rowley, figured this out in the early 1980s, when he convinced a few local restaurants to sign up for air shipments of the river's first run. At the time, half the appeal lay in the fact that the Copper River kings started their run in mid-May, about two weeks earlier than other Alaskan salmon. They were the season's first fresh salmon, and restaurants not just in Seattle but all along the West Coast proved eager to get some.

Once fresh, cheap, farmed salmon became available twelve months of the year, Rowley had to find another way to sell the Copper River's seasonality. So in 1997, with help from Alaska Airlines and some helicopters, he invented a race to get the first catches down to Seattle. Although the first year he and a friend were the only team racing, the media loved it. So did the city's upscale Metropolitan Market grocery chain, which burnished its own reputation by joining the race then feting the fish's arrival with barbecues and in-store chef's demos. By 2005 the welcoming of the Copper River salmon had become, as one local food columnist put it, "as much a Seattle thing as the Running of the Bulls is a Pamplona thing."[54] It had also become more than a "Seattle thing," as retailers and restaurants across the nation found that certain clientele would pay almost any price for a taste of the first Copper River king, delivered fresh from an Alaska Airlines cargo hold.

The commercial success of the Copper River salmon perhaps testifies to our longing for the seasonality long celebrated, with considerably more reverence, in the "first salmon" ceremonies of the North Pacific's indigenous peoples. Certainly the marketing made the most of the fish's own evolutionary success in the wild. It suggested that consumers would taste the fish's vitality—the vitality that helped each generation survive the 350-mile migration to its spawning grounds—in every bite of its brilliant red flesh. This was especially true if it had been pulled from the river only a day or two before. This aspect of the Copper River salmon's wildness appealed to consumers on a more visceral level than information about, say, the sustainable management of the region's fisheries, or the cleanness of the waters, or even the reportedly superior Omega-3 content of wild salmon over farmed. And by the early 2000s, salmon from elsewhere in Alaska were challenging the Copper River brand on these very grounds. Companies selling salmon from the Yukon had an especially strong case, because its fish had to migrate nearly 2,000 miles to their spawning grounds, and therefore boasted an even higher fat content (on average 25 percent versus the Copper River's 18 percent). But the Yukon run starts later in the summer and in a much more remote location. Indeed, the Yukon's very wildness makes it a much tougher and costlier place from which to air-ship fresh salmon than the Copper River—which, despite all the wild imagery, is actually only a few minutes by plane from the city of Cordova, itself not far from Anchorage.

Not all the new wild Alaska salmon brands aimed at the same sky-high market as the Copper River. Not all of them even aimed to sell their salmon fresh. The counter-trend to the Copper River mania was what might be called artisanal blast-freezing. For remote communities, this was a less risky approach than relying on Alaska Airlines flights that might not have space for their cargo. For the smallest operators—families, twosomes—this method allowed them to avoid the almost impossible logistics of shipping fresh catch in the middle of an open-sea fishing season. For consumers it promised fish indistinguishable from fresh, available year-round. But consumers had heard this before. So the challenge for fishermen attempting to direct-market their frozen fish was twofold. First, they had to master the skills and tools that would really deliver fish "fresher than fresh," even from the most remote locations. Second, they had to win over the doubters.

Fisherman Rick Oltman appeared to have succeeded on both fronts when I met him in Port Townsend, Washington, in late December 2006. He was selling frozen salmon and lingcod off his boat, which he does a couple of times a week during the long off-season. He started selling his own fish in 1998, when the market was awash with farmed salmon, because "it was either that or go out of business." First, though, he had to install a blast freezer on his half-century-old, forty-six-foot-long wooden troller, the *Cape Cleare*. With a freezer on board he could go farther out to sea and stay there longer. He could freeze fish within an hour of landing them, much faster than fish caught on most factory boats. The problem was that most blast-freezers were themselves factory-sized, so Oltman, with no background in refrigeration engineering, had to build his own. In particular, he had to figure out how to keep several tons of fish at forty below without also freezing the boat's hydraulic system.

I got a sense of the results after a few minutes of standing on deck. On a mild sunny day, it felt like Antarctica was seeping up through the bottom of my shoes. The hold itself was almost too cold to peer into. And the fish were, well, very frozen. A foursome of tourists wandered by and looked skeptical. But Oltman's regulars swore by his salmon. One customer, a former vegan with a heart condition, said it saved his life. Back in town, the Wild Coho restaurant showcases Oltman's fish, calling it "'better than fresh' Alaskan salmon." At the Saturday farmers market, the Cape Cleare grilled salmon sandwich cart draws a steady lunchtime crowd.

Selling to locals is relatively easy in a small neighborly place like Port Townsend. But it doesn't earn enough to keep a boat running or to keep tens of thousands of pounds of fish in deep cold storage for several months a year. Oltman is the first to admit that even fishing by hook and line from an old trawler is an energy- and gear-intensive business. So like many of the area's other direct-marketing fishermen, Oltman sells to restaurants in Seattle and ships orders across the country. Like them, he has a website describing how his salmon are caught and processed (in particular, how freezing preserves freshness) complete with a gallery of photos of Alaskan fish and wilderness.

This formula worked for many small-scale sellers of frozen wild salmon in the early 2000s, and *National Fisherman* predicted that it would work for others.[55] At one level, the strong market reflected demand for more care and technical mastery in the handling of a food that had so often disappointed. Retailers, chefs, and ordinary consumers were more willing to trust frozen fish if they felt they could trust the people behind it. As *Frozen Food Age* said in a 2004 article, "in today's seafood market it is finally okay to be a very cold fish." Indeed, in some foodie circles, especially on the West Coast, frozen salmon with the right name attached to it had distinction. Frozen had become the new fresh.[56]

At another level, the premiums earned by small-boat salmon fishermen like Oltman reflected the cachet of salmon that came, as he put it, "with a story." This wasn't unique to fish, of course. Peoples' hunger for food they could connect to a place and a face fed the popularity of farmers markets from the early 1990s onward. But unlike the farmers at those markets, fishermen could not tell a *local* story about their Alaskan salmon. Instead they could tell about the pristine wildness of the salmon's home, and how directly it had come from there. Consumers' appetite for this type of story—about fish untamed and untainted by industrial civilization—was not new. But the negative publicity about farmed salmon helped, by all accounts, to revitalize it.

It also boosted the sales of fishermen like Pete Knutson, who started direct-marketing his fish long before most, originally just to avoid working for a corporate processor. He never did freeze his catch at sea. He still doesn't believe it's necessary in the relatively proximate wilds of southeast Alaska, as long as the fish are otherwise promptly and carefully handled. For a long while he shipped his salmon on ice from southeast Alaska to Seattle on Alaska Airlines. But now he packs them in insulated totes of refrigerated seawater and ships them south by truck and ferry to a blast-freezer near Seattle. That's his fish story. It's a relatively low-tech and low-cost approach that lacks the drama of the Copper River. Nonetheless, by the early 2000s Knutson had won renown as not just a very political fisherman but also one whose salmon appeared in some of Seattle's most esteemed restaurants.

This renown eventually caught the attention of the Sub-Zero refrigerator company, which recruited him for an advertisement featuring "portraits" of small farmers and fisherpeople. The company wanted to show Knutson holding aloft a wild salmon as a symbol of the high-quality fresh foods that potential Sub-Zero customers would supposedly appreciate. But the film crew arrived during the off-season, when all of Knutson's own fish was frozen solid. The company did not think an ice-crusted salmon would look right in an ad about fresh foods. So someone on the film crew was dispatched to Pike Place Market to pick up a fresh one. Knutson laughs at the end of that story. The salmon that Sub-Zero found him was indeed very fresh. It was also farmed.

The Live Trade

For most of Alaska's fishermen, what Seattle consumers want ultimately matters much less than what Asia wants. Historically, Asia meant Japan. For decades, Japan was the world's largest and most demanding market for many kinds of seafood. Fishermen from Ketchikan to Gloucester to Sidney catered to its tastes. Since Japan's national airline, JAL, first began carrying fresh Canadian bluefin tuna in the 1970s, Japanese companies have led the way in developing technologies for shipping seafood around the globe. The technique of "superfreezing" down to negative 65 degrees Celsius (negative 85 degrees Fahrenheit) turns a half-ton tuna into a brick of top-grade sashimi, with a storage life of two years. Fish frozen on board in the Mediterranean now supplies sushi restaurants in Moscow and Los Angeles.[57]

Japan's standards for seafood freshness are now as familiar

across the world as sushi itself. But as a market for the world's fish (fresh or otherwise), Japan is increasingly overshadowed by China, a country with its own ancient—but very different—fish foodways. Sushi restaurants have of course sprung up in the major cities, and younger generations have grown fond of seafood varieties once scorned for their tough oily flesh. Traditionally, though, uncooked fish was usually considered unwholesome as well as inappropriate for social occasions. Rapid steaming was and remains the favored preparation, especially in the Cantonese traditions of southern China and Hong Kong.[58]

In these places, freshness is not merely an important quality in seafood but the defining one. Seafood restaurants and market stalls commonly advertise live seafood in characters that translate as "vital delicious freshness from the sea." As in Japan, wild-caught fish usually costs more than farmed. It is considered cleaner, more "natural," and stronger. But whereas sushi bar diners assess the quality of fish in pieces—looking at the color and texture and cut—Chinese diners want to see their fish whole and swimming. They want the whole fish because the head and tail both taste good and give symbolic completeness to the meal. And they want to see swimming because it proves that a fish is still full of delicious life. This vitality is especially precious in a food that itself symbolizes wealth and well being. Served whole and steamed as a traditional banquet dish, fish even sounds the same as "abundance" in both Cantonese and Mandarin.[59]

Chinese demand for tail-flipping freshness feeds a trade that has long been centered in Hong Kong, even though more and more of the products now end up in mainland markets. Once reliant on the coral reef fisheries of the South China Sea, Hong Kong merchants now import live seafood from all across the South Pacific, as well as from Anchorage, Johannesburg, and Boston. And while shipping fish alive was once the cheapest and most reliable way to keep them fresh in tropical regions, these days the most expensive species travel by air. After all, in this business the greatest threat to freshness—and the money it's worth—is not warmth but death. This fact alone makes it unlike any other fresh food trade on earth.

In the United States, the biggest markets for live seafood (besides shellfish) are, not surprisingly, the cities with the largest Chinese populations. Not so long ago, though, the live-fish trade was much more extensive. The 1899 Bulletin of the United States Fish Commission called it "one of the most satisfactory methods of marketing marine foods, not only because of the superior quality of the product, but also because it avoids costly processes of preservation." In some places fish were simply caught and kept alive offshore until they could be sold for a higher price. This was the case of the Detroit River's once-plentiful whitefish. But large quantities of fish were shipped live by boat and rail. In the mid-nineteenth century, some New England halibut and cod fishermen specialized in live catch. They fished with hook and line, kept the fish below deck in well-smacks, and sold them to merchants who themselves had tanks for live products. The careful handling paid off; live halibut sold for four times as much as dead. By the turn of the century, rail cars equipped with aerated tanks were carrying live carp from lakes and rivers up to 1,500 miles to New York and Philadelphia. Most ended up in markets in Jewish neighborhoods, where the *Fishing Gazette* described consumers as "extremely particular" about the freshness of their fish. In 1930, the two cities together received about six million pounds of live carp.[60]

Eventually, fishing boats and fish markets in the United States got rid of their smacks and most of their tanks. Neither ice nor mechanical refrigeration preserved freshness as effectively as life, but they preserved much more *efficiently*. Dead fish took up less space, needed no air, and required less supervision than live fish. They could be sold through the same channels and pre-packaged forms as other refrigerated and frozen perishables. And Americans increasingly preferred to buy them that way. They were willing to trade just-caught freshness—which many American consumers had never tasted anyway—for greater convenience.

In Hong Kong, growing affluence had the opposite effect on demand for live fish. Before and shortly after World War Two, the only Hongkongers who regularly ate fish fresh out of the water were the island's boat people. A handful of floating restaurants also kept pens full of live seafood. Otherwise, fishermen salted much of their catch, for ice was a rare commodity. By the 1960s, demand for live fish was growing along with the colony's economy. The best places to find it were in coastal towns such as Aberdeen and Castle Peak Bay, where restaurant diners could choose from pens holding everything from prawns to parrotfish. In the city's wet markets, live seafood was still relatively rare but much appreciated. "The Cantonese will buy a live fish if they possibly can," said the anthropologist Eugene Anderson of the island's Chinese population, "since freshness is of critical importance in cooking."

Anderson spent years among Hong Kong's boat people, observing how their lives changed as the island filled up with mainland immigrants. On one hand, if they could catch fish alive, they could sell it for a huge premium. "The residents of Hong Kong are almost without exception gourmets in regard to seafood," Anderson wrote. "Even a poor man will pay extremely high prices."[61] On the other hand, the island's growing prosperity was decimating local stocks. By the 1970s motorized boats were becoming more common; the use of dynamite to stun catches was illegal but still practiced. Pollution was increasing, and land reclamation was shrinking harbor habitats. High-rise housing developments, meanwhile, were crowding out fishponds in Hong Kong's remaining rural areas. Once again, an industrializing society was destroying its nearest and dearest food supply.[62]

As Hongkongers grew wealthier, though, they came to expect live fish whether or not it came from nearby. After all, localness had no effect on the freshness of live fish; if anything, the obvious pollution in Victoria Harbor raised questions about the wholesomeness of local catch. But bringing fish back alive from more distant waters was beyond the means of many Hong Kong fishermen. As Anderson predicted, many were forced ashore and into menial wage labor. It took cash and connections to get far in the live-fish trade.

For those who had both, Asia's coral reefs teemed with profitable, edible life. By the mid-1970s, Hong Kong vessels were traveling hundreds of miles to reefs across the South China Sea. A few had reached the Philippines. The smallest live-catch fishermen installed simple wells in their traditional Hong Kong–style wooden boats. At fifty or sixty feet long, they couldn't go far or carry much cargo, since each ton of fish needed anywhere from eight to fourteen tons of water. Smaller boats soon gave way to specially built live-fish transport vessels (LFTVs) equipped with skiffs, portable holding pens, and teams of hired fishermen, many of them Filipino or mainland Chinese. Although these carriers could hold up to twenty tons of fish and stay at sea for up to a month—enough time to reach many islands in the South Pacific, as well as Indonesia—longer voyages meant higher mortality rates. Often 40 percent of the cargo died on the way to Hong Kong. Careful handling and clean water improved fishes' chances of survival, but some hazards were beyond any crew's control. And the losses were enormous, for the wholesale price for a single live "plate-sized" humphead wrasse (this was considered the ideal size, at about one kilogram) could run up to seventy-five dollars. If it died en route to market, it was worthless.[63]

By the late 1980s, the most valuable live fish traveled to Hong Kong on passenger flights.[64] Companies based in the Philippines, Indonesia, Australia, and other exporting countries prepared each fish for its journey. First it needed to be starved for twenty-four hours, to minimize the chance that it would vomit. Then it had to be anesthetized and chilled (to minimize respiration), packed in a plastic bag pumped full of water and oxygen, sealed into a Styrofoam cooler, and finally rushed to the airport. The process was not cheap—each fish needed three times its own weight in water—but it brought mortality down to around 5 percent. For exporters of high-value fish, the math was obvious. Soon Australia and Malaysia shipped most of their live seafood by air. Meanwhile the LFTVs continued to call at ports with no direct flights to Hong Kong, such as the Seychelles, Fiji, and the Maldives.

Back in Hong Kong itself, what was once a luxury soon seemed commonplace. People hardly noticed the wooden-topped blue trucks that raced buckets of fish through the city's blaring traffic. They were unfazed by the splashing and spraying that made the wet markets so wet. They trained their Filipino and Indonesian maids to cook fish "Hong Kong style": steamed, with ginger and green onion.[65] They perused aquariums like menus. Just as the Copper River salmon race became a "Seattle thing," selecting one's meal from a tank became a Hong Kong thing. To get a taste of local tradition, guidebooks recommended visits to Lamma Island or the village of Sai Kung, where some eateries offered twenty-five different kinds of live seafood, most of it from foreign countries. To impress clients, businessmen opted for in-town restaurants where a single humphead wrasse might cost three hundred dollars.

Seventy-something Cheung Tim remembers a time when not even rich people ate much live fish. Like many members of her generation, she came to Hong Kong from the mainland just after the end of World War Two. She grew up in a poor inland village, where seafood of any kind was a luxury. But her first home on Hong Kong Island was in Aberdeen, a port town full of boat people. She bought their fish—it was dead, she recalls—and learned from them how to steam it. It was a useful skill for her first job, working as a maid for a wealthy family from Shanghai. Later she married a chef and had three sons. Her husband is now retired and ill, and when she cooks fish it is mostly for her sons' families. But she always buys it live; she enjoys the routine even if cleaning the fish at home smells up

Hongkonger Cheung Tim looks for the liveliest of live fish at a wet market in Kowloon. (Author's photograph)

the tiny kitchen. She takes the bus to the wet market, chats with her favorite vendors, and watches the tanks a while before choosing. She looks for the fish that swim and flip the most. The more energetic, she explains, the better they'll taste.

The market is in a working-class neighborhood, and many of the vendors sell farmed freshwater fish from the Pearl River. As live seafood goes, it's cheap but also suspect in the eyes of many Hongkongers. They know about the mainland's pollution (plenty of it drifts toward Hong Kong), as well as its poor food-safety record. Farmed fish contaminated with malachite green, a carcinogenic pesticide, was just one of many products recalled from Hong Kong markets in the early 2000s.[66]

Some vendors also display fish on ice, which they sell for one-half or even one-third the price of their live stocks. But these are no good for steaming, Tim says, only frying. Other shoppers agree. Frying hides the taste of staleness. You fry fish if you cannot afford better. Or, some will add, if you are a Westerner and don't know any better.

The wet-market shoppers all describe the taste of a just-killed fish as *tsim,* which means both sweet and savory, like monosodium glutamate (and fresh seafood is full of glutamic acid).[67] They all have opinions about steaming techniques. But as with consumers in any urban industrial society, their knowledge about food thins out farther from the kitchen.[68] They may have heard that the prized humphead wrasse was recently listed as an endangered species, and that other wild fish may contain ciguatera, a naturally occurring poison.[69] They may not know that a synthetic poison, sodium cyanide, is often used to catch wild live fish. Squirted from a spray bottle or mixed with chopped chum salmon to make a pasty bait, cyanide stuns fish, making it easy for divers to pull them out of their hiding places in coral reefs. The chemical poses no known threat to consumers, because it does not concentrate in muscle tissue and breaks down fairly quickly. But it kills smaller species as well as coral. For tropical reef habitats, it is more lethal than dynamite. Most countries supplying the Hong Kong market have banned cyanide fishing. In most of those countries, the bans don't work. Cyanide is too efficient, undetectable, and available, since it is used by gold and silver mines across Southeast Asia. And the short-term profits are too tempting, both for the fishermen themselves and for the various officials who look the other way.[70]

For many coastal communities in Southeast Asia and the South Pacific, the prices offered for live catch far exceed whatever they could earn from ordinary fishing or, for that matter, any other line of work. One study in Indonesia estimated that live-fish divers earned up to ten times more than other fishermen, and up to three times more than university lecturers. And it doesn't cost much to get started. Live fish, after all, don't need refrigeration, and the big buyers typically supply on credit all the other necessary gear, from spray bottles to motorboats. The boats alone make the trade attractive to the young men usually hired as divers. But they take on both debt and dangerous work. Poorly trained and minimally equipped, live-fish divers risk injury and death from the bends (otherwise known as decompression sickness). As the fish grow scarcer and divers go deeper, the risks increase. Cyanide poisoning is also a danger, both to the divers and to anyone who eats the small fish killed by the divers' chemical spray.[71]

The live-fish trade has introduced more than cyanide into fishing communities. Like some other extractive industries—tropical timber, diamonds, gold, in some places oil—it generates wealth that tends to be both short-lived and socially corrosive. It makes some community members richer and leaves others with fewer fish; it heightens conflicts over rights to fishing grounds; it drives men to neglect farming in favor of diving. Eventually the reef is depleted and the traders move on, leaving behind a poorer local food supply.

Live-catch fishing is not inherently more ecologically and socially unsustainable than other kinds of fishing. Cyanide may be the easiest way to get fish out of the water alive, but it's not the only way. Properly managed, the live-fish trade can help fishermen earn a decent income from relatively small but high-value catches. This has happened in Australia, where hook-and-line fishers earn much more than they did when selling their catches to chilled and frozen seafood processors. And the fact that live fish don't need ice or electric refrigeration still makes the trade attractive to coastal communities throughout the South Pacific.[72] The problem is not the aliveness and wildness of the fish *per se* but rather that these qualities are highly valued by the rich yet supplied, across great distances, mainly by the poor. Under these circumstances the trade in precious freshness resembles the diamond trade; it defies "proper management."

For this reason some countries banned live reef fishing altogether in the 1990s and early 2000s. Hong Kong's marine biologists and environmentalists, meanwhile, called for regulation, consumer education, and the development of sustainable aquaculture. They pointed out that in blind taste tests most people couldn't tell the difference between wild and farmed grouper. They appealed to consumers' self-interest in avoiding ciguatera, a poison found only in wild stocks (and much more dangerous than the pesticides detected in some farmed fish). They called on restaurants and supermarkets and fishmongers to help raise awareness.[73] One government fishery official was optimistic that younger generations would not insist on buying endangered live species because they had been raised on fast food and did not care much about live fish, period. But changing the habits of status-conscious consumers was tough. "They don't know what the taste is," the official said, "but if it says in the magazines that it's expensive, that it's good stuff—they will go for it. They can pay for it."

The View from Seafood Street

Andy Vik knew all about the environmentalists' campaigns. A live-fish importer-exporter and the owner of a small chain of seafood restaurants, he was also vice president of the Hong Kong Seafood Merchants Association. Unlike many of the other Hong Kong fish merchants, he was not from a fishing family; he worked in the hotel business until he saw opportunities in the live-fish trade in the mid-1990s. His fluent English helped him carve out a niche in products from Australia, South Africa, and even the United States, where he bought Maine lobster and Boston geoduck clams. Much of the controversy surrounding cyanide and tropical species depletion didn't directly concern his dealings in coldwater shellfish. But he was affected by broader shifts in demand for live fish—shifts that arguably made Hong Kong the wrong focus for the environmentalists' awareness campaigns.

"Business in Hong Kong has dropped," Vik said, describing a market that was wealthy but also small, aging, and increasingly omnivorous. "The people who can spend a lot, they go to the hotel, they try the most expensive Japanese food, the French food, whatever they want." People who cannot spend a lot have no interest in hundred-dollar lobsters. On the mainland, though, the newly rich would gladly pay that price. So more and more of the live fish brought into Hong Kong was re-exported to the mainland. Some studies suggested that the figure was as high as 60 or 70 percent. Exactly how much that amounted to in volume or value was impossible to say. Hong Kong was, after all, a free-trade city, and the trade in live fish was freer than most. By law, only airfreight fish had to be reported. Even government officials admitted that a large proportion of the live seafood arriving in Hong Kong was smuggled into the mainland on fishermen's boats. Like the above-board shippers, the smugglers usually sent their cargo to the port of Yantien, on the outskirts of the Shenzhen.

Vik and I went to Yantien one afternoon in January 2007. He wanted to see how a client, a Mr. Long, handled his shipments of South African lobster. The lobsters were flown direct to Hong Kong in Styrofoam coolers, then ferried by boat to Yantien, and finally driven by truck to Long's warehouse on Seafood Street. We went by train and taxi. When we arrived in the early evening, Long was drinking tea in his upstairs office. The lobsters were late, he said, stuck in Shenzhen traffic. He proposed dinner down the street, at a waterfront restaurant. The neighborhood was full of smallish warehouses like his, dimly lit and lined with tanks. A few shops sold rubber boots, plastic baskets, scoop-nets. A few skinny dogs scrounged around a rice vendor. Seafood Street didn't look like much, but Vik said that huge quantities of live fish passed through here on route to China's big-city markets.

The restaurant was enormous, with a warren of windowless meeting rooms on the second floor. Not that the view was so great —the banquet floor overlooked a couple of late-model black Audi sedans in the parking lot, and a harbor full of container ships shrouded in smog. The menu swimming in the indoor tanks was more impressive. Our meal included Mexican geoduck, steamed yellow croaker, and Indonesian lobster. All very *tsim*. Long said he had started importing lobsters some fifteen years ago, after China's own stocks grew scarce and costly. It was good money then, but the competition had grown fierce; a lot of people had gotten into the seafood import business.

Midway through the second round of beers, Long's phone rang: the lobsters had arrived. Back at the warehouse, workers hurried coolers off the truck. They were migrants from the countryside, Vik explained—much cheaper than Hong Kong labor. His South African lobsters made up only part of the shipment. Others came from Australia, Indonesia, India, and Sri Lanka. Some were giant, others prawn-sized. Some traveled in wood shavings; the Indian crustaceans came packed in crumpled newspaper. The workers untangled and sorted them quickly, throwing aside the occasional dead one, putting the rest into tanks. Watching them work, I was not sure whether I was more impressed by the effort and expense that went into keeping sea creatures alive or the fact that these particular creatures had survived their globe-spanning voyage. Vik was more matter-of-fact. The market demanded this quality called life; he supplied it. And, he noted approvingly, the mortality rate of his South African lobsters was less than 5 percent.

The lobsters' journey was not quite over. They'd rest overnight in Long's warehouse, then be repacked for one last trip to his wholesale shop in Guangzhou, a city at the center of South China's industrial revolution. Plenty of people there had money and would appreciate the lobsters' meaty size and cold-water origins. Indeed, Seafood Street's tanks held a wealth of fish they wanted to eat and to be seen eating. China's newly rich couldn't get enough of delicious freshness from the sea. In more ways than one, they were hungry for the good life.

But how long would the planet's waters support this life? Back at the Boston seafood show, participants at a session on "international sourcing" were both excited about China as a market and worried about its future as a supplier of 40 percent of the world's farmed fish, much of it cheap and frozen. A panel member predicted that the demand of China's growing middle class would eventually drive up prices of farmed species like shrimp and tilapia, and that the audience of retailers and processors might need to find new bargains in countries such as Vietnam. He did not predict that the United States would begin blocking shipments of China's farmed fish only a few months later, after inspectors found samples contaminated with chemicals such as malachite green. This was old news to China's own consumers, as well as to those in Hong Kong. They had heard plenty about the Pearl River Delta's pesticide and antibiotic-ridden fish farms. It was one reason investors saw opportunities for purer, pricier aquaculture in various sites around Hong Kong itself, including one on the fourteenth story of a high-rise.[74]

In the United States, the contaminated fish episode fueled calls for both tighter controls on imported foods and more support for local suppliers. It should also have reminded us of what our own industrial revolution did to local fish supplies. All too often, our last wild food has proven not just a symbol of wealth and well being, but also one of its casualties.

Epilogue

In our days, everything seems pregnant with its
contrary . . . The newfangled sources of wealth, by
some strange weird spell, are turned into sources of
want.

—Karl Marx, *The People's Paper,* April 19, 1856

When I first started to work on this book, the word "locavore" didn't exist. By late 2007 it had been chosen *Oxford American Dictionary*'s Word of the Year. Locavore's rise from the fringes of the San Francisco foodie scene to national "it" word reflects partly the increasing cultural authority of foodies themselves. American presidents once told citizens to plant wartime gardens. Today the chef Alice Waters gives presidents the same advice, arguing that a White House garden would show a commitment to the war against unhealthful eating. Fellow Berkeleyite and best-selling author Michael Pollan has gone from writing about his backyard vegetable garden to authoring *Newsweek* articles on solutions to global hunger.

Not since the golden age of French gastronomy have chefs and food writers commanded so much attention. But whereas gastronomes like Escoffier and Brillat-Savarin mostly wrote about cooking and table manners, today's foodies are weighing in on the entire food chain. And their calls to shorten the chain—to bring the food supply "back home"—have reached an avid audience. The French chefs of the nineteenth century (and the twenty-first) would agree with Waters about the superior freshness of local greens, picked just hours before serving. Today's local food movement, though, is not just about freshness. It's about finding good-tasting solutions to the growing list of problems caused by global food. You could say it's the quest for perfect food.

This quest drove *Time* magazine's 2007 cover story "Eating Better Than Organic," which chronicles a journalist's search for the perfect apple. Unhappy with the two options available at his Manhattan neighborhood grocery—a New York apple likely sprayed with something toxic, or an organic apple hauled from California—he finds what he's looking for on a local family farm (though the apple on *Time*'s cover looks suspiciously like a waxed Washington State Red Delicious). Watching the farmer's son head barefoot into the chicken coop, the journalist said, "I realized I had never before felt so connected to my food." The connection, he decides, more than makes up for the farm's lack of organic certification.

Pollan, too, seeks perfection in his wildly popular book *Omnivore's Dilemma,* which ends with the author dining on a wild pig he shot himself. As Pollan emphasizes, what made the meal perfect was not just the tastiness of the pork, braised with locally gathered morels, nor just the pleasure of eating with friends and family. It was also the meal's "nearly perfect transparency":

> Scarcely an ingredient in it had ever worn a label or bar code or price tag, and yet I knew almost everything there was to know about its provenance and its price. I knew and could picture the very oaks and pines that had nourished the pigs and the mushrooms that were nourishing us. And I knew the true cost of this food, the precise sacrifice of time and energy and life it had entailed . . . it was cheering to realize just how little this preindustrial and mostly preagricultural meal had diminished the world. My pig's place would soon be taken by another pig.[1]

Of course, not everyone who wants to eat local can or wants to shoot their own boar. Conveniently, even Wal-Mart began promoting local food in 2006 with its "Salute to American Farmers" program. Every month it chose a few growers around the country and featured their produce at nearby Supercenters. "As Californians search for the perfect pumpkin this Halloween season," a Wal-Mart press release announced in October 2007, the store would feature pumpkins from a third-generation Central Valley family farm. In Oregon and Minnesota, Wal-Mart shoppers could buy their own states' melons and apples; Alabamans could buy local sweet potatoes. The retailer pointed out that in addition to buying fresh produce from existing local growers, it encouraged new production near its regional distribution centers. Cilantro once shipped from California to New York would now be sourced in southern Florida: "Not only will these efforts save food miles, but they will provide our customers with fresher products."[2]

The most committed locavores would never be caught food shopping in Wal-Mart, and not just because it defines "local" far more expansively than they do. The headlines are full of reasons to pull one's diet and dollars out of Wal-Mart and the global food system more generally. Apart from the risks to personal health (whether in the form of Chinese farmed shrimp, tainted California spinach, or just too much high fructose corn syrup), the threats to the health of the planet are many and serious. The purveyors of global food are racking up food miles (despite Wal-Mart's claims to the contrary), fostering destructive farming practices, and, perhaps worst of all, encouraging the rest of the world to eat like Americans.

Supporting the local food movement, it seems, offers an obvious way both to protest and to stay protected from global food. If the borders of the locavores' 100- or 150-mile diets are admittedly arbitrary, the point of drawing them is not. Locavores want their foodsheds to sustain everything that Wal-Mart and its ilk have already threatened: small farms and businesses, open space, a sense of community, and plenty of distinctive, delicious, healthful, and *genuinely* fresh food, grown by people they know and trust.

It is a very attractive vision. It is also not new. In the late nineteenth century, a similar dream inspired all kinds of utopian experiments, both fictional and real-life. Nineteenth-century Americans gobbled up novels about perfect societies, and some, including recent immigrants, tried to start their own. Taking advantage of cheap lands out west, they took on names such as New Harmony, Heaven Colony, Fruitlands, True Inspirationists, Perfectionists. By the century's end, the search for a fresh start fueled the creation of more than 300 planned communities in the United States.[3]

Feeding utopia worked better on paper than in real life. Fictional utopians enjoyed pastoral bliss and sci-fi convenience. They walked streets lined with fruit-laden trees, ate plentiful, perfectly balanced meals, and never washed dishes. If they used money, all food prices were fair; if they ate meat, it was clean and cruelty-free. Since most authors of such books were city types, they tended to devote less attention to food production than to distribution and preservation. In their imagined worlds, no one went hungry, no food went to waste, and technology kept fruits and vegetables "in their natural state" longer and better than refrigeration.[4]

Members of real-life communities, by contrast, faced the real-life challenges of weather, sickness, pests, and just getting along. Some went hungry; some didn't like the communal cooking.[5] But even the best-fed communities, with a few exceptions, faced a sustainability problem that had nothing to do with the soil they farmed. At a time of rapid, confusing change, they struggled to maintain order and stability. Founded to show the world a different, better way to live, many just focused on keeping worldly influences at bay. Some communities descended into conflict and totalitarianism before falling apart. More often, they just went stale.[6]

I doubt that many of today's locavores would call themselves utopians. After all, even hardnosed economists and public health experts would agree that much about local food makes sense. High fuel prices could turn locally grown, low-input foodstuffs into relative bargains.[7] Less transport and storage of fresh foods saves vitamins as well as energy. Local food markets also allow for efficient communication: consumers can express their preferences directly to producers ("can you bring more red chard to next week's market?") and also learn what to reasonably expect ("no, the season's finished"). As Pollan observes in *Omnivore's Dilemma,* this kind of "perfect knowledge" about food—its provenance and seasons and costs—is easier to come by at the local farm or farmers market than in the aisles of any supermarket, even Whole Foods. Ideally such knowledge will empower consumers to make better (if not necessarily perfect) choices. The choice of local food, according to this line of thought, isn't just fresher, safer, and more sustainable; it's altogether more enlightened.

The history of modern freshness is full of people whose ideas of an enlightened diet included an abundance of fresh food. Among them were the refrigeration engineers like Charles Tellier. It would be hard to call them the heroes of the story, given that their inventions are partly responsible for much that we dislike and distrust about our mass-produced, long-haul supermarket perishables. Still, it's a story that's helpful for thinking about our current food quandaries, and for understanding what's missing from the local foodshed's perfect knowledge.

This story reminds us of all those who've come to depend on the global fresh-food economy for their livelihoods, whether as farmers or fishermen or as employees of larger operations. This dependence isn't by itself a reason to maintain the status quo any more than arms industry jobs justify going to war, especially since many of these livelihoods, whether in Iowa packhouses or Indonesia's live-fish trade, are grueling and dangerous. The point here is simply that the same larger forces that have created prosperous local foodsheds in some parts of the world have undermined them in others. It is easier to be a locavore in Berkeley than in Burkina Faso, and not just because the weather's milder. The health of a local food economy depends to a large degree on its wealth—not only in terms of household spending power, but also (and arguably more importantly) in the public resources available for everything from road maintenance to irrigation to community gardens. The basic infrastructure of locavorism, in other words, can't be taken for granted everywhere.

At the same time, the history of modern freshness shows that even with the best resources and intentions, the utopian vision of an unchanging local food economy really is a fiction. Refrigeration engineers dreamed of machines that could preserve this steady state, whether in steamships or in home kitchens. They ended up with machines that, by slowing spoilage, set off all kinds of other changes in food culture and commerce. The history of the fridge illustrates (to reverse the old truism) that the more you try to keep things the same, the more they are apt to change.

This is certainly apparent in my own foodshed around Dartmouth College. The area has long profited from the nostalgia of both Dartmouth alum (many of whom retire here, or never leave) and country inn–hopping tourists. Over the past several years, though, artisanal cheesemakers and other old-fashioned farm enterprises have sprouted up all over the Connecticut River Valley. Now the tourists and second–home owners come to eat as well as to leaf-peep. But as in other foodie destinations—New York's Hudson River Valley, California's Sonoma County—agricultural land values have climbed skyward. This makes it a hard place to run an ordinary dairy farm or apple orchard. Farmers who actually need to live off their land face a choice: sell out or add value. And that means that much of the local food is priced out of reach of the area's less-wealthy locals.[8]

Could local food be made more equitable? Sure. So could global food, and for that matter the nation's food. I'm finishing this book at a time when rarely a day passes without news about crises caused by food injustices, from obesity to famine and food riots.[9] The history of freshness offers no simple solutions. On the contrary, it suggests that any route that seems perfectly simple is probably no solution at all, especially if it proposes going back to an imagined past. Local food has much to offer, at least in some localities. But eating it will not undo history. A tour of the modern fridge reveals a world of interdependencies and inequalities, forged through trade, conquest, and politics. It is a world of sharp contradictions between marketed ideals and industrial realities. Nothing is as pure or natural as we'd like, but there's no shutting the door on this world. Anyone who owns a fridge knows that ignoring the contents for too long is never a good idea. On the bright side, the same tour also shows how humans' hunger for better food has inspired both creativity and courage. More than ever, this kind of freshness is worth keeping.

Notes

Bibliography

Acknowledgments

Index

Notes

Introduction

1. U.S. Food and Drug Administration, Public Meeting on Use of the Term "Fresh" on Foods Processed with Alternative Technologies, Chicago, July 21, 2000. A transcript of the meeting is available at www.cfsan.fda.gov/~dms/flfresh.html.
2. U.S. Food and Drug Administration, "Subpart F—Specific Requirements for Descriptive Claims That Are Neither Nutrient Content Claims nor Health Claims, Sec. 101.95 'Fresh,' 'Freshly Frozen,' 'Fresh Frozen,' 'Frozen Fresh,'" *Code of Federal Regulations, Title 21*, 2006.
3. On the history of food preservation methods, see Sue Shepard, *Pickled, Potted, and Canned: How the Art and Science of Food Preserving Changed the World* (New York: Simon and Schuster, 2000); C. Anne Wilson, *Waste Not, Want Not: Food Preservation from Early Times to the Present Day* (Edinburgh: Edinburgh University Press, 1991).
4. Marion Nestle, *What to Eat* (New York: North Point Press, 2006).
5. Richard O. Cummings, *The American and His Food: A History of Food Habits in the United States* (Chicago: University of Chicago Press, 1941); Jean Louis Flandrin, "Seasoning, Cooking and Dietetics in the Late Middle Ages," in *Food: A Culinary History from Antiquity to the Present*, ed. J. L. Flandrin, M. Montanari, and A. Sonnenfeld (New York: Penguin, 1999), pp. 313–327.
6. On early twentieth-century nutritional thought and diet fads, see

Harvey Levenstein, *Revolution at the Table: The Transformation of the American Diet* (Berkeley: University of California Press, 2003); Hillel Schwartz, *Never Satisfied: A Cultural History of Diets, Fantasies and Fat* (New York: Free Press, 1986).

7. Michael French and Jim Phillips, *Cheated Not Poisoned? Food Regulation in the United Kingdom, 1875–1938* (Manchester: Manchester University Press, 2000).

8. Marvin Harris and Eric B. Ross, eds., *Food and Evolution: Toward a Theory of Human Food Habits* (Philadelphia: Temple University Press, 1987).

9. L. K. Wright, *The Next Great Industry: Opportunities in Refrigeration and Air Conditioning* (New York: Funk and Wagnalls, 1939), p. 30.

10. Sustain, *Eating Oil: Food Supply in a Changing Climate*, a Sustain/Elm Farm Research Centre Report, London, December 2001; John Humphrys, *The Great Food Gamble* (London: Hodder and Stoughton, 2002). The relationship between food miles and fuel consumption is far from straightforward, especially once the energy costs of food production in different regions are included in the calculations. According to one controversial study, it is more energy-efficient for Britain to import lamb by boat from New Zealand than to raise it domestically. See Caroline Saunders, Andrew Barber, and Greg Taylor, *Food Miles—Comparative Energy/Emissions Performance of New Zealand's Agriculture Industry*, Lincoln University Research Report 285 (July 2006).

11. For example: Michael Pollan, *The Omnivore's Dilemma: A Natural History of Four Meals* (New York: Penguin Press, 2006); Alisa Smith and J. B. Mackinnon, *Plenty: One Man, One Woman, and a Raucous Year of Eating Locally* (New York: Harmony, 2006); Barbara Kingsolver, with Steven L. Hopp and Camille Kingsolver, *Animal, Vegetable, Miracle: A Year of Food Life* (New York: HarperCollins, 2007); Michael Pollan, *In Defense of Food: An Eater's Manifesto* (New York: Penguin Press, 2008).

12. The most comprehensive account of refrigeration's technological development and economic impacts in the United States remains Oscar E. Anderson, *Refrigeration in America: A History of a New Technology and Its Impact* (Princeton, N.J.: Princeton University Press, 1953).

13. William G. Sickel, "Refrigeration on Ocean Steamships," in *Premier congrès international du froid Paris, 5 au 12 octobre 1908*, ed. Inter-

national Congress of Refrigeration, under the direction of J. de Loverdo (Paris: Secrétariat général de l'Association internationale du froid, 1908), p. 766.

14. Woods Hutchinson, "The Physical Basis of Brain-Work," *North American Review* 146 (1888) 522–531; T. J. Jackson Lears, *No Place of Grace: Antimodernism and the Transformation of American Culture, 1880–1920* (Chicago: University of Chicago Press, 1994). On the literati's views of rural life, see Leo Marx, *The Machine in the Garden: Technology and the Pastoral Ideal in America* (New York: Oxford University Press, 1964).

15. Lears, *No Place of Grace.*

16. It has been suggested that if Louis Pasteur had not discovered bacteria, someone would have had to invent them, because they became such a convenient focus for late nineteenth–century anxieties about urban hygiene. By the 1920s the same could be said of vitamins, except that they played the role of the invisible saviors rather than the villains. By the 1930s, vitamin pills and fortified foods offered a less fresh form of salvation. Bruno Latour, *The Pasteurization of France* (Cambridge, Mass.: Harvard University Press, 1988), pp. 23, 25; Rima D. Apple, *Vitamania: Vitamins in American Culture* (New Brunswick, N.J.: Rutgers University Press, 1996). On the changes in late nineteenth– and early twentieth–century nutritional thought, see Levenstein, *Revolution at the Table.*

17. Advertisement sponsored by the Western Growers' Protective Association, *Ladies' Home Journal* (March 1930).

18. Steven Stoll, *The Fruits of Natural Advantage: Making the Industrial Countryside in California* (Berkeley: University of California Press, 1998); Douglas Cazaux Sackman, *Orange Empire: California and the Fruits of Eden* (Berkeley: University of California Press, 2005).

19. Quoted in T. J. Jackson Lears, *Fables of Abundance: A Cultural History of Advertising in America* (New York: Basic Books, 1994), p. 179.

20. Michael Schudson, *Advertising, the Uneasy Persuasion: Its Dubious Impact on American Society* (New York: Basic Books, 1984). On popular insecurities about modernity, see Robert H. Wiebe, *The Search for Order, 1877–1920* (New York: Hill and Wang, 1967). The classic work on advertisers' role in selling modernity to ambivalent consumers is Roland Marchand, *Advertising the American Dream: Making Way for Modernity, 1920–1940* (Berkeley: University of California

Press, 1985). See also J. T. Jackson Lears, "American Advertising and the Reconstruction of the Body, 1880–1930," in *Fitness in American Culture: Images of Health, Sport, and the Body, 1830–1940,* ed. Kathryn Grover (Amherst: University of Massachusetts Press, 1989), pp. 47–66; T. J. Jackson Lears, "From Salvation to Self-Realization: Advertising and the Therapeutic Roots of the Consumer Culture, 1880–1930," *Advertising & Society Review* 1 (2000).

21. Marx, *Machine in the Garden,* p. 3.

22. Edna Ferber, "Maymeys from Cuba," in *Buttered Side Down: Stories,* ed. Edna Ferber (New York: Grosset and Dunlap, 1911).

1. Refrigeration

1. Oscar E. Anderson, *Refrigeration in America: A History of a New Technology and Its Impact* (Princeton, N.J.: Princeton University Press, 1953), pp. 5–6; H. J. Teutenberg, "History of Cooling and Freezing Techniques and Their Impact on Nutrition in Early Twentieth Century Germany," in *Food Technology, Science, and Marketing: European Diet in the Twentieth Century,* ed. Adel P. den Hartog (East Linton: Tuckwell Press, 1995), p. 52; Barry Donaldson and Bernard Nagengast, *Heat and Cold: Mastering the Great Indoors* (Atlanta: American Society of Heating, Refrigerating and Air-Conditioning Engineers, 1994), p. 18.

2. Letter to cousin, December 10, 1805. "Frederic Tudor: Ice King," *Journal of the Business Historical Society* 6, 4 (1932): 1–8.

3. Carl Seaburg and Stanley Paterson, *The Ice King: Frederic Tudor and His Circle* (Boston: Massachusetts Historical Society, 2003). For a fuller account of Tudor's life and the trade he started, see Gavin Weightman, *The Frozen Water Trade* (New York: Hyperion, 2004).

4. Weightman, *The Frozen Water Trade,* p. 144.

5. Ibid., p. 155.

6. H. Pearson, "Frederic Tudor, Ice King," *Proceedings of the Massachusetts Historical Society* 65 (1933): 183.

7. Donaldson and Nagengast, *Heat and Cold,* p. 46.

8. Anderson, *Refrigeration in America,* p. 21.

9. "Ice and the Ice Trade," *Hunt's Merchants' Magazine* (1855): 175; Anderson, *Refrigeration in America,* p. 27.

10. Richard O. Cummings, *The American and His Food: A History of*

Food Habits in the United States (Chicago: University of Chicago Press, 1941), p. 34.

11. Actually, Moore's icebox probably worked better than those manufactured later because he put the ice compartment on top, allowing for better circulation of the cold air. Most later models stored ice on the bottom, where it melted more slowly. Anderson, *Refrigeration in America,* p. 23.

12. Cummings, *The American and His Food,* p. 39. The refrigeration industry continued to recommend two iceboxes even in the 1890s because "milk and butter, especially, acquire an unpleasant odor and flavor . . . and are said by physicians to be rendered unwholesome, if not poisonous, by the presence of the meat." Current Opinion, *Ice and Refrigeration* 4 (1893).

13. Weightman, *The Frozen Water Trade,* p. 172; Anderson, *Refrigeration in America,* p. 27.

14. Anderson, *Refrigeration in America,* p. 22.

15. "Ice: How Much of It Is Used, and Where It Comes From," *De Bow's Review* 19 (1955): 709.

16. J.C.C., "Ice Made by Mechanical Power," *Scientific American,* September 22, 1849, p. 3.

17. On the history of refrigeration invention, see Donaldson and Nagengast, *Heat and Cold.*

18. Anderson, *Refrigeration in America,* p. 86; George Briley, "A History of Refrigeration," *ASHRAE Journal* (2004): S31–34.

19. Anderson, *Refrigeration in America,* p. 110.

20. Ibid., p. 113.

21. Thomas A. Bird, "The Ice Man as an Advertiser," *Ice and Refrigeration* 38 (1910): 103.

22. David A. Brown, "Advertising Ice," *Ice and Refrigeration* 38 (1910): 212–213.

23. Ibid.

24. Bird, "The Ice Man as an Advertiser."

25. Census figures cited in Anderson, *Refrigeration in America,* pp. 114–115.

26. Ibid.

27. Emile Bonnechaux, "L'industrie du froid en Asie, Afrique, Australie, et aux Etats-Unis," in *Premier congrès international du froid Paris, 5 au 12 octobre 1908,* ed. International Congress of Refrigeration (Paris:

Secrétariat général de l'Association internationale du froid, 1908). In some places, local people actively avoided ice. One French official said this of Arab café owners in colonial Mozambique (formerly a Portuguese colony): "Although they provide ice for their customers, they do not want to eat or drink anything that has touched ice or been preserved by ice, meaning that their diet is reduced to dried meat and vegetables." "Les entrepôts frigorifiques sous les tropiques," *La Glace et les industries du froid* 4 (1907).

28. "Ice in Europe," *Ice and Refrigeration* 3 (1892): 359–362.

29. H. D. Renner, *The Origin of Food Habits* (London: Faber and Faber, 1944), p. 221.

30. On trade norms at Les Halles, see Kyri Claflin, "Culture, Politics and Modernization in Paris Provisioning, 1880–1920," Ph.D. diss., Boston University, 2006.

31. Guy Chemla, *Les ventres de Paris: les Halles, la Villette, Rungis: l'histoire du plus grand marché du monde* (Grenoble, Glénat, 1994), p. 213, n. 146. Six other fruit sellers in the Ile-de-France region also owned cold storage chambers, including one in Montreuil. J. de Loverdo, *Monographie sur l'etat actuel de l'industrie du froid en France* (Paris: Association Francaise du Froid, 1910), Appendix.

32. "Le Frigoriphobie," *L'Industrie Frigorifique* 11 (1913): 126.

33. H. Brun, "Les entrepôts frigorifiques," *L'Industrie Frigorifique* 1 (1903): 17–26.

34. Ibid.

35. "L'étiquetage des fruits conservés par le froid et le phobie du froid," *La Revue Générale du Froid* 3 (1911): 536–538.

36. "La réfrigération des fruits et les préjugés courants," *La Revue Générale du Froid* 2 (1910): 154–155.

37. W. Weddel and Company, *Review of the Chilled and Frozen Meat Trade* (London, 1922), pp. 4, 6. See also Claflin, "Culture, Politics and Modernization in Paris Provisioning, 1880–1920."

38. The Food Administration did not discourage the consumption of all meats, just beef and pork. It also encouraged consumers to get more of their protein from dry beans and legumes. Harvey A. Levenstein, *Revolution at the Table: The Transformation of the American Diet* (Berkeley: University of California Press, 2003), chapter 11.

39. C. Houston Goudiss and Alberta M. Goudiss, *Foods That Will Win the War and How to Cook Them* (New York: Forecast Publishing, 1918), p. 33.

40. "Storing Your Vegetables: Your Government Tells You Exactly How to Do It," *Ladies' Home Journal* (October 1917) ; J. Shumway, "The Woman and the War," *Ladies' Home Journal* 34 (September 1917): 28, 85–88, 99; Levenstein, *Revolution at the Table,* p. 139.

41. M. Neil, "The Vegetables You Have Grown," *Ladies' Home Journal* (October 1917): 30; Mary Swartz Rose, *Everyday Foods in Wartime* (New York: Macmillan, 1918).

42. Levenstein, *Revolution at the Table,* chapters 4, 8–10.

43. Ibid., p. 141.

44. The diets of the "leaf-eating" Japanese and Chinese were considered from this perspective. On these ideas, see Elmer V. McCollum, Elsa Orent-Keiles, and Harry G. Day, *The Newer Knowledge of Nutrition* (New York: Macmillan, 1939); several editions of this book were published. Elmer V. McCollum, *A History of Nutrition; the Sequence of Ideas in Nutrition Investigations* (Boston: Houghton Mifflin, 1957).

45. Levenstein, *Revolution at the Table,* p. 141; Grace Aspinwall, "The Joys of Raw Food," *Good Housekeeping* 50 (January 1910): 110–112; George J. Drews, *Unfired Food and Hygienic Dietetics: For Prophylactic (Preventive) Feeding and Therapeutic (Remedial) Feeding: (Treatise on Food in the Cause, Prevention and Cure of Disease)* (Chicago: G. J. Drews, 1909); Eugene Christian and Molly Griswold Christian, *Uncooked Foods & How to Use Them* (New York: The Health-Culture Company, 1904). For more on food fads, see Hillel Schwartz, *Never Satisfied: A Cultural History of Diets, Fantasies and Fat* (New York: Free Press, 1986).

46. Rose, *Everyday Foods in Wartime,* pp. 40, 46.

47. C. Frederick, "Your Health Depends on Your Eating—So Why Not Get Better Acquainted with Fruit?" *Ladies' Home Journal* 36 (February 1919): 53.

48. Lewis Edwin Theiss, "What Shall We Eat to Be Well?" *Good Housekeeping* (July–August 1920): 157.

49. "Nation-wide Newspaper Campaign Now Underway," *Western Grower and Shipper* 1 (February 1930): 12–13.

50. Elmer Verner McCollum and Nina Simmonds, *The American Home Diet: An Answer to the Ever Present Question, What Shall We Have for Dinner* (Detroit: Frederick C. Mathews, 1920).

51. "The National Educational and Goodwill Advertising Campaign for the Ice Industry," *Ice and Refrigeration* 73 (1927): 158–161.

52. J. A. Harlan, "Selling Refrigeration," *Ice and Refrigeration* 72 (1927): 449–450.

53. For more detail see Levenstein, *Revolution at the Table*.

54. Ibid., p. 194.

55. Anderson, *Refrigeration in America*, p. 209.

56. E. Whitehorne, "Household Refrigeration," *The House Beautiful* (September 1921).

57. R. T. Frazier, "The Household Ice Refrigerator," *Refrigerating Engineering* 18 (1929).

58. Anderson, *Refrigeration in America*, p. 210.

59. Peter Neff, "Domestic Refrigerating Machine," *Ice and Refrigeration* 49 (1915): 143–144.

60. Lisa Mae Robinson, "Safeguarded by Your Refrigerator: Mary Engle Pennington's Struggle with the National Association of Ice Industries," in *Rethinking Home Economics: Women and the History of a Profession*, ed. Sarah Stage and Virginia B. Vincenti (Ithaca, N.Y.: Cornell University Press, 1997), p. 255; David E. Nye, *Electrifying America: Social Meanings of a New Technology, 1880–1940* (Cambridge, Mass.: MIT Press, 1990), p. 276.

61. Donaldson and Nagengast, *Heat and Cold*, pp. 205–207; Joel Connolly, "Difficulties Encountered in the Control of Mechanical Refrigeration," *American Journal of Public Health* 20 (1930): 252–256; Anderson, *Refrigeration in America*, p. 93.

62. Jonathan Rees, "'I Did Not Know . . . Any Danger Was Attached': Safety Consciousness in the Early American Ice and Refrigeration Industries," *Technology and Culture* 46 (2005): 542, 553.

63. Ruth Schwartz Cowan, "How the Refrigerator Got Its Hum," in *The Social Shaping of Technology*, ed. Donald Mackenzie and Judy Wajcman (Philadelphia: Open University Press, 1985), p. 206.

64. Ibid.

65. The electric grid spread fastest between 1918 and 1928, when two million homes a year were wired annually. Most rural households, however, did not get electricity until after 1935, under the auspices of Tennessee Valley Authority and Rural Electrification Administration. Nye, *Electrifying America*, pp. 239, 265, 299.

66. Ruth Schwartz Cowan, *More Work for Mother* (New York: Basic Books, 1983), pp. 136–137.

67. James Cullen, "To Ice!" *Ice and Refrigeration* 73 (1927): 162.

68. Anderson, *Refrigeration in America*, p. 216; "National Educational Publicity for Ice," *Ice and Refrigeration* 70 (1926): 546–547.

69. M. E. Pennington, *Journeys with Refrigerated Foods: Eggs* (Chicago: National Association of Ice Industries, 1928), p. 2.

70. Robinson, "Safeguarded by Your Refrigerator," 259, 263.

71. M. Kingsley, "Household Refrigeration Bureau," *Ice and Refrigeration* 70 (1926): 376–377.

72. "Attractive and Effective Newspaper Advertisements of Ice," *Ice and Refrigeration* 71 (1926): 159–160.

73. "Plans for Advertising the Ice Industry in 1928," *Ice and Refrigeration* 73 (1927): 211–218.

74. *Ladies Home Journal* (July 1929).

75. Some refrigeration propaganda called fifty degrees the scientifically accepted "danger point"; other sources put it at sixty degrees. In fact, scientists already knew that different foods keep best at different temperatures; certainly the most perishable foods, like fresh milk and fish, will not keep long even at "several degrees below fifty."

76. *Good Housekeeping* (September 1925): 114; quoted in Peter R. Grahame, "Objects, Texts and Practices: The Refrigerator in Consumer Discourses between the Wars," in *The Socialness of Things: Essays on the Social-Semiotics of Objects*, ed. S. H. Riggins (New York: Mouton de Gruyter, 1994), p. 289.

77. Frigidaire, *Frigidaire Recipes: Prepared Especially for Frigidaire Automatic Refrigerators Equipped with the Frigidaire "Cold Control"* (Dayton, Ohio: Frigidaire Corporation, 1929); Kelvinator, *New Delights from the Kitchen* (Detroit, Mich.: 1930).

78. M. J. Crosby, "Refrigeration Cookery by Electricity," *Ladies' Home Journal* (1927): 129.

79. Alice Bradley, *Electric Refrigerator Menus and Recipes. Recipes Prepared Especially for the General Electric Refrigerator* (Cleveland, Ohio: General Electric Co., 1927), p. 18; Frigidaire, *Frigidaire Recipes*, p. 84.

80. Gove Hambidge, "This Age of Refrigeration," *Ladies' Home Journal* (August 1929): 93+; Rees, "'I Did Not Know . . . Any Danger Was Attached,'" p. 556. In 1929, 840,000 electric refrigerators sold, compared with 1,053,000 iceboxes; in 1930, 850,000 electric models sold, but only 419,000 iceboxes. In 1934, electric refrigerator sales had climbed to 1,390,000, and icebox sales were down to 276,000. United

States National Resources Committee, Science Committee, headed by W. F. Ogburn, *Technological Trends and National Policy, Including the Social Implications of New Inventions, June 1937* (Washington, D.C.: U.S. Government Printing Office, 1937), p. 317.

81. Cowan, *More Work for Mother*, p. 196.
82. According to census figures, an estimated 80 percent of U.S. households had refrigerators by 1950; by 1965 estimated ownership was more than 90 percent. W. T. Pentzer, "The Giant Job of Refrigeration," in *Yearbook of Agriculture: Protecting Our Food*, ed. J. Hayes (Washington, D.C.: U.S. Department of Agriculture, 1966), pp. 123–138.
83. On American food history from the 1930s onwards, see Harvey A. Levenstein, *Paradox of Plenty: A Social History of Eating in Modern America* (New York: Oxford, 1993); Sandy Isenstadt, "Visions of Plenty: Refrigerators in America around 1950," *Journal of Design History* 11 (1998): 311–321. Crosley Home Appliances refrigerator models promised to keep "bushels of fresh things," while International Harvester sold 7-Climate Refrigerators.
84. Hambidge, "This Age of Refrigeration."
85. Leonora Baxter, "The New Ice Age," *Golden Book Magazine* (January 1931): 83–86.
86. Hambidge, "This Age of Refrigeration."
87. Levenstein, *Paradox of Plenty*, p. 27.
88. Already by 1919, Del Monte advertised its fruits and vegetables as "canned the day they are picked, with all their natural freshness and flavor." *Ladies' Home Journal* (July 1919): 106.

2. Beef

1. José Bové and François Dufour, trans. Anna de Casparis, *The World Is Not for Sale: Farmers against Junk Food* (New York: Verso, 2001), p. 53.
2. F. C. Parrish, Aging of Beef (Chicago: National Cattlemen's Beef Association); www.askthemeatman.com/dry_aged_beef.htm.
3. Finbar McCormick, "The Distribution of Meat in a Hierarchical Society: The Irish Evidence," in *Consuming Passions and Patterns of Consumption*, ed. P. Miracle and N. Milner (Cambridge, England: McDonald Institute for Archaeological Research, 2002), pp. 25–31; Pauline Wilson Wiessner and Wulf Schiefenhövel, *Food and the Status*

Quest: An Interdisciplinary Perspective (Providence, R.I.: Berghahn Books, 1996).

4. Caroline Skeel, "The Cattle Trade between Wales and England from the Fifteenth to the Nineteenth Centuries," *Transactions of the Royal Historical Society, 4th Series,* 9 (1926): 135–158; I. Blanchard, "The Continental European Cattle Trades, 1400–1600," *Economic History Review* 39 (1986): 427–460; Kristoff Glamann, "The Cattle Trades," in *The Cambridge Economic History of Europe,* ed. M. M. Postan and H. J. Habakkuk (Cambridge, England: Cambridge University Press, 1966), pp. 232–240.

5. Jack Cecil Drummond and Anne Wilbraham, *The Englishman's Food: A History of Five Centuries of English Diet* (London: J. Cape, 1958), p. 53; Jeremy Rifkin, *Beyond Beef: The Rise and Fall of the Cattle Culture* (New York: Penguin, 1992), pp. 53–54.

6. Virginia DeJohn Anderson, *Creatures of Empire: How Domestic Animals Transformed Early America* (New York: Oxford University Press, 2004), p. 155.

7. Karen Friedmann, "Victualling Colonial Boston," *Agricultural History* 47 (1973): 195; Sarah F. McMahon, "A Comfortable Subsistence: The Changing Composition of Diet in Rural New England, 1620–1840," *The William and Mary Quarterly* 42 (1985): 26–65.

8. Jason W. Moore, "The Modern World-System as Environmental History? Ecology and the Rise of Capitalism," *Theory and Society* 32 (2003): 307–377.

9. Drummond and Wilbraham, *The Englishman's Food,* p. 350.

10. Mark Finlay, "Early Marketing of the Theory of Nutrition: The Science and Culture of Liebig's Extract of Meat," in *The Science and Culture of Nutrition, 1840–1940,* ed. Harmke Kamminga and Andrew Cunningham (Atlanta: Rodopi, 1995), pp. 48–76.

11. Friedrich Engels, *The Condition of the Working Class in England* (New York: Penguin, 1967 [1887]), p. 104.

12. Richard Perren, *The Meat Trade in Britain, 1840–1914* (London: Routledge and Kegan Paul, 1978), pp. 108–114.

13. Simon Gabriel Hanson, *Argentine Meat and the British Market: Chapters in the History of the Argentine Meat Industry* (Stanford, Calif.: Stanford University Press, 1938), p. 19. The number of patents taken out for mechanical refrigeration methods in Britain increased from eleven in the 1850s to thirty in the 1860s and fifty-six in the years

1870–1874. James Critchell and Joseph Raymond, *A History of the Frozen Meat Trade* (London: Constable and Company, 1912), p. 435.

14. Charles Elliot, "The Preservation of Food," *Journal of the Society of Arts* 9 (1861): 95–97.

15. Perren, *The Meat Trade in Britain,* pp. 115, 158; also I. M. Greg and S. H. Towers, *Cattle Ships and Our Meat Supply* (London: The Humanitarian League's Publications, no. 15, 1894).

16. W. Bridges Adams, "Letter on the Preservation of Food," *Journal of the Society of Arts* 13 (1865): 339.

17. Drummond and Wilbraham, *The Englishman's Food,* pp. 316–317; B. C. Steet, "On the Preservation of Food, Especially Fresh Meat and Fish, and the Best Form for Import and Provisioning Armies, Ships and Expeditions," *Journal of the Society of Arts* 13 (1865): 309–315.

18. Steet, "On the Preservation of Food," p. 315.

19. Charles Tellier, *Histoire d'une invention moderne: le frigorifique* (Paris: C. Delagrave, 1910).

20. Robert Lesage, *Charles Tellier, le père du froid* (Paris: A. Giraudon, 1928), p. 166.

21. Tellier, *Histoire d'une invention moderne,* p. 35.

22. Alfred Massé, *Le troupeau français et la guerre: viande indigène, viande importée* (Paris: Librairie Agricole de la Maison Rustique, 1915), p. 7.

23. Tellier, *Histoire d'une invention moderne,* p. 200; Charles Tellier, "Conservation de la viande et autres substances alimentaires," *Usine frigoforique d'Auteuil* (1871):10–11.

24. "Le 'Frigorifique' á Buenos-Ayres," *L'Illustration,* March 10, 1877, p. 151; Charles Tellier, "Communication aux actionnaires de la société fondatrice pour la conservation de la viande fraiche par le froid" (1877).

25. Tellier, *Histoire d'une invention moderne,* p. 323.

26. E. Menalque, "Les frigorifiques dans les abbatoirs," *Industrie Frigorifique* 2 (1904): 273–278.

27. Massé, *Le troupeau français et la guerre.*

28. Menalque, "Les frigorifiques dans les abbatoirs." Cookbooks devoted to these meats explained what they were and how to treat them. Le Frigo: comment le parer, comment le cuire, comment le préparer, 1919.

29. Georges Lafond, *L'industrie frigorifique Argentine et la crise de "la vie chere"* (Paris: Société d'Editions Internationales, 1912), p. 4.

30. "Reprint of 'Conseil General de la Seine Paris, seance du 21 Juin,'" *La Revue Générale du Froid* 3 (1911): 373–385.
31. Lafond, *L'industrie frigorifique Argentine et la crise de "la vie chere,"* p. 32.
32. "Manifestion du 15 fevrier en l'honneur du Charles Tellier," *Industrie Frigorifique* 11 (1913): 67–73; G. Le Roy, *La mort de Charles Tellier: ses obseques* (Paris: Association Francaise du Froid, 1913).
33. See also Willis Raymond Woolrich, *The Men Who Created Cold: A History of Refrigeration* (New York: Exposition Press, 1967).
34. The most important consumer groups at the turn of the century were cooperatives. They generally favored the import of foreign refrigerated meat because it would lower prices. J.-J. Meusy, ed., *La Bellevilloise (1877–1939): un page de l'histoire de la coopération et du mouvement ouvrier français* (Paris: Créaphis, 2001); Ellen Furlough and Carl Strikwerda, *Consumers against Capitalism?: Consumer Cooperation in Europe, North America, and Japan, 1840–1990* (Lanham, Md.: Rowman and Littlefield, 1999).
35. Alfred Dupont Chandler, *The Visible Hand: The Managerial Revolution in American Business* (Cambridge, Mass.: Belknap Press of Harvard University Press, 1977).
36. William Cronon, *Nature's Metropolis: Chicago and the Great West* (New York: W. W. Norton, 1991), pp. 249–250.
37. Rudolf A. Clemen, *The American Livestock and Meat Industry* (New York: Ronald Press, 1923), pp. 225.
38. John H. White, *The Great Yellow Fleet: A History of American Railroad Refrigerator Cars* (San Marino, Calif.: Golden West Books, 1986), p. 32.
39. J. Ogden Armour, *The Packers, the Private Car Lines and the People* (Philadelphia: Henry Altemus, 1906), p. 20; Clemen, *The American Livestock and Meat Industry*, p. 220.
40. Louis Franklin Swift, *The Yankee of the Yards: The Biography of Gustavus Franklin Swift* (Chicago: A. W. Shaw, 1927), p. 29.
41. Perren, *The Meat Trade in Britain, 1840–1914*, pp. 100, 129.
42. Ibid., p. 130; George Putnam, *Supplying Britain's Meat* (London: George Harrap, 1923), p. 68.
43. January 10, 1876. Quoted in Hanson, *Argentine Meat and the British Market*, p. 42.
44. James Macdonald, *Food from the Far West* (London: W. P. Nimmo, 1878), pp. 263, 273.

45. Edward Everett Dale, *Cow Country* (Norman: University of Oklahoma Press, 1965), pp. 101–103; Richard Graham, "The Investment Boom in British-Texan Cattle Companies 1880–1885," *Business History Review* 34 (1960): 421–445; W. Turrentine Jackson, "British Interests in the Range Cattle Industry," in *When Grass Was King,* ed. Maurice Frink (Denver: University of Colorado Press, 1956), pp. 135–334.

46. Clemen, *The American Livestock and Meat Industry;* James Whitaker, *Feedlot Empire: Beef Cattle Feeding in Illinois and Iowa, 1840–1900* (Ames: Iowa State University Press, 1975).

47. Clemen, *The American Livestock and Meat Industry,* p. 68.

48. Swift, *The Yankee of the Yards,* pp. 206–207.

49. Clemen, *The American Livestock and Meat Industry,* pp. 283.

50. "Dressed Beef," *Ice and Refrigeration* 5 (1894): 397–398.

51. Swift, *The Yankee of the Yards,* pp. 206–207.

52. Cronon, *Nature's Metropolis,* p. 242. On butchers' opposition in Britain, see Perren, *The Meat Trade in Britain, 1840–1914,* pp. 127–128.

53. Charles Edward Russell, *The Greatest Trust in the World* (New York: Ridgway-Thayer, 1905), p. 175.

54. Cronon, *Nature's Metropolis,* p. 244; Russell, *The Greatest Trust,* p. 177.

55. Cronon, *Nature's Metropolis,* p. 237.

56. George Vest, *Report of the Select Committee on the Transportation and Sale of Meat Products, Senate Report No. 829,* 51st Congress, 1st Session, United States Senate, 1890; Armour, *The Packers, the Private Car Lines and the People,* pp. 476–477.

57. Chandler, *The Visible Hand,* p. 397; James R. Barrett, *Work and Community in the Jungle: Chicago's Packinghouse Workers, 1894–1922* (Urbana: University of Illinois Press, 1987), p. 16.

58. Mary Yeager Kujovich, "The Refrigerator Car and the Growth of the American Dressed Beef Industry," *Business History Review* 44 (1970): 460–482.

59. Ibid., p. 480.

60. "A Model Abattoir," *Ice and Refrigeration Illustrated* 8 (1895).

61. Vest, *Report of the Select Committee,* p. 363. See also Philip Armour's testimony, pages 426–427.

62. Quoted in Cronon, *Nature's Metropolis,* p. 250.

63. Critiques of the meatpacking industry are abundant, and some are more polemical than others. Rifkin, *Beyond Beef;* Donald D. Stull and

Michael J. Broadway, *Slaughterhouse Blues: The Meat and Poultry Industry in North America* (Belmont, Calif.: Thomson/Wadsworth, 2004); Danielle Nierenberg, *Happier Meals: Rethinking the Global Meat Industry,* WorldWatch paper (Washington, D.C.: WorldWatch Institute, 2005).

64. Bessie Louise Pierce and Joe Lester Norris, *As Others See Chicago: Impressions of Visitors, 1673–1933* (Chicago: University of Chicago Press, 2004).

65. J. de Loverdo, *Le froid artificiel et ses applications industrielles, commerciales et agricoles* (Paris: Dunod, 1903), p. 278.

66. Hanson, *Argentine Meat and the British Market,* p. 150.

67. George Rommel, "Notes on the Animal Industry of Argentina," *Twenty-fifth Annual Report of the Bureau of Animal Industry* (Washington, D.C.: Government Printing Office, 1908), pp. 315–334.

68. In 1869 Argentina's population was 1.7 million; over the next forty years, immigration added another 2.6 million people. Peter H. Smith, *Politics and Beef in Argentina: Patterns of Conflict and Change* (New York: Columbia University Press, 1969), p. 14; Henry S. Ferns, *Britain and Argentina in the Nineteenth Century* (Oxford: Clarendon Press, 1960); Silvio R. Baretta and John Markoff, "Civilization and Barbarism: Cattle Frontiers in Latin America," *Comparative Studies in Society and History* 20 (1978): 587–620.

69. Critchell and Raymond, *A History of the Frozen Meat Trade,* pp. 208, 250; Vest, *Report of the Select Committee,* testimony of P. D. Armour, p. 424.

70. *Buenos Aires Standard,* May 30, 1907; *The Financial Times,* September 6, 1908; *Review of the River Plate,* Oct. 24, 1908; all quoted in Hanson, *Argentine Meat and the British Market,* pp. 142–143, 148.

71. *La Prensa,* November 24, 1908; quoted in Hanson, *Argentine Meat and the British Market,* p. 144.

72. Ibid., pp. 155–156; Smith, *Politics and Beef in Argentina,* p. 35.

73. Hanson, *Argentine Meat and the British Market,* pp. 158–163.

74. In the 1920s, estimated per capita consumption of red meat in Buenos Aires was around 140 kilograms, including 118 kilograms of beef. Juan E. Richelet, *A Defence of Argentine Meat* (Buenos Aires, 1930), p. 89.

75. Michael Johns, "Industrial Capital and Economic Development in Turn of the Century Argentina," *Economic Geography* 68 (1992): 188–204.

76. John Barrett, "Buenos Ayres: The Paris of America," *New York Times,* August 19, 1906.

77. Emilio Frers, June 27, 1913, cited in Smith, *Politics and Beef in Argentina,* p. 66.

78. Ibid., p. 43; Michael Johns, "The Antinomies of Ruling Class Culture: The Buenos Aires Elite, 1880–1910," *Journal of Historical Sociology* 6 (1993): 74–101.

79. Roger Gravil, "State Intervention in Argentina's Export Trade between the Wars," *Journal of Latin American Studies* 2 (1970): 161.

80. Smith, *Politics and Beef in Argentina;* Gravil, "State Intervention in Argentina's Export Trade between the Wars," p. 259.

81. Jorge Schvarzer, "The Argentine Riddle in Historical Perspective," *Latin American Research Review* 27 (1992): 169.

82. J. Kandell, "Argentine Cattle Industry in Crisis as Herds Dwindle," *New York Times,* December 28, 1974, p. 1; Monte Reel, "Many Incredulous at Call to Eat Less Meat in Bid to Curb Inflation," *Washington Post,* March 20, 2006.

83. The account of the Vesteys draws on Phillip Knightley, *The Rise and Fall of the House of Vestey: The True Story of How Britain's Richest Family Beat the Taxman and Came to Grief* (London: Warner Books, 1993). Quote p. 27.

84. P. Locamus, *Madagascar et ses richesses* (Paris: Augustin Challamel, 1896), p. 32; P. Locamus, *Madagascar et l'alimentation Européenne: céréales et viandes* (Paris: Augustin Challamel, 1896), p. 74; Georges Foucart, *Le commerce et la colonisation à Madagascar* (Paris: Augustin Challamel, 1894); M. Moussu, "L'exploitation de la richesse en gros bétail dans nos colonies Africaines (Madagascar, Ouest Africaine) par l'industrie frigorifique," *Industrie Frigorifique* 12 (1914): 3–8; Gwyn Campbell, *An Economic History of Imperial Madagascar, 1750–1895* (New York: Cambridge University Press, 2005).

85. Locamus, *Madagascar et l'alimentation Européenne,* pp. xii, xiii.

86. Joret, "Le transport des produits coloniaux au moyens du froid," *L'Industrie Frigorifique* 1 (1903): 27–30.

87. Virginia Thompson and Richard Adloff, *The Malagasy Republic* (Stanford, Calif.: Stanford University Press, 1965), p. 413.

88. *Press Coloniale,* May 10, 1922; cited in Antoine Gener, "De l'elevage du gros bétail à Madagascar," Faculté de Droit, Université d'Alger, 1927, p. 205; Maurice Gontard, *Madagascar pendant la Première*

Guerre Mondiale (Tananarive: Imprint Société malgache d'édition, 1969), p. 89.

89. M. Geoffroy, "Le froid à Madagascar," *Revue Générale du Froid* 12 (1931): 261–263.

90. Gener, "De l'elevage du gros bétail à Madagascar," p. 188; Frederick Winslow Taylor, *The Principles of Scientific Management* (New York: Harper, 1911).

91. Maurice Piettre, *Les bases d'un grand elevage colonial* (Paris: Association Colonies Sciences, 1929), pp. 7–8; Maurice Piettre, *L'avenir des industries animales aux colonies* (Centre de Perfectionnement Technique, 1944), p. 3; Gener, "De l'elevage du gros bétail à Madagascar," p. 163.

92. Gener, "De l'elevage du gros bétail à Madagascar," p. 163.

93. Compte Rendu du tournée dans le district de Vohemar (Majunga 22–26 Mai 1954 par delegue du Gouveneur General, Cl. Cozanet) (MAD PM 574): Correspondence, affaires economiques 1954–1956 (MAD PM 710); Compte-Rendu de la session du Conseil Superieur de l'Elevage, 16 Octobre 1954 (MAD GGM 5 [14)] D). From Archives d'Outre Mer, Aix-en-Provence.

94. Critchell and Raymond, *A History of the Frozen Meat Trade*, p. 394.

95. S. Sanderson, "The Emergence of the 'World Steer': Internationalization and Foreign Domination in Latin American Cattle Production," in *Food, the State, and International Political Economy: Dilemmas of Developing Countries*, ed. F. L. Tullis and W. L. Hollist (Lincoln: University of Nebraska Press, 1986), pp. 123–147; Carol Ryan Dumas, "Brazilian Meat Company to Purchase Swift, Industry Differs on Sale," *Tri-State Neighbor,* July 5, 2007.

96. Upton Sinclair, *The Jungle* (New York: New American Library, 1906), p. 377.

97. Knightley, *The Rise and Fall of the House of Vestey;* Phillip Knightley, "Curse of the Spam Clan," *Daily Mail,* August 14, 1999; Sibilla Brodzinsky, "Squatters Sit Tight as Land Revolution Is Put to the Test in Venezuela," *The Guardian,* January 25, 2005, p. 15; Kathy Marks, "Four Decades after Wave Hill, Aborigines in Renewed Battle for Land Rights," *The Independent,* August 19, 2006.

98. Roger Horowitz, *Putting Meat on the American Table: Taste, Technology, Transformation* (Baltimore: Johns Hopkins University Press, 2006), pp. 149–150.

99. Thomas Friedman, "Iowa Beef Revolutionized Packing Industry," *New York Times,* June 2, 1981; M. J. Broadway, "Following the Leader: IBP and the Restructuring of Canada's Meatpacking Industry," *Culture & Agriculture* 18 (1996): 3–8; Bill Ganzel, "IBP, Boxed Beef and a New 'Big Four,'" Wessels Living History Farm, www. livinghistoryfarm.org/farminginthe50s/money_17.html.

100. Eric Schlosser, *Fast Food Nation* (New York: Harper Perennial, 2005), p. 5.

3. Eggs

1. Edward Wentworth, "Mystery of the Egg and Chick," in *Eggs,* ed. P. Mandeville (Chicago: Progress Publications, 1933), p. 78.

2. Mary E. Pennington, "What Is an Egg?" in *Eggs,* ed. P. Mandeville, pp. 37–38.

3. Harold McGee, *On Food and Cooking: The Science and Lore of the Kitchen* (New York: Scribner, 2004), p. 81.

4. Pete and Gerry interview, October 2005. Eggs may be stored for a week on the farm, and then two weeks en route to the supermarket.

5. Daniel J. Browne, *The American Poultry Yard* (New York: C. M. Saxton, 1850). The silver and golden Hamburghs were among the "everlasting layers."

6. Page Smith and Charles Daniel, *The Chicken Book* (Boston: Little, Brown, 1975), p. 176.

7. January receipts accounted for 25 percent of the annual total, whereas May receipts accounted for 18 percent. New York City wholesale prices for a dozen white eggs ranged from $0.36 in April to $0.83 in November. M. A. Jull et al., "The Poultry Industry," in U.S. Department of Agriculture, *Yearbook of the Department of Agriculture* (Washington, D.C.: Government Printing Office, 1925), pp. 385–386, 404.

8. William V. Cruess, *Home and Farm Food Preservation* (New York: Macmillan, 1918), p. 149; Dora Morrell Hughes, *Thrift in the Household* (Boston: Lothrop, Lee and Shepard, 1918), pp. 82–86.

9. Alvin Wood Chase, *Dr. Chase's Recipe Book* (Detroit: F. B. Dickerson Company, 1891). See also Harry R. Lewis, *Productive Poultry Husbandry* (Philadelphia: J. B. Lipincott, 1919), p. 441. The search for techniques to keep eggs "as good as" fresh was not limited to the

United States. For example, "Assurez la conservation parfaite des oeufs," *Vie à la Campagne* (1911): 283.

10. Lewis, *Productive Poultry Husbandry,* p. 25; Smith and Daniel, *The Chicken Book,* p. 232. Even in 1930, half the farms with chickens reported owning fewer than fifty birds. For these farms the average flock size was twenty-three. W. O. Wilson, "Housing," in *American Poultry History 1823–1973,* ed. O. A. Hanke, J. L. Skinner, and J. H. Florea (Madison, Wisc.: American Poultry Historical Society, 1974), p. 218.

11. Domestic chickens did by then lay considerably more eggs than their wild ancestors, who might not produce more than thirty a year. Milo Hastings, *The Dollar Hen* (Blodgett, Oreg.: Norton Creek Press, 1909 [2003]), p. 19. On the longer history of the chicken see Smith and Daniel, *The Chicken Book.*

12. Lewis, *Productive Poultry Husbandry,* p. 29.

13. W. B. Morehouse, quoted in James Howard Florea, "Education," in *American Poultry History 1823–1873,* p. 75. See also Frances Willard, *Occupations for Women* (New York: Success Company, 1897).

14. Jull et al., "The Poultry Industry," p. 283.

15. Charles Engel, "Cold in Modern Life: The Importance of Refrigeration" *Scientific American Supplement* 68 (1909): 168; "Editorial Notes," *Ice and Refrigeration* 11 (1896): 172; J. de Loverdo, *Le froid artificiel et ses applications industrielles, commerciales et agricoles* (Paris: Dunod, 1903).

16. J. Frederic Thorne, *Chinese Eggs: Conditions of Egg Production and Competition in Pacific Coast Markets Prices, Imports Exports* (Eugene: University of Oregon Press, 1916).

17. Petaluma Chamber of Commerce, *Petaluma, Sonoma County, California: The Largest Poultry Center in the World* (Petaluma: City of Petaluma, 1916), p. 11; Lewis, *Productive Poultry Husbandry,* p. 11. On Petaluma's poultry history, see Kenneth L. Kann, *Comrades and Chicken Ranchers: The Story of a California Jewish Community* (Ithaca, N.Y.: Cornell University Press, 1993).

18. Smith and Daniel, *The Chicken Book,* pp. 252–253.

19. Oscar E. Anderson, *Refrigeration in America: A History of a New Technology and Its Impact* (Princeton, N.J.: Princeton University Press, 1953), pp. 46–47; Mary E. Pennington, "Fifty Years of Refrigeration in the Egg and Poultry Industry," *Ice and Refrigeration* 101 (1941): 43–48.

20. "Advertising Cold Storage," *Ice and Refrigeration* 73 (1927): 347–349; "Ice in Europe," *Ice and Refrigeration* 3 (1892): 359–362.

21. L. Houllevigue, "Causerie scientifique: le congrès du froid," *Journal du Syndicat de la Boucherie de Paris* (1912): 2.

22. Anderson, *Refrigeration in America*, p. 70.

23. "Eggs Have Fruity Flavor: Forty Thousand in Boston Found to Have a Novel Taste," *New York Times*, August 31, 1900.

24. *Ice and Refrigeration* 6 (1894): 112; "Refrigeration Abroad," *Ice and Refrigeration* 26 (1904).

25. Letter: "Danger of Eating Cold Storage Food," *New York Times*, August 4, 1906.

26. "Cold Storage Meats Good Three Months," *New York Times*, January 30, 1907; Arthur L. Nason, *Report of the Commission to Investigate the Subject of the Cold Storage of Food and of Food Products Kept in Cold Storage* (Boston: Wright and Potter Printing, 1912), p. 272; *Journal of the American Medical Association*, quoted in "Cold Storage Prejudice Declining," *Ice and Refrigeration* 43 (1912): 56–57.

27. As the name implies, candling simply means holding an egg up to a bright light in order to detect the volume and viscosity of its contents. The less fresh the egg, the more watery the white is, and the larger the airspace.

28. I. C. Franklin, "The Service of Cold Storage in the Conservation of Foodstuffs," in U.S. Department of Agriculture, *Yearbook of Agriculture* (Washington, D.C.: Government Printing Office, 1917), p. 366.

29. "The Senate Committee Hearings," *Ice and Refrigeration* 38 (1910): 385–387.

30. Philip B. Hawk, *What We Eat and What Happens to It: The Results of the First Direct Method Ever Devised to Follow the Actual Digestion of Food in the Human Stomach* (New York: Harper and Brothers, 1919), p. 76.

31. Miriam Dexter, "The Housekeeping Club," *Good Housekeeping* 50 (1910): 263–267.

32. Walter E. Clark, *The Cost of Living* (Chicago: A. C. McClurg, 1915); Alfred W. McCann, *Thirty Cent Bread: How to Escape a Higher Cost of Living* (New York: George H. Doran, 1917), pp. 73, 77.

33. These arguments appear repeatedly in *Ice and Refrigeration*, the main trade journal of the American refrigeration industry, from around 1906 through the 1920s. See, for example, Mary E. Pennington, "Better Food for the Masses," *Ice and Refrigeration* 75 (1928): 33–35. Egg

production increased from approximately 450 million dozen in 1880 to 1.9 billion dozen in 1907. S. S. Van deer Vaart, "Growth and Present Status of the Refrigerating Industry in the United States," in J. de Loverdo, ed., *Premier congrès international du froid Paris, 5 au 12 octobre 1908,* vol. 3 (Paris: Secrétariat général de l'Association internationale de froid, 1908), p. 341.

34. "Got 'em in the Ice Box," *Los Angeles Times,* April 23, 1902, p. 5; "Ruined by Trust," *Boston Daily Globe,* April 24, 1902, p. 1; "Corner in Eggs," *The Hartford Courant,* April 19, 1902, p. 1; "Food Combine May Come," *New York Times,* April 18, 1902, p. 1.

35. "The Hen as a Trust Buster," *Los Angeles Times,* May 7, 1905, p. 114; "Hens Happy," *Boston Daily Globe,* January 20, 1906, p. 1.

36. These warehouses were usually privately owned but were open for public use (unlike the meatpackers' stores). Most cold storage companies were relatively small and their ownership distributed among stockholders. Boston's Quincy Market, for example, had 228 stockholders in 1910. These companies also typically did not own the goods they stored; these belonged to a wide range of wholesalers, some bigger than others. For these reasons, all state and federal investigations into a possible "cold storage monopoly" concluded that it did not exist. Massachusetts Commission on Cold Storage of Food, *Report of the Commission to Investigate the Subject of the Cold Storage of Food and of Food Products Kept in Cold Storage* (Boston, 1912), pp. 94–96; Anderson, *Refrigeration in America,* p. 134.

37. Letter: "A Trust in Food?" *The Washington Post,* December 11, 1909, p. 6.

38. "Cold Storage Ordinance," *Ice and Refrigeration* 31 (1906).

39. Anderson, *Refrigeration in America,* p. 138.

40. "Got 'em in the Ice Box," p. 5.

41. Frank A. Horne, "Legislation Affecting Cold Storage and Cold Stored Products," *Ice and Refrigeration* 41 (1911): 180.

42. Quoted in "Anti Cold Storage Agitation," *Ice and Refrigeration* 38 (1910): 104–106.

43. Thomas A. Bird, "The Ice Man as an Advertiser," *Ice and Refrigeration* 38 (1910). "To Dine on Embalmed Food: Produce Merchants Invite City Officials to Cold Storage Meal," *Chicago Daily Tribune,* September 27, 1911.

44. Pennsylvania Representative J. Hampton Moore, in hearings for HR 27173: "A bill to regulate the storage of food supplies in the district of

Columbia." Quoted in "Cold Storage Legislation," *Ice and Refrigeration* 39 (1910): 52–53.

45. "Senate Committee Hearings," *Ice and Refrigeration* 38 (1910): 385–387; "Hearings on Senate Bill 136," *Ice and Refrigeration* 41 (1911): 1–11.
46. Hearings, June 9, 1910, quoted in "Hearings on Senate Bill 136."
47. Hearings, May 18, 1910, quoted in ibid.
48. Barbara Heggie, "Ice Woman: Dr. Mary Engle Pennington," *New Yorker,* September 6, 1941. For further biographical information on Pennington, see Rima Apple, "Science Gendered: Nutrition in the United States, 1840–1940," in *The Science and Culture of Nutrition, 1840–1940,* ed. Harmke Kamminga and Andrew Cunningham (Atlanta: Rodopi, 1995), pp. 129–154.
49. Mary E. Pennington, "Relation of Cold Storage to the Food Supply and the Consumer," *Annals of the American Academy of Political and Social Science* 48 (1913): 154–155.
50. Constance Carruthers, "Women Solve Problem of High Living Cost," *Los Angeles Times,* December 12, 1912.
51. "Clubwomen and Cheap Egg Sale," *Chicago Tribune,* December 16, 1912, p. 3; "War of Housewives for Cheaper Food," *New York Times,* December 16, 1912, p. 1.
52. "Cold Storage Prejudice Declining," *Ice and Refrigeration* 43 (1912): 56–57; American Warehousemen's Association, "Proceedings of the 25th Annual Meeting of the American Warehousemen's Association" (New York: 1915), p. 235.
53. M. E. Pennington et al., "The Egg and Poultry Demonstration Car Work in Reducing Our $50,000,000 Waste in Eggs," in U.S. Department of Agriculture, *Yearbook of Agriculture* (Washington, D.C.: U.S. Government Printing Office, 1914): 363–379.
54. "Cold Storage Eggs Good," *Ice and Refrigeration* 40 (1911): 35–36; "Advertising Cold Storage," *Ice and Refrigeration* 73 (1927): 347–349.
55. American Warehousemen's Association, "Proceedings of the 25th Annual Meeting of the American Warehousemen's Association," pp. 236–237.
56. M. A. Jull et al., "The Poultry Industry," p. 388; M. E. Pennington, "Demonstrating Cold Storage Foods at Health Exposition," *Ice and Refrigeration* 64 (1923): 333–334.

57. Quoted in W. M. O'Keefe, "Cold Storage Division A. W. A.," *Ice and Refrigeration* 78 (1930): 513–515.

58. M. E. Pennington, "Address to National Convention of the United Master Butchers of America," *Ice and Refrigeration* 59 (1920): 98.

59. O'Keefe, "Cold Storage Division A. W. A."; "Doom of Cold-Storage Eggs," *Business Week,* March 21, 1936, p. 28; William Jasper, "Marketing," in *American Poultry History 1823–1873,* ed. O. A. Hanke, J. L. Skinner, and J. H. Florea (Madison, Wisc.: American Poultry Historical Society, 1974), p. 312; D. M. Rutherford, "Getting Fresh with Our Fresh Eggs," *Pacific Rural Press,* December 29, 1934. On contemporary federal egg standards, see www.ams.usda.gov/poultry/standards/AMS-EggSt-1995.htm.

60. Hastings, *The Dollar Hen,* p. 178.

61. Walter Hogan, "The Call of the Hen: Or the Science of the Selection and Breeding of Poultry," Petaluma, Calif.: *Petaluma Daily Courier,* 1913; Lewis, *Productive Poultry Husbandry;* Smith and Daniel, *The Chicken Book,* p. 247.

62. Jull et al., "The Poultry Industry," p. 388; Petaluma Chamber of Commerce, *Petaluma, Sonoma County, California: The Largest Poultry Center in the World* (Petaluma Chamber of Commerce, 1916), quoted in Smith and Daniel, *The Chicken Book,* p. 253.

63. Dexter, "The Housekeeping Club," pp. 263–267.

64. Browne, *The American Poultry Yard.*

65. Frank Platt, "Poultry Keeping: An Art, a Science, an Industry," in *Eggs,* ed. P. Mandeville (Chicago: Progress Publications, 1933), p. 81.

66. Kathy Cooke, "From Science to Practice, or Practice to Science? Chickens and Eggs in Raymond Pearl's Agricultural Breeding Research," *Isis* 88 (1997): 62–86.

67. Ernest Cobb, *The Hen at Work: A Brief Manual of Home Poultry Culture* (New York: G. P. Putnam's Sons, 1919).

68. *Leghorn World* (February 1931): 339.

69. Quoted in Smith and Daniel, *The Chicken Book,* p. 29.

70. Lewis, *Productive Poultry Husbandry,* p. 274.

71. Donald Bell, "Forces That Have Helped Shape the U.S. Egg Industry," *Poultry Tribune* (1995): p. 36. Other products that claimed to help winter egg production included Exadol, a poultry feed supplement, and the Gizzard Capsule, a de-worming pill (advertised in *American Poultry Journal* and *Leghorn World*).

72. Spanish farmers may have made this discovery well before Edison's invention, according to Francisco Dieste's book *Tratado economico dividido en tres discursos,* first published in 1781. As a 1936 article in *Nature* noted, the basic assumption in the eighteenth century, as in the twentieth, was that if hens received more light, they would eat more and thus lay more. "The keeper during winter would disturb the hens in their sleep, and make them go to the trough at which there should be lights or torches of wood or other material so that the birds could see the food. The hens grew accustomed within a week to eat at that hour, and 'come running as soon as they saw the light.'" John Randal Baker, "Increasing Winter Egg-Production in Spain More Than a Hundred Years Ago," *Nature* 143 (1936): 477; Francisco Dieste y Buil, *Tratado economico dividido en tres discursos. I. Crianza de gallinas, y considerables utilidades, que producen à su dueño. II. Compra de primales para venderlos al año siguiente por carneros. III. Modo de procurar la extincion de fieras perjudiciales al ganado, y aves domésticas, y que las de rapiña lo sean memos* (Madrid: Benito Cano, 1803).

73. Smith and Daniel, *The Chicken Book,* p. 264; Platt, "Poultry Keeping: An Art, a Science, an Industry," pp. 135–136.

74. "Electric Light and Egg Production," *Scientific American* 120 (1919): 272.

75. Y. P. Bhosale, "How to Secure More Eggs in Winter," *Leghorn World* 11 (1926): 53+.

76. "Interest in Forced Egg Production Waning," *The Hen Coop* 6 (1922): 1.

77. D. C. Kennard and V. D. Chamberlin, *All-Night Light for Layers,* Ohio Agricultural Experiment Station, Bulletin 476 (1931): 1.

78. Ibid., pp. 8, 11. The study period was between October 2 and March 4.

79. H. V. Tormohlen, "Do Not Short Change the Pullets," *Leghorn World* 14 (1930): 42.

80. "Eggs from China: Millions Frozen for Shipment," *Literary Digest* 122 (1936): 38.

81. "Use of Winter Lighting," *American Egg and Poultry Review* 19 (1940). Advertised on the back cover of the April 1942 edition of *American Poultry and Egg Review.*

82. "Rate of Production Mounting," *American Egg and Poultry Review* 106 (1941); "Egg Lay Rate Holds Record High," *American Egg and Poultry Review* 106 (1941): 262.

83. "Doom of Cold-Storage Eggs," *Business Week,* March 21, 1936, p. 28.

84. E. B. Heaton, managing director of the American Institute of Poultry Industries, quoted in "Seek Change in Terminology," *American Egg and Poultry Review* (April 1940): 146–147.

85. R. H. Switzler, "Refrigerated Warehousing over the Years," in *Proceedings of the American Warehousemen's Association,* p. 69.

86. Katherine Jellison, *Entitled to Power: Farm Women and Technology, 1913–1963* (Chapel Hill, N.C.: University of North Carolina Press, 1993), pp. 156–157.

87. U.S. Department of Agriculture, Economic Research Service. On wartime rationing: Amy Lynn Bentley, "Eating for Victory: United States Food Rationing and the Politics of Domesticity during World War Two" (Ph.D. diss., University of Pennsylvania, 1992), pp. ix, 288; Stephanie A. Carpenter, *On the Farm Front: The Women's Land Army in World War II* (DeKalb: Northern Illinois University Press, 2003).

88. Jellison, *Entitled to Power,* pp. 156–157; Kahn, *Comrades and Chicken Farmers,* p. 166.

89. E. Smith Kimball, "Characteristics of U.S. Poultry Statistics," *Journal of Farm Economics* 22 (1940): 361.

90. Gordon Sawyer, *The Agribusiness Poultry Industry* (New York: Exposition Press, 1971), p. 218.

91. William Jasper, *Poultry Farm Practices and Egg Quality,* U.S. Department of Agriculture Production and Marketing Administration, Marketing Research Report 22 (1952); William J. Stadelman and Owen J. Cotterill, *Egg Science and Technology* (New York: Food Products Press/Haworth Press, 1995), ch. 1.

92. www.aeb.org; Donald Bell, "Forces That Have Helped Shaped the U.S. Egg Industry," *Poultry Tribune* (1995): 32–33.

93. On the development of antibiotic-resistant bacteria in eggs, see Nicols Fox, *Spoiled: The Dangerous Truth about a Food Chain Gone Haywire* (New York: Basic Books, 1997).

94. Starre Vartan, "Happy Eggs: 'Free Range,' 'Cage Free,' 'Organic'— What's the Story?—Eating Right," *The Environmental Magazine* (May–June 2003).

95. According to Cal-Maine, "In fiscal 2004, specialty shell eggs were estimated to represent approximately 8.2% of the Company's shell egg dollar sales." www.calmainefoods.com; Marion Nestle, *What to Eat* (New York: North Point Press, 2006), p. 262.

96. See, for example, the websites of the following producers: www.peteandgerrys.com; www.nestfresh.com.

97. David Joachim, "Egg-Vertising by CBS: An Idea Not for Poaching," *International Herald Tribune,* July 18, 2006. Interview, Bradley Parker, Sept. 2007.

98. Sarah F. Gale, "Radlo Foods Hatches a High-Tech Egg Safety Plan," *Food Safety Magazine* 11 (2005/6): 70–74. Interview, David Radlo, Sept. 2007.

99. Robert Weisman, "Born Free Eggs Etches Data on Its Fragile Wares," *Boston Globe,* September 19, 2005.

4. Fruit

1. Andrew Fearne and David Hughes, "Success Factors in the Fresh Produce Supply Chain: Insights from the U.K.," *British Food Journal* 102 (2000): 760–772.

2. Harold McGee, *On Food and Cooking: The Science and Lore of the Kitchen* (New York: Scribner, 1984), p. 138. See also Adam Gollner, *The Fruit Hunters: A Story of Nature, Adventure, Commerce, and Obsession* (New York: Scribner, 2008).

3. John Robson, "Fruit in the Human Diet: Fruit in the Diet of Prehistoric Man and of the Hunter-Gatherer," *Journal of Human Nutrition* 32 (1978): 19–26.

4. McGee, *On Food and Cooking,* pp. 163–166.

5. S. Peneau et al., "Importance and Consumer Perception of Freshness of Apples," *Food Quality and Preference* 17 (2006): 9–19.

6. Dana G. Dalrymple, "The Development of an Agricultural Technology: Controlled-Atmosphere Storage of Fruit," *Technology and Culture* 10 (1969): 35–48; Eugene Kupferman, "The Early Beginnings of Controlled Atmosphere Storage," *Post Harvest Pomology Newsletter* 7 (1989): 3–4.

7. See, for example, the comments by Stephen Woods, the owner of Poverty Lane Orchards in Lebanon, New Hampshire, in David Karp, "Apples with Pedigrees Selling in Urban Edens," *New York Times,* October 20, 2004.

8. McGee, *On Food and Cooking,* p. 360; Xavier Rocques, *Les industries de la conservation des aliments* (Paris: Gautier-Villars, 1906); D. B. Bremer, "Fruit in Cold Storage," *Ice and Refrigeration* 3 (1892): 33–34.

9. Marcus Cato, *On Agriculture*, Loeb Classical Library, vol. 283 (Cambridge, Mass.: Harvard University Press, 1935).

10. Frank Browning, *Apples* (London: Allan Lane, Penguin Press, 1998), p. 59; Bridget Ann Henisch, *Fast and Feast: Food in Medieval Society* (University Park: Pennsylvania State University Press, 1976), pp. 113–117; *Le trésor de santé de la vie humaine* (1607), quoted in Jean Louis Flandrin, "Seasoning, Cooking and Dietetics in the Late Middle Ages," in *Food: A Culinary History from Antiquity to the Present*, ed. J. L. Flandrin, M. Montanari, and A. Sonnenfeld (New York: Penguin, 1999), pp. 313–327; ibid., p. 324.

11. Jean Louis Flandrin, "The Early Modern Period," in *Food*, pp. 349–373; L. F. Newman, "Some Notes on Foods and Dietetics in the Sixteenth and Seventeenth Centuries," *The Journal of the Royal Anthropological Institute of Great Britain and Ireland* 76 (1946): 39–49.

12. H. Frederic Janson, *Pomona's Harvest: An Illustrated Chronicle of Antiquarian Fruit Literature* (Portland, Oreg.: Timber Press, 1996), p. 118.

13. Florent Quellier, *Des fruits et des hommes: l'arboriculture fruitière en Ile-de-France* (Rennes: Presses Universitaires de Rennes, 2003), quoted on p. 60.

14. Charles-Jean de Combles and Antoine Boudet, *Traité de la culture des pêchers* (Paris: Chez Antoine Boudet, 1745), quoted in Quellier, *Des Fruits et des hommes*, p. 209.

15. Ralph Austen, "A Treatise of Fruit Trees," printed by William Hall for Amos Curteyne, 1665. Available at EEBO, Early English Books Online, p. 33.

16. Gordon Dunthorne, *Flower and Fruit Prints of the 18th and Early 19th Centuries, Their History, Makers and Uses, with a Catalogue Raisonné of the Works in Which They Are Found* (Washington, D.C.: author, 1938); Pierre Joseph Redouté and J. A. Guillemin, *Choix des plus belles fleurs: prises dans différentes familles du ráegne végétal et de quelques branches des plus beaux fruits* (Paris: Chez l'auteur, 1827).

17. Jean-Roger Schabol and Marie-Rose Simoni-Aurembou, *Parlers et jardins de la banlieue de Paris au Xviiie siècle* (Paris: Klincksieck, 1982), p. 27.

18. Pierre Jean-Baptiste Legrand d'Aussy, *Histoire de la vie privée des Français: depuis l'origine de la nation jusqu'à nos jours* (Paris: P. D.

Pierres, 1782); Schabol and Simoni-Aurembou, *Parlers et jardins de la banlieue de Paris au Xviiie siècle*, p. 151.

19. C.A. de Bengy-Puyvallée, *Mémoire sur la culture du pêcher* (Paris: Librairie Agricole de la Maison Rustique, 1860), p. 16. Banks quoted in Janson, *Pomona's Harvest*, p. 226.

20. U. P. Hedrick, *A History of Horticulture in America to 1860* (New York: Oxford University Press, 1950); Mahlon Stacy, quoted in ibid., p. 78.

21. Ibid., pp. 100, 57.

22. J. Hector St. John de Crèvecoeur, *Letters from an American Farmer: Describing Certain Provincial Situations, Manners, and Customs, and Conveying Some Idea of the Late and Present Interior Circumstances of the British Colonies in North America* (London: T. Davies, 1782), quoted in Hedrick, *A History of Horticulture in America to 1860*, p. 60.

23. Quoted in ibid., p. 59.

24. Hedrick, *A History of Horticulture in America to 1860*, p. 332.

25. Richard O. Cummings, *The American and His Food: A History of Food Habits in the United States* (Chicago: University of Chicago Press, 1941), p. 34.

26. Hedrick, *A History of Horticulture in America to 1860*, pp. 235–236.

27. William A. Taylor, "The Influence of Refrigeration on the Fruit Industry," in *Yearbook of Agriculture*, ed. U.S. Department of Agriculture (Washington, D.C.: Government Printing Office, 1900), p. 562.

28. As late as 1849, city councils banned sales of fresh fruit during cholera outbreaks. Cummings, *The American and His Food*, p. 43.

29. Guadeloupe Vallejo, "Gold Hunters of California. Ranch and Mission Days in Alta California," 41 (December 1890): 183–192.

30. Edward J. Wickson, *The California Fruits and How to Grow Them* (San Francisco: Pacific Rural Press, 1926), p. 47; *Country Gentleman* 10 (July 2, 1857); *Country Gentleman* 20 (December 18, 1862): 394.

31. Wickson, *The California Fruits and How to Grow Them*, p. 49.

32. "California Fruits," *The Horticulturist* 14 (1859): 23–25.

33. Ellen Liebman, *California Farmland: A History of Large Agricultural Landholdings* (Totowa, N.J.: Rowman and Allanheld, 1983); George L. Henderson, *California and the Fictions of Capital* (New York: Oxford University Press, 1999).

34. C. F. Dowsett, *A Start in Life: A Journey across America: Fruit Farming in California* (London: Dowsett, 1891), p. 86.

35. Solano County Board of Supervisors, "Solano County: Land of Fruit, Grain and Money," *Vallejo Evening Chronicle*, 1905; David Vaught, *Cultivating California: Growers, Specialty Crops and Labor, 1875–1920* (Baltimore: Johns Hopkins University Press, 1999), p. 50; Edward J. Wickson, *The Vacaville Early Fruit District of California* (San Francisco: California View Publishing, 1888); Sabine Goerke-Shrode, "Publication Showcased Vacaville to the Rest of the World," http://www.solanoarticles.com/history/index.php/weblog/more/ publication_showcased_vacaville_to_the_rest_of_the_world/.

36. William R. Nutting, *California Views in Natural Colors* (San Francisco: California View Publishing, 1889), pp. 7, 10, 26; N. P. Chipman, *Report upon the Fruit Industry of California* (California State Board of Trade, 1889), p. 35.

37. Nutting, *California Views in Natural Colors*, p. 5. See also Vaught, *Cultivating California*, especially ch. 2; Don Mitchell, *The Lie of the Land: Migrant Workers and the California Landscape* (Minneapolis: University of Minnesota Press, 1996).

38. In central California's Placer County, the cost of foothill land leapt thirtyfold in thirteen years, from $10 per acre in 1880 to $300 in 1893. By 1917, an acre was worth $1,500. Land values climbed equally fast in orange-growing country. Lawrence J. Jelinek, *Harvest Empire: A History of California Agriculture* (San Francisco: Boyd and Fraser, 1982), p. 53; Vaught, *Cultivating California*, p. 80; Richard Randall, "Making Markets for 1925 Crops of Prunes and Apricots," *Western Advertising* 1 (1920): 7+; Ronald Tobey and Charles Wetherell, "The Citrus Industry and the Revolution of Corporate Capitalism in Southern California, 1887–1944," *California History* (1995): 6–21. Cheap mortgages provided an added incentive to switch to orchard crops. On the importance of credit in the rise of orchard production—which requires a significant up-front capital investment and then at least three years' lag time before the orchards yield fruit—see Paul W. Rhode, "Learning, Capital Accumulation and the Transformation of California Agriculture," *Journal of Economic History* 55 (1995): 773–800; George L. Henderson, "Nature and Fictitious Capital: The Historical Geography of an Agrarian Question," *Antipode* 30 (1998): 73–118.

39. Taylor, "'The Influence of Refrigeration on the Fruit Industry."

40. Citrus shipments increased from 1,000 carloads in 1886 to 25,000 in 1900. Donald J. Pisani, *From the Family Farm to Agribusiness: The Irrigation Crusade in California and the West, 1850–1931* (Berkeley:

University of California Press, 1984), pp. 449–450; Jelinek, *Harvest Empire*, p. 58.

41. H. Vincent Moses, "G. Harold Powell and the Corporate Consolidation of the Modern Citrus Enterprise, 1904–1922," *Business History Review* 69 (1995): 119–155; Archibald Shamel, "The Esthetic Side of Orange Growing in the Southwest," *California Citrograph* (January 1928): 79, quoted in Anthea Hartig, "'In a World He Has Created:' Class Collectivity and the Growers' Landscape of the Southern California Citrus Industry, 1890–1940," *California History* (1995): 101.

42. Massachusetts, Report of the Commission on the Cost of Living, 1910, p. 255.

43. Steven Stoll, *The Fruits of Natural Advantage: Making the Industrial Countryside in California* (Berkeley: University of California Press, 1998), pp. 64–70. For this section, Stoll's chapter "Organize and Advertise" has been extremely helpful.

44. J. A. Filcher, California State Board of Trade Manager, quoted in ibid., p. 73; original quotation from *Pacific Rural Press* 55 (January 1, 1898).

45. E. Woehlke, "In the Orange Country: Where the Orchard Is a Mine— The Human Factor among the Gold-bearing Trees of California," *Sunset* 26 (March 1911): 263.

46. Tobey and Wetherell, "The Citrus Industry and the Revolution of Corporate Capitalism in Southern California, 1887–1944," p. 8.

47. Don Francisco, "The Plans Behind Sunkist Advertising," *California Citrogragh* (1920): 9.

48. Gordon T. McClelland and Jay T. Last, *California Orange Box Labels: An Illustrated History* (Beverly Hills, Calif.: Hillcrest Press, 1985).

49. Francisco, "The Plans Behind Sunkist Advertising," p. 9.

50. Don Francisco, "How the 'Sunkist' Campaign Was Sold to the Association," *Western Advertising* 1 (1920): 29.

51. Don Francisco, *Cooperative Advertising: A Social Service as Well as a Powerful Sales Force* (Los Angeles: California Fruit Growers Exchange, 1920).

52. The CFGE's publicity cost growers 2.5 cents per box. Ibid.

53. Don Francisco, "The Advertising of Agricultural Specialties," *American Cooperation* 2 (1928): 137. The increase in orange eating may have been more dramatic than Francisco's number suggest. The Department of Commerce estimated per capita consumption of 6.7 pounds in 1899, 13.8 pounds in 1919, and 25.4 pounds in 1931. E. G.

Montgomery and C. H. Kardell, *Apparent Per Capita Consumption of Principal Foodstuffs in the United States,* Department of Commerce, Domestic Commerce Series 38 (Washington, D.C., 1930), pp. 2–4.

54. Douglas Cazaux Sackman, "'By Their Fruits Ye Shall Know Them': Nature Cross Cultural Hybridization and the California Citrus Industry, 1893–1939," *California History* (1995): 83–99. See also Harvey A. Levenstein, *Revolution at the Table: The Transformation of the American Diet* (Berkeley: University of California Press, 2003), p. 154.

55. J. Walter Thompson Company, "Presentation on California Fresh Bartlett Pears" (Giannini Foundation for Agricultural Economics, University of California, Berkeley, 1936), pp. 36, 52.

56. Frank Swett, "Collective Selling by Pear Growers," *Pacific Rural Press,* January 1, 1921.

57. Francisco, *Cooperative Advertising,* pp. 18–19.

58. George Holmes, "Consumers' Fancies," in *1904 Yearbook of Agriculture,* U.S. Department of Agriculture (Washington, D.C.: Government Printing Office, 1905), pp. 417–434.

59. Ibid., p. 434.

60. Paul Armstrong, "Sunkist Advertising—How It Sells California Oranges and Lemons," *California Citrograph* (1923): 5.

61. F. W. Read, "The Objectives of Standardization," *Blue Anchor* (1928): 9.

62. F. W. Read, "'Blue Anchor' Grade Specifications, Season 1927," *Blue Anchor* 4 (1927): 7+.

63. Stoll, *The Fruits of Natural Advantage,* p. 98.

64. Ibid., p. 113.

65. Moses, "G. Harold Powell and the Corporate Consolidation of the Modern Citrus Enterprise, 1904–1922," p. 37.

66. Richard C. Sawyer, *To Make a Spotless Orange: Biological Control in California* (Ames: Iowa State University Press, 1996), p. 17; Stoll, *The Fruits of Natural Advantage,* pp. 118–119.

67. Whorton, *Before Silent Spring,* ch. 4.

68. Growers in the Pacific Northwest faced some of the same pests and responded to them in similar ways. In fact, it was Britain's discovery of arsenic-tainted apples from Washington state that unleashed the next round of controversy in 1925. Ibid., ch. 5.

69. "Battle of the Bugs," *Pacific Rural Press,* February 27, 1926, quoted in Stoll, *The Fruits of Natural Advantage,* p. 122.

70. "Better Fruit Essential," *Pacific Rural Press,* March 17, 1923.

71. A. Prilleray, "La refrigeration et l'approvisionnement en fruits du marché de Londres," *Industrie Frigorifique* 2 (1904); A. Prilleray, "Transports frigorifiques des fruits coloniaux," *Industrie Frigorifique* 3 (1905): 123–127.

72. Charles Baltet, *Les fruits populaires* (Paris: Bibliotheque du Jardin, 1889); E-A. Carrière, *Montreuil-Aux-pêches* (Paris, 1890), p. 99; C. A. de Bengy-Puyvallée, *Mémoire sur la culture du pêcher* (Paris: Librairie Agricole de la Maison Rustique, 1860).

73. Hippolyte Langlois, *Le livre de Montreuil-aux-pêches* (Paris: Librairie Firmin-Didot, 1875). "Un cultivateur de Montreuil tue ses arbres sous lui comme un ardent cavalier ses chevaux surmenés. Mais avec la précaution toujours prise d'avoir sous la main des sujets jeunes qui succèdent aux arbres épuisés . . . La terre s'use vite ici, presque aussi vite que les arbres" (pp. 97–98).

74. Ibid., p. 94.

75. "La vente de pêches aux Halles," *Journal du syndicat des cultivateurs de département de la Seine,* September 1, 1900.

76. Sources on the nineteenth-century history of Montreuil include Carrière, *Montreuil-aux-pêches;* Fernand Bournon, *Etat des communes à la fin du Xix siècle: Montreuil* (Paris: Conséil General, 1906); Schabol and Simoni-Aurembou, *Parlers et jardins de la banlieue de Paris au Xviiie siècle;* Jacques Brunet and Nicole Savard, *Les Savards: histoires des vies* (Paris: Valette Editions, 2006).

77. Léon Loiseau, *De l'ensachage des fruits* (Paris: Librairie et imprimerie horticoles, 1903), p. iii.

78. The best accounts of women's work in Montreuil can be found in the interviews collected in Brunet and Savard, *Les Savards.*

79. Conference agricole à Bry-sur-Marne, *Journal de Syndicat des Cultivateurs,* Montreuil, 1895.

80. Leon Nerdeux, "Les nouvelles installations dans l'alimentations et l'industrie fruitiere," *Industrie Frigorifique* 4 (1906): 244–245; Léon Loiseau, "Du role des appareils frigorifiques dans la conservation des fruits," *Industrie Frigorifique* 4 (1906): 334–338.

81. Société Régionale d'Horticulture de Montreuil, Reunion mensuelle 18 juillet 1914.

82. Quelques lettres de noblesse, *Bulletin de la Société Régionale d'Horticulture de Montreuil* (1975); Philippe Schuller, Société Régionale d'Horticulture de Montreuil, http://pschuler.club.fr/srhm/.

83. Interview, André Patereau, July 15, 2006.
84. One of the earliest references can be found in the twelfth-century text Ibn al-Awwām, Y.i.M., *Le livre de l'agriculture d'Ibn al-Awam* (Paris: A. Franck, 1864), p. 606. I am indebted to Antoine Jacobsohn for this reference.
85. Loiseau, *De l'ensachage des fruits.*
86. The peaches sold for three to four francs each; the average daily wage for a Parisian worker was eight francs. Harvest workers in Montreuil earned only three francs a day. P. Schuller, personal communication.
87. Philippe Schuller, "La Famille Vassout, le sens de la tradition," *Bulletin de la Société Régionale d'Horticulture de Montreuil* (2004): 9–15. At Les Halles in the 1930s, Montreuil peaches continued to cost several times more than bulk peaches from Spain or southern France. Edmond Garnier, *L'agriculture dans le département de la Seine et le marché Parisien du point de vue ravitaillement alimentaire* (Poitiers: Imprimerie l'Union, 1939), p. 274.
88. According to Garnier, Montreuil garden land cost 300 francs per hundred square meter before World War One and 3,000–10,000 francs in the late 1930s. Garnier, *L'agriculture dans le département de la Seine et le marché Parisien du point de vue ravitaillement alimentaire.*
89. Already in the 1930s, French horticultural experts urged growers to adopt American standardization and packing practices. Albert Maumené, "Ni fardage, ni maquillage: la beauté vraie," *Vie à la Campagne,* June 15, 1934, pp. 4, 7.
90. Jelinek, *Harvest Empire,* p. 89.
91. Julie Guthman, *Agrarian Dreams? The Paradox of Organic Farming in California* (Berkeley: University of California Press, 2004).
92. The British journalist Joanna Blythman coined this term. Joanna Blythman, "Strange Fruit," *The Guardian,* September 7, 2002.
93. Peneau et al., "Importance and Consumer Perception of Freshness of Apples."
94. Bee Wilson, "In the Pink," *New Statesman,* February 12, 2001.
95. Jon Mooallem, "Twelve Easy Pieces," *New York Times Magazine,* February 12, 2006.

5. Vegetables

1. Harold McGee once again offers one of the best layperson's explanations of how vegetables live and die. Harold McGee, *On Food and*

Cooking: The Science and Lore of the Kitchen (New York: Scribner, 2004), chs. 6 and 7.

2. For more information on how fresh produce ages, see N. A. M. Eskin, "Biochemical Changes in Raw Foods: Fruits and Vegetables," in *Biochemistry of Foods*, ed. N. A. M. Eskin (San Diego: Academic Press, 1990), pp. 69–165; Vijay K. Mishra and T. V. Gamage, "Post-Harvest Physiology of Fruit and Vegetables," in *Handbook of Food Preservation*, 2nd ed., ed. M. Shafiur Rahman (Taylor and Francis, 2007), pp. 19–48.

3. Sidney Mintz, *Sweetness and Power: The Place of Sugar in Modern History* (New York: Penguin, 1986); C. Anne Wilson, "From Garden to Table: How Produce Was Prepared for Immediate Consumption," in *The Country House Kitchen Garden, 1600–1950: How Produce Was Grown and How It Was Used*, ed. C. A. Wilson (London: Sutton Publishers in association with the National Trust, 1998), pp. 144–161.

4. Ibid., p. 149.

5. Nicolas de Bonnefons, *Les délices de la campagne. suite du jardinier François, ou est enseigné à preparer pour l'usage de la vie tout ce qui croist sur la terre, & dans les eaux* (Paris: Par la Compagnie des marchands libraires du Palais, 1665); May Woods, *Glass Houses: A History of Greenhouses, Orangeries and Conservatories* (London: Aurum Press, 1988), p. 114.

6. T. Tryon, *A pocket-companion, containing things necessary to be known by all that values their health and happiness being a plain way of nature* (1694), p. 18.

7. T. Tryon, *The good house-wife made a doctor* (London, 1692), p. 224.

8. Johann Heinrich von Thünen and Peter Geoffrey Hall, *Isolated State; an English Edition of Der Isolierte Staat* (Oxford: Pergamon Press, 1966). See also D. Block and E. M. du Puis, "Making the Country Work for the City: Von Thünen's Ideas in Geography, Agricultural Economics and the Sociology of Agriculture," *American Journal of Economics and Sociology* 60 (2001): 79–98.

9. *Hovey's American Gardener* (1935); Richard O. Cummings, *The American and His Food: A History of Food Habits in the United States* (Chicago: University of Chicago Press, 1941), p. 34. On rural families' use of root cellars and other means to store vegetables, see Sarah F. McMahon, "'All Things in Their Proper Season': Seasonal Rhythms of

Diet in Nineteenth-Century New England," *Agricultural History* 63 (1989): 130.

10. Thomas F. de Voe, *The Market Assistant: Containing a Brief Description of Every Article of Human Food Sold in the Public Markets of the Cities of New York, Boston, Philadelphia, and Brooklyn* (New York: Orange Judd, 1866), p. 321.

11. "Our Culinary Vegetables," *Hovey's American Gardener* 31, 8 (1865): 225–233.

12. A. E. Prugh, "Why Western Vegetables Are Popular," *Western Grower and Shipper* 1 (August 1930): 8+. For a fuller discussion of early twentieth-century changes in nutritional opinion and their effects on American eating habits, see Harvey Levenstein, *Paradox of Plenty: A Social History of Eating in Modern America* (New York: Oxford, 1993); Harvey Levenstein, *Revolution at the Table: The Transformation of the American Diet* (Berkeley: University of California Press, 2003). For this section of the chapter I am especially indebted to Levenstein's chapter on "the newer nutrition" in *Revolution at the Table*. Other useful sources include Hillel Schwartz, *Never Satisfied: A Cultural History of Diets, Fantasies and Fat* (New York: Free Press, 1986); Harmke Kamminga and Andrew Cunningham, eds., *The Science and Culture of Nutrition, 1840–1940* (Atlanta: Rodopi, 1995); W. B. Gratzer, *Terrors of the Table: The Curious History of Nutrition* (Oxford: Oxford University Press, 2005).

13. Cummings, *The American and His Food*, p. 127; W. O. Atwater, "Pecuniary Economy of Food," *The Century* 25 (1888): 437–446.

14. Atwater, "Pecuniary Economy of Food," p. 445; C. F. Langworthy, "Food and Diet in the United States," *Yearbook of Agriculture*, U.S. Department of Agriculture (Washington, D.C.: Government Printing Office, 1907) p. 376.

15. On Kellogg, Fletcher, and other "food faddists," see Levenstein, *Revolution at the Table*, pp. 86–97; also Horace Fletcher, *The New Glutton or Epicure* (New York: Frederick A. Stokes, 1906); L. Margaret Barnett, "Every Man Is His Own Physician: Dietetic Fads, 1890–1914," in *The Science and Culture of Nutrition, 1840–1940,* ed. H. Kamminga and Andrew Cunningham (Atlanta: Rodopi, 1995), pp. 155–178. Older texts on raw food include Eugene Christian and Mollie Griswold Christian, *Uncooked Foods & How to Use Them* (New York: Health-Culture Company, 1904); George J. Drews, *Unfired Food and Hygienic Dietetics: For Prophylactic (Preventive) Feeding and Thera-*

peutic (Remedial) Feeding: (Treats on Food in the Cause, Prevention and Cure of Disease) (Chicago: G. J. Drews, 1909); Grace Aspinwall, "The Joys of Raw Food," *Good Housekeeping* 50 (January 1910): 110–112.

16. J. M. Booher, ed., *Scientific Weight Control: An Improved System for Reducing or Increasing Weight, Together with an Explanation of the Benefits to Be Gained from Weighing Daily* (Chicago: Chicago Continental Scale Works, 1925), p. 42.

17. On the "Oriental" diets, see Elmer Verner McCollum, Elsa Orent-Keiles, and Harry G. Day, *The Newer Knowledge of Nutrition,* 5th ed. (New York: Macmillan, 1939), pp. 512–517 (several editions of this book were published). Note that not all assessments were so favorable; one popular guidebook told readers that "Earth's little peoples, like the Japanese, are dwarfed, the scientists declare, because they don't get enough milk and milk's fat-soluble vitamines." C. Houston Goudiss, *Eating Vitamines: How to Know and Prepare the Foods That Supply These Invisible Life-Guards, with Two Hundred Tested Recipes and Menus for Use in the Home* (New York: Funk and Wagnalls, 1922), p. 20. See also Elmer Verner McCollum, *A History of Nutrition; the Sequence of Ideas in Nutrition Investigations* (Boston: Houghton Mifflin, 1957).

18. McCollum, Orent-Keiles, and Day, *The Newer Knowledge of Nutrition,* p. 564.

19. Goudiss, *Eating Vitamines,* pp. 3, 20; Elmer Verner McCollum, "What to Teach the Public Regarding Food Values?" *Journal of Home Economics* 10 (1918): 202.

20. Arnold Shircliffe, *The Edgewater Beach Hotel Salad Book* (Chicago: Hotel Monthly Press, 1926), pp. 68, 99. Other salad cookbooks that refer to the health benefits include Elizabeth O. Hiller, *The Calendar of Salads: 365 Answers to the Daily Question: "What Shall We Have for Salad?"* (New York: P. F. Volland, 1916); H. J. Heinz Company, *Heinz Book of Salads* (Pittsburgh, 1925).

21. Ross H. Gast, "History of California Vegetables," *Western Grower and Shipper* 17 (1945): 27+; Edward J. Wickson, *The California Vegetables in Garden and Field; a Manual of Practice, with and without Irrigation, for Semitropical Countries* (San Francisco: Pacific Rural Press, 1910); E. J. Ryder, "The New Salad Crop Revolution," in *Trends in New Crops and New Uses,* ed. J. Janick and A. Whipley (Alexandria, Va.: ASHS Press, 2002), pp. 408–412.

22. Burton Anderson, *America's Salad Bowl: An Agricultural History of the Salinas Valley* (Salinas, Calif.: Monterey County Historical Society, 2000), pp. 105–106.

23. W. H. Winterrowd, "Design and Construction of Refrigeration Cars," *Ice and Refrigeration* 63 (1922): 144–148; A. A. Tavernetti, "The Salinas Deal in the Roaring Twenties," *Western Grower and Shipper* 21 (1950): 18+; William H. Friedland, Amy E. Barton, and Robert J. Thomas, *Manufacturing Green Gold: Capital, Labor, and Technology in the Lettuce Industry* (New York: Cambridge University Press, 1981).

24. Anderson, *America's Salad Bowl*, p. 114; Betty Doty, "Grower-Shipper Assn Celebrates Golden Year," *Salinas Californian Weekender* (1980): 35–36.

25. Of the 55,636 railcars that shipped in 1930, 37,473 carloads, or 67 percent, were from California. H. A. Jones, *The Head-lettuce Industry of California* (Berkeley, Calif., 1932). Figures for New York railcar receipts of lettuce demonstrate iceberg's growing popularity. Iceberg accounted for approximately 80 percent of the U.S. commercial lettuce crop by 1926, versus 38 percent in 1917. H. R. Wellman, *Lettuce* (Berkeley, Calif., 1926).

26. Helen B. Lamb, "Industrial Relations in the Western Lettuce Industry," Ph.D. diss., Harvard University, 1942, p. 99; Gabriella Petrick, "'Like Ribbons of Green and Gold': Industrializing Lettuce and the Quest for Quality in the Salinas Valley, 1920–1965," *Agricultural History* 80 (2006): 269–295.

27. *Ladies' Home Journal* (March 1930): 155.

28. Vincent E. Rubatzky and Mas Yamaguchi, *World Vegetables: Principles, Production, and Nutritive Values* (New York: Chapman and Hall, 1997); Ryder, "The New Salad Crop Revolution."

29. "Selling Western Sunshine," *Western Grower and Shipper* 1 (December 1929): 7+.

30. Daniel R Hodgdon, "Vegetable Salads Are Praised as Health Food Par Excellence," *The Washington Post*, May 20, 1928; Iva McFadden, "Taking the 'Happenstance' Out of Marketing," *Los Angeles Times*, February 10, 1929; Alma Whitaker, "Beauty Diets Still Require Freak Menus," *Los Angeles Times*, January 12, 1936.

31. Christine Frederick, "Improving Our Restaurants," *The American Restaurant* (1928): 68, cited in Levenstein, *Revolution at the Table*, p. 167.

32. Lamb, "Industrial Relations in the Western Lettuce Industry," p. 16.

33. Ruth Schwartz Cowan, *More Work for Mother* (New York: Basic Books, 1983), pp. 121–122. For a fuller discussion of the relationship between the decline of domestic service and the rising demand for foods that required less cooking, see Levenstein, *Revolution at the Table*, ch. 5.

34. Cowan, *More Work for Mother*; B. J. Fox, "Selling the Mechanized Household: 70 Years of Ads in *Ladies Home Journal*," *Gender and Society* 4, 1 (1990): 25–40.

35. George A. Sweet, "A Grower Looks at the Future," *Western Grower and Shipper* 3 (1932): 8.

36. Levenstein, *Revolution at the Table*, p. 161.

37. E. G. Montgomery, *Points Brought out in the Canned Food Survey* (Washington, D.C.: U.S. Department of Commerce, 1926), cited in Levenstein, *Revolution at the Table*, p. 251.

38. C. F. Langworthy, "Green Vegetables and Their Uses in the Diet," *Yearbook of Agriculture*, U.S. Department of Agriculture (Washington, D.C.: Government Printing Office, 1911), p. 452.

39. Laura Shapiro, *Perfection Salad: Women and Cooking at the Turn of the Century* (New York: Farrar, Straus, and Giroux, 1986), p. 96.

40. Olive Green, *One Thousand Salads* (New York: G. P. Putnam's Sons, 1909); Hiller, *The Calendar of Salads*; culinary pamphlets: H. J. Heinz Company, *Heinz Book of Salads; The Salad Bowl* (Best Foods, 1929); *Today . . . What Salad . . . What Dessert? Jell-O Brings Dozens of Answers* (Jell-O, 1928).

41. "When the Man of the House Makes the Salad," *Good Housekeeping* (September 1927): 76.

42. Ross H. Gast, "Salads as a First Course an Old California Custom," *Western Grower and Shipper* (1938): 11.

43. "Our Changing Diet," *The Washington Post*, January 16, 1932, p. 6.

44. Ibid.

45. William O. Jones, "The Salinas Valley: Its Agricultural Development," Ph.D. diss., Stanford University, 1947.

46. Tavernetti, "The Salinas Deal in the Roaring Twenties."

47. O. D. Miller, "'All Work and No Play—'", *Western Grower and Shipper* 1 (1930): 14.

48. Jones, "The Salinas Valley," p. 281. Other sources on the early history and organization of the California lettuce industry include Lamb, "In-

dustrial Relations in the Western Lettuce Industry"; William O. Jones, "A Case Study in Risk Distribution: The California Lettuce Industry," *Journal of Farm Economics* 33 (1951): 235–241; Judith C. Glass, "Conditions Which Facilitate Unionization of Agricultural Workers: A Case Study of the Salinas Valley Lettuce Industry," Ph.D. diss., University of California, Los Angeles, 1966.

49. Ross H. Gast, "My Own Page," *Western Grower and Shipper* (1931). On the power of the big growers see Petrick "'Like Ribbons of Green and Gold'"; Friedland, Barton, and Thomas, *Manufacturing Green Gold;* Margaret Fitzsimmons, "The New Industrial Agriculture: The Regional Integration of Specialty Crop Production," *Economic Geography* 62 (1986): 334–353.

50. F. H. Higgens, "A 'Programmed' Agriculture," *Western Grower and Shipper* 1 (1930): 12+; "Power Cuts Growing Costs," *Western Grower and Shipper* (1932): 9. Discourses on "power farming" are discussed in D. K. Fitzgerald, *Every Farm a Factory: The Industrial Ideal in American Agriculture* (New Haven: Yale University Press, 2003), pp. 75–105.

51. On the challenges of harvesting iceberg, see Friedland, Barton, and Thomas, *Manufacturing Green Gold;* Petrick, "'Like Ribbons of Green and Gold.'"

52. Workers with long arms and torsos relative to their legs were considered better suited to stooping. Jones, "The Salinas Valley," p. 282. On the history of racial stereotypes in California agricultural labor relations, see also Don Mitchell, *The Lie of the Land: Migrant Workers and the California Landscape* (Minneapolis: University of Minnesota Press, 1996).

53. Lamb, "Industrial Relations in the Western Lettuce Industry," pp. 189–190.

54. Between 1930 and 1938, eighty firms joined the GSVA, but ninety-five firms withdrew from an industry that had an average of fifty to sixty firms active at any one time. Jones, "A Case Study in Risk Distribution," p. 239.

55. Howard A. DeWitt, "The Filipino Labor Union: The Salinas Lettuce Strike of 1934," *Amerasia Journal* 5 (1978): 1–21; "Valley Strike Brings Death," *Los Angeles Times,* September 1, 1934, p. 2; "Filipino Labor Camp Burned in California," *New York Times,* September 23, 1934, p. 6.

56. John Steinbeck, "Dubious Battle in California," *The Nation* 143 (1936): 302–304.

57. Glass, "Conditions Which Facilitate Unionization," pp. 77–78, emphasis added.

58. "The Real Issue," *Western Grower and Shipper* (1936): 5.

59. Steinbeck, "Dubious Battle in California."

60. "Free Salinas Streets of Lettuce Strikers," *New York Times,* September 18, 1936, p. 48.

61. The fatal blow came when Filipino field workers, who did not belong to the same union as the shed workers, voted not to join the strike. They received sizable (though temporary) wage raises during the strike. The shed workers had previously shown little interest in organizing field workers, partly because of language barriers, partly because of the field workers' migratory status, and partly, no doubt, because of racism. Glass, "Conditions Which Facilitate Unionization."

62. For examples of the press coverage of the strike see "Women Clubbed in Lettuce Strike, Mediation Near," *Christian Science Monitor,* September 18, 1936; "Citizen 'Army' Routs Strikers; Troops Wait Call to Salinas," *Los Angeles Times,* September 17, 1936; "Loyal Workers Battle Strikers in Salinas," *Los Angeles Times,* October 5, 1936; "Leader Warns Coast Strikers of 'Death Trap,'" *The Washington Post,* September 18, 1936; Charleton Williams, "Salinas Torn by Guerrilla Warfare," *Los Angeles Times,* September 26, 1936. For a retrospective analysis of the photographic coverage see R. S. Street, "The 'Battle of Salinas': San Francisco Bay Area Press Photographers and the Salinas Valley Lettuce Strike of 1936," *Journal of the West* 26, 2 (1987): 41–51.

63. "Novels, Plays, Books, Radio Spotlight California's Agricultural Labor Problems," *Western Grower and Shipper* (1939): 14+; Carey McWilliams, *Factories in the Field: The Story of Migratory Farm Labor in California* (Boston: Little, Brown and Company, 1939); John Steinbeck, *The Grapes of Wrath* (New York: Viking Press, 1939).

64. Paul Ortiz, "From Slavery to Cesar Chavez and Beyond: Farmworker Organizing in the United States," in *The Human Cost of Food: Farmworkers' Lives, Labor, and Advocacy,* ed. C. D. Thompson and M. Wiggins (Austin: University of Texas Press, 2002), pp. 249–276; Susan Ferriss and Ricardo Sandoval, *The Fight in the Fields: Cesar Chavez and the Farmworkers Movement* (New York: Harcourt Brace, 1997).

65. Robert J. Thomas, *Citizenship, Gender, and Work: The Social Organization of Industrial Agriculture* (Berkeley: University of California Press, 1985), p. 88; M. Zahara, S. Johnson, R. Garrett, "Labor Requirements, Harvest Costs and the Potential for Mechanical Harvest of Lettuce," *Hortscience* 99, 6 (1974): 535–537; Anderson, *America's Salad Bowl,* p. 122.

66. Dave Stildolph, "Vacuum Cooling Continues to Grow," *Fruit and Vegetable Review* (1955): 30–32. Sources on vacuum packing were consulted at the Monterey County Historical Society in February 2007. See also Petrick, "'Like Ribbons of Green and Gold'"; Glass, "Conditions Which Facilitate Unionization."

67. "Quick Chill for Salinas Lettuce," *P.G.& E Progress* 32, 9; B. A. Friedman, "Vacuum Cooling Upheld in Tests," *Western Grower and Shipper* 23 (July 1952): 21+.

68. Stildolph, "Vacuum Cooling"; C. B. Moore, "'Evolution' of Packing Industry," *Western Grower and Shipper* (July 1953).

69. Ernesto Galarza, *Merchants of Labor: The Mexican Bracero Story; an Account of the Managed Migration of Mexican Farm Workers in California, 1942–1960* (Charlotte, Calif.: McNally and Loftin, 1964).

70. Friedland, Barton, and Thomas, *Manufacturing Green Gold,* p. 66.

71. C. B. Moore, "Labor Trends," *Western Grower and Shipper* (1950): 25+.

72. Glass, "Conditions Which Facilitate Unionization," p. 128.

73. Petrick, "'Like Ribbons of Green and Gold,'" p. 281.

74. Stildolph, "Vacuum Cooling"; Doty, "Grower-Shipper Assn Celebrates Golden Year."

75. Glass, "Conditions Which Facilitate Unionization," p. 128; Harland Padfield and William Edwin Martin, *Farmers, Workers and Machines: Technological and Social Change in Farm Industries of Arizona* (Tucson: University of Arizona Press, 1965), pp. 284–285; D. Stildolph, "Vacuum Cooling"; Doty, "Grower-Shipper Assn Celebrates Golden Year"; D. Vera, "Critical Problems Dictate Industry Harmony," *Western Grower and Shipper* (November 1960): 52–53.

76. "It's a Process—Not a Fad," *Western Grower and Shipper* (1946): 8–9.

77. Ibid.; Helen E. Goodrich, "Processed Foods Gain Vital Issue," *Western Grower and Shipper* 22 (1951): 25+.

78. Ross H. Gast, "Competition Comes in Cans, Too," *Western Grower*

and Shipper (1937): 8; "Here's What Radio Says about Iceberg Head Lettuce," *Western Grower and Shipper* (1937): 19; "Iceberg Lettuce on the Air," *Western Grower and Shipper* (1937): 9.

79. "It's a Process—Not a Fad."

80. J. H. Collins, "Vegetables Will Be Dressed Up," *Western Grower and Shipper* 17 (December 1945): 31+; A. L. Martin, "Who'll Do the Pre-Packaging?" *Western Grower and Shipper* 17 (1946): 88+.

81. "Packaging: Who Wants What?" *Western Grower and Shipper* (1955): 14+.

82. Barbara Tellus, "Salad Month Western Style," *Western Grower and Shipper* (1966): 13+.

83. C. B. Moore, "Consumer Packaging and Research," *Western Grower and Shipper* (1947): 22+.

84. "A New Look for Lettuce," *Western Grower and Shipper* 31 (1962): 11+.

85. "Perishable Produce," *Progressive Grocer Associates,* May 1, 1997; Gary Lucier et al., *Fruit and Vegetable Backgrounder,* Economic Research Service, U.S. Department of Agriculture, VGS-313-012222 (2006), p. 12.

86. Sherry Frey, "New Rules for Perishables," *FMI Show,* 2007.

87. "What's Fresh in Fresh-cut Packaging?" *Brand Packaging* 9, 1 (2005).

88. "Spotlight: Product Shelf Life Extension," *Fresh Cut* (March 2005), www.freshcut.com/mar2005/productshelflife.htm.

89. U.S. Department of Agriculture Economic Research Service, "U.S. Lettuce: Per Capita Use, 1960–2005" (U.S. Department of Agriculture, 2006).

90. For one of the first portrayals of "foodies," see Ann Barr and Paul Levy, *The Official Foodie Handbook: Be Modern—Worship Food* (New York: Timbre Books, 1984). On the role of Chez Panisse in promoting baby lettuce, see Julie Guthman, "Fast Food/Organic Food: Reflexive Tastes and the Making of 'Yuppie Chow,'" *Social and Cultural Geography* 4 (2003): 45–58; Burkhard Bilger, "Salad Days: How a Lowly Leaf Became a High-End Delicacy," *New Yorker,* September 6, 2004, pp. 136+.

91. Bilger, "Salad Days."

92. Carol Ness, "Earthbound Farm: Backyard Farmers Emerge as Top Organic Produce Brand," *San Francisco Chronicle,* May 3, 2006; Michael Rosenwald and Yian Q. Mui, "California Farm Firm Linked to Tainted Spinach," *Washington Post,* September 16, 2006.

93. Georgia Dullea, "What's Really Big Now? Tiny Vegetables," *New York Times,* June 15, 1985; Regina Schrambling, "Forever Young: Many of the Springtime Table's Most Appealing Delicacies Are Still Glowing with the Tender Blush of Babyhood," *Los Angeles Times,* May 5, 2004; M. Brindley, "Baby Veg May Be Popular But Are They Any Good for Farmers or the Environment?" *Western Mail and Echo,* August 5, 2005, p. 5.

94. Jock O'Connell, Bert Mason, and John Hagen, *The Role of Air Cargo in California's Agricultural Export Trade* (Fresno: Center for Agricultural Business, California State University, 2005), pp. 80–81.

95. Susanne Freidberg, *French Beans and Food Scares: Culture and Commerce in an Anxious Age* (New York: Oxford University Press, 2004).

96. Ibid.

97. Edward F. Fischer and Peter Benson, *Broccoli and Desire: Global Connections and Maya Struggles in Postwar Guatemala* (Stanford, Calif.: Stanford University Press, 2006).

98. Susanne Freidberg, "Cleaning up Down South: Supermarkets, Ethical Trade and African Horticulture," *Social and Cultural Geography* 4 (2003): 353–368.

6. Milk

1. In 2002 Vermont, by far the largest dairy producer in New England, recorded $342,440,000 in sales of milk and dairy products. In 2004–2005, there were 1,259 dairy farms (most employing fewer than 5 people) and 27 off-farm processors, which together provided 1,752 jobs. Ben and Jerry's accounted for many of these. In 2005, tourists to the state spent $1.57 billion and supported, both directly and indirectly, 36,250 jobs. Dairy Task Force Report, *Vermont State Agricultural Overview 2005* (National Agricultural Statistics Service, U.S. Department of Agriculture); Vermont Department of Tourism and Marketing, *The Travel and Tourism Industry in Vermont* (Montpelier, 2006).

2. Valerie Essex Cheke, *The Story of Cheese-Making in Britain* (London: Routledge and K. Paul, 1959); Patricia Lysaght, *Milk and Milk Products from Medieval to Modern Times: Proceedings of the Ninth International Conference on Ethnological Food Research, Ireland, 1992* (Edinburgh: Canongate Academic Press, 1994).

3. J. D. Burks, "Clean Milk and Public Health," *Annals of the American*

Academy of Political and Social Science 37 (1911): 192; Jacqueline Wolf and Leslie C. Frank, "A State-Wide Milk Sanitation Program: Commentary," *Public Health Reports* 121 (2006 [1924]): 174–189; M. J. Rosenau, *The Milk Question* (Boston: Houghton Mifflin, 1912), p. 6.

4. Lysaght, *Milk and Milk Products from Medieval to Modern Times*, Introduction.

5. Lore Alford Rogers, *Fermented Milks*, Circular (U.S. Department of Agriculture Bureau of Animal Industry, 1911). For a review of the scientific literature on the potential health benefits of dairy products cultured with "probiotic" bacteria, see S. C. Leahy, D. G. Higgins, G. F. Fitzgerald, and D. van Sinderen, "Getting Better with Bifidobacteria," *Journal of Applied Microbiology* 98 (2005): 1303–1315.

6. Cheke, *The Story of Cheese-Making in Britain*, pp. 84, 103.

7. J. Twamley, *Dairying Exemplified, or the Business of Cheese-Making* (Providence: Carter and Wilkinson, 1796); Cheke, *The Story of Cheese-Making in Britain*, p. 12.

8. Elinor Oakes, "A Ticklish Business: Dairying in New England and Pennsylvania, 1750–1812," *Pennsylvania History* 47 (1980): 195–212; Sarah F. McMahon, "'All Things in Their Proper Season': Seasonal Rhythms of Diet in Nineteenth-Century New England," *Agricultural History* 63 (1989): 130; Judith Moyer, "From Dairy to Doorstep: The Processing and Sale of New Hampshire Dairy Products, 1860s to 1960s," *Historical New Hampshire* 58 (2003): 101–122.

9. Joshua Johnson, *The Art of Cheese Making Reduced to Rules, and Made Sure and Easy, from Accurate Observation & Experience, Published for the Help of Dairy Women* (Albany, N.Y.: Charles R. and George Webster, 1801), quoted in Andrew F. Smith, "The Origins of the New York Dairy Industry," in *Milk Beyond the Dairy: Proceedings of the Oxford Symposium on Food and Cookery 1999*, ed. H. Walker (Totnes, Devon, England: Prospect Books, 2000), p. 319.

10. References to maggot-filled cheese were common in the press, as were suggestions for how to avoid this problem: "Red pepper, so called, is a complete antidote against flies impregnating cheese so as to produce maggots. Take one and put it into a delicate piece of linen, moisten it with a little fresh butter, and rub your cheese frequently. It not only gives a very fine color to your cheese, but it is so pungent that no fly will touch it." "Rubbing Cheese with Red Pepper Preserves It Against

Mites," *Vermont Gazette* (1827), 1; "On Cheese Making," *Rutland Herald* (1795), p. 4. On early cheesemaking in northern New England, see Paul Kindstedt, *American Farmstead Cheese* (White River Junction, Vt.: Chelsea Green, 2005), ch. 2.

11. An estimated 8,500 cows lived in and around London in 1800, including hundreds in now-densely populated neighborhoods such as Islington, Tottenham Court Road, and Paddington. G. E. Fussell, *The English Dairy Farmer, 1500–1900* (New York: A. M. Kelley, 1966), p. 304.

12. Frances Milton Trollope, *Domestic Manners of the Americans* (New York: Penguin Books, 1997 [1832]), p. 32, quoted in Fussell, *The English Dairy Farmer, 1500–1900*, p. 317.

13. E. Melanie DuPuis, *Nature's Perfect Food: How Milk Became America's Drink* (New York: New York University Press, 2002), p. 30.

14. Jacqueline Wolf, *Don't Kill Your Baby: Public Health and the Decline of Breastfeeding in the 19th and 20th Centuries* (Columbus: Ohio State University Press, 2001), ch. 1.

15. James Flexner, "The Battle for Pure Milk in New York City," in *Is Loose Milk a Health Hazard? The Report of the Commission Appointed by Dr. Shirley W. Wynne, Commissioner of Health of the City of New York, to Study the Public Health Aspects of the Sale of Loose Milk in New York City and to Make Recommendations* ed. E. F. Brown (New York: The Commission, 1931), pp. 161–196.

16. *Frank Leslie's Illustrated Newspaper* coverage of swill dairies began on May 8, 1858. Andrea G. Pearson, "'Frank Leslie's Illustrated Newspaper' and 'Harper's Weekly': Innovation and Imitation in Nineteenth-Century American Pictorial Reporting," *Journal of Popular Culture* 23 (1990): 81; DuPuis, *Nature's Perfect Food*, p. 40.

17. Such practices were neither new nor uniquely American. See Fussell, *The English Dairy Farmer, 1500–1900*, pp. 304–306. On early views of "railroad milk," see Horatio Newton Parker, *City Milk Supply* (New York: McGraw-Hill, 1917), p. 203.

18. John L. Bernardi, "A Century and a Quarter of Milk Contracting in New England: The Hood Company, 1846–1970," Ph.D. diss., University of Pennsylvania, 1971; H. P. Hood Company, *The Hood Story: A Century of Progress in the New England Dairy Industry*, H. P. Hood Collection, Box 2, Archives of Historic New England, Boston.

19. Bernardi, "A Century and a Quarter of Milk Contracting in New England," p. 17.

20. L. B. Bacon, "Institutional Factors Affecting the Marketing of Milk in Boston," Ph.D. diss., Harvard University, 1934, p. 16.

21. Harold F. Wilson, *The Hill Country of Northern New England: Its Social and Economic History, 1790–1930* (New York: AMS Press, 1967); T. D. Seymour Bassett, "500 Miles of Trouble and Excitement: Vermont Railroads, 1848–1861," *Vermont History* 49 (1981): 98, 101–102, 133.

22. A. F. Sanborn, "The Future of Rural New England," *Atlantic Monthly* (1897): 74–84.

23. On creamery butter: Howard Russell, *A Long Deep Furrow: Three Centuries of Farming in New England* (Hanover, N.H.: University Press of New England, 1976), p. 424; Moyer, "From Dairy to Doorstep," p. 104. On factory cheese: Loyal Durand, Jr., "The Migration of Cheese Manufacture in the United States," *Annals of the Association of American Geographers* 42 (1952): 263–282; Loyal Durand, Jr., "The Historical and Economic Geography of Dairying in the North Country of New York State," *Geographical Review* 57 (1967): 24–47; Kindstedt, *American Farmstead Cheese,* p. 30.

24. Wilson, *The Hill Country of Northern New England,* pp. 208, 312.

25. Quoted in Ibid., 300.

26. Jan Albers, *Hands on the Land: A History of the Vermont Landscape* (Cambridge, Mass.: MIT Press, 2000), p. 246. See also Dona Brown, *Inventing New England: Regional Tourism in the Nineteenth Century* (Washington, D.C.: Smithsonian Institution Press, 1995).

27. Parker, *City Milk Supply,* p. 217.

28. Joseph Scott MacNutt, *The Modern Milk Problem in Sanitation, Economics, and Agriculture* (New York: Macmillan, 1917), pp. 128–133; Robert Milham Hartley, *An Historical, Scientific, and Practical Essay on Milk as an Article of Human Sustenance* (New York: Arno Press, 1977); Henry E. Alvord and Raymond A. Pearson, *The Milk Supply of Two Hundred Cities and Towns* (Washington, D.C.: U.S. Department of Agriculture Bureau of Animal Industry, 1903). See also DuPuis, *Nature's Perfect Food.*

29. Burks, "Clean Milk and Public Health," p. 443; Wolf, *Don't Kill Your Baby,* ch. 2.

30. W. H. Park and L. E. Holt, "Report upon the Results with Different

Kinds of Pure and Impure Milk in Infant Feeding in Tenement Houses and Institutions of New York City: A Clinical and Bacteriological Study," *Archives of Pediatrics* 20 (1903): 881–910; Rosenau, *The Milk Question;* Harvey A. Levenstein, "'Best for Babies' or 'Preventable Infanticide'? The Controversy over Artificial Feeding of Infants in America, 1880–1920," *Journal of American History* 70 (1983): 75–94; Rosenau, *The Milk Question,* p. 256.

31. S. Sharwell, "Lessons to be Learned from an Inspection that Follows Milk from the Cow to the Consumer," in *Fourth Annual Report of the International Association of Dairy and Milk Inspectors,* ed. I. Weld (Washington, D.C., 1915), pp. 194–197.

32. Rosenau, *The Milk Question,* p. 259.

33. *Milk-borne Infectious Disease,* Mrs. William Lowell Putnam Papers, 1887–1935, Schlesinger Library, Radcliffe Institute for Advanced Study, Harvard University, Cambridge, Mass. (hereafter Putnam Papers), Box 6, Folder 103.

34. Hollis Godfrey, *The Health of the City* (Boston: Houghton Mifflin, 1910), p. 41.

35. Alexis Bernstein's thesis provides an excellent analysis of these two philanthropists' relationships with scientists and other milk reformers. Alexis Bernstein, "The Land of Milk and Money: Philanthropy, Milk and Infant Mortality, 1893–1914," B.A. Honors thesis, Harvard University, 2005.

36. Lina Gutherz Straus, *Disease in Milk: The Remedy, Pasteurization; The Life Work of Nathan Straus* (New York: Dutton, 1917); Julie Miller, "To Stop the Slaughter of the Babies: Nathan Straus and the Drive for Pasteurized Milk," *New York History* 73 (1993): 159–184; Bernstein, "The Land of Milk and Money."

37. Miller, "To Stop the Slaughter of the Babies," pp. 177, 179.

38. Rosenau, *The Milk Question,* pp. 226–228. More than a century after the debates about pasteurization's effects on the nutritional value of milk began, there remains no scientific consensus. Pasteurization is known to kill the healthy bacteria as well as vitamin C, which milk contains little of anyway. Calcium and protein are unaffected. As of 2007, there was growing evidence that farm children who drank raw milk were less likely to get allergies or asthma, but it was not entirely clear whether the immunity was due to the milk itself or to the children's early exposure to farm animals. M. R. Perkin, "Unpasteurized

Milk: Health or Hazard?" *Clinical & Experimental Allergy* 37 (2007): 627–630. On the campaign for raw milk, see www.realmilk.com.

39. Manfred J. Waserman, "Henry L. Coit and the Certified Milk Movement in the Development of Modern Pediatrics," *Bulletin of the History of Medicine* 46 (1972): 359–390.

40. Samuel Prescott, "The Milk Supply of Boston," *Science Conspectus* 2 (1911): 22–28.

41. Bernstein, "The Land of Milk and Money," ch. 2.

42. "Few bills in the history of the legislature have been so strongly endorsed as the Ellis Clean Milk Bill," Putnam Papers, Box 7, Folder 139. Letter from Arthur Barber, March 31, 1911, "Complaints of treatment of farmers," Putnam Papers, Box 6, Folder 85.

43. Letter from C. W. Carpenter, March 30, 1911, ibid.

44. Boston Chamber of Commerce, *Investigation and Analysis of the Production, Transportation, Inspection and Distribution of Milk and Cream in New England* (Boston: Chamber of Commerce, 1915), p. 37; M. Putnam, "Talk to mothers," mimeo, Putnam Papers, Box 27, Scrapbook 464v.

45. M. Putnam, "Consumers plead for Massachusetts Milk," Address delivered before the Farmers' Institute, Marlboro, Mass., March 27, 1912, Putnam Papers, Box 4, Folder 49.

46. Putnam Papers, Box 5, Folder 78. "The enactment of the consumers' milk bill a tremendous benefit to milk producers," Putnam Papers, Box 4, Folder 49.

47. Letter to the editor, *The Union*, March 4, 1916, Putnam Papers. Scrapbook 464v, Box 27.

48. Bernstein, "The Land of Milk and Money," p. 62.

49. Arthur W. Gilbert, *The Food Supply of New England* (New York: Macmillan, 1924), p. 47.

50. DuPuis, *Nature's Perfect Food*, ch. 4; Daniel Block, "Protecting and Connecting: Separation, Connection and the U.S. Dairy Economy, 1840–2002," *Journal for the Study of Food and Society* 6 (2002): 22–30.

51. Gary Wheelock, New England Dairy Council Promotion Board, personal communication.

52. For more detailed discussions, see Alden Coe Manchester, *The Public Role in the Dairy Economy: Why and How Governments Intervene in the Milk Business* (Boulder, Colo.: Westview Press, 1983); Kenneth

W. Bailey, *Marketing and Pricing of Milk and Dairy Products in the United States* (Ames: Iowa State University Press, 1997); Alden C. Manchester and Don P. Blayney, *Milk Pricing in the United States,* Agriculture Information Bulletin 761 (Market and Trade Economics Division, Economic Research Service, U.S. Department of Agriculture, 2001).

53. George Max Beal and Henry Harrison Bakken, *Fluid Milk Marketing* (Madison, Wisc.: Mimir Publishers, 1956), pp. 101–102; Wilson, *The Hill Country of Northern New England,* p. 307.

54. Manchester, *The Public Role in the Dairy Economy,* p. 30.

55. Ibid., p. 39; T. J. Kriger, "Syndicalism and Spilled Milk: The Origins of Dairy Farmer Activism in New York State, 1936–1941," *Labor History* 38 (1997): 266–286.

56. Kriger, "Syndicalism and Spilled Milk"; *New England Homestead,* May 7, 1910, p. 678; Wilson, *The Hill Country of Northern New England,* pp. 339–340.

57. "Milk Strike Threatened," *Los Angeles Times,* September 8, 1938, p. 1; "Here Are Foods to Substitute for Fresh Milk," *Chicago Tribune,* May 1, 1940, p. 16.

58. "Wider 'Milk Shed' to Combat Strike: Dr. Wynne Declares He Will Import Product from West to Supply City's Needs," *New York Times,* May 9, 1933, p. 19; "Milk Strike Cuts City's Supply 37%; Troopers on Guard," *New York Times,* August 18, 1939, p. 1.

59. "Milk Strike Pickets and Police Fight: Illinois Street Battle Scene," *Los Angeles Times,* October 14, 1935; "Milk Strike Threatened," *Los Angeles Times,* September 8, 1938; "Deputy Killed in Milk Strike Clash in East," *Chicago Tribune,* July 4, 1941.

60. The first of these was the Agricultural Adjustment Act of 1933; federal milk marketing regulations were revised and elaborated in the Agricultural Marketing Agreement Act of 1937. E. Erba and A. M. Novakovic, *The Evolution of Milk Pricing and Government Intervention in Dairy Markets* (Ithaca, N.Y.: Department of Agricultural, Resource, and Managerial Economics, Cornell University, 1995); Bailey, *Marketing and Pricing of Milk and Dairy Products in the United States.*

61. Paul S. McComas, "The New England Dairy Industry, with Special Reference to Inter-Regional Competition," Ph.D. diss., Harvard University, 1947.

62. Andrew Novakovic, "Pricing Milk in the U.S.—Spatial Values of Plant

Pay Prices and Price Received by Dairy Farmers," National Workshop for Dairy Economists and Policy Analysts, Charleston, S.C., 2007; Block, "Protecting and Connecting."

63. Robert O. Sinclair,"The Economic Effects of Bulk Milk Handling on the Dairy Industry of Vermont," M.S. thesis, University of Vermont, 1955.

64. Gordon J. Fielding, "Dairying in Cities Designed to Keep People Out," *Professional Geographer* 14 (1962): 12–17. See also DuPuis, *Nature's Perfect Food*, ch. 9.

65. Bailey, *Marketing and Pricing of Milk and Dairy Products in the United States;* E. Melanie DuPuis and Daniel Block, "Sustainability and Scale: U.S. Milk-Market Orders as Relocalization Policy," *Environment and Planning* 40 (2008): 1987–2005.

66. Per capita fluid milk consumption peaked in 1945 at forty-five gallons per year. By 1970 it had fallen to thirty-one gallons, and by 2001 to less than twenty-three gallons. Until the mid-1980s, Americans drank more whole milk than lower-fat varieties; by 2001, they drank on average only eight gallons of whole milk per year (out of the total twenty-three). Judy Putnam and Jane Allshouse, "Trends in U.S. Per Capita Consumption of Dairy Products, 1909–2001," *Amber Waves* 1 (2003): 12–13.

67. Andrew Malcolm, "Dairy Output Rises Despite Efforts to Curb It," *New York Times*, 1983; Erba and Novakovic, *The Evolution of Milk Pricing and Government Intervention in Dairy Markets;* Vermont Dairy Promotion Council, "Dairy Farm Numbers," www.vermont-dairy.com/dairy_industry/farms/numbers (accessed July 19, 2007).

68. On differences in regional milk-production costs, see Economics Research Service, "Milk Costs and Returns" (Washington, D.C.: U.S. Department of Agriculture, 2005). On the mid-century history of the New England dairy market, see McComas, "The New England Dairy Industry."

69. Adrian E. Logan, "Dairying in Vermont: Farming and the Changing Face of Agriculture, 1945–1992," Ph.D. diss., University of Vermont, 1998, pp. 87–88; R. Kelly Myers, *Survey Report of Current Use Practices in New Hampshire* (Durham, N.H.: Institute for Policy and Social Science Research, 1993).

70. "Milk and Political Kindness," *Boston Globe*, April 8, 2007.

71. Ronald W. Cotterill et al., *Toward Reform of Fluid Milk Pricing in Southern New England: Farm Level, Wholesale and Retail Prices in*

the Fluid Milk Marketing Channel: 2003–2006 (Storrs: Food Marketing Policy Center, University of Connecticut, 2007).

72. Albers, *Hands on the Land,* p. 202.

73. Interview, July 2007. See also www.locavores.com.

74. Joe Drape, "Should This Milk Be Legal?" *New York Times,* August 8, 2007; Nathanael Johnson, "The Revolution Will Not Be Pasteurized: Inside the Raw Milk Underground," *Harper's* (April 2008): 71–78. See also "Why a Campaign for Real Milk?" Weston Price Foundation, www.realmilk.com (accessed June 8, 2008).

75. Peter Hirschfeld, "Senate Doubles Raw Milk Threshold," *Rutland Herald,* April 10, 2008.

76. Gabriel León y Saúl Maldonado, "Pactan Productores de Leche Diálogo con el Gobierno," *La Jornada,* January 30, 2008.

77. D. R. Strobel and C. J. Babcock, *Recombined Milk: A Dependable Supply of Fluid Milk Far from the Cow* (Washington, D.C.: U.S. Department of Agriculture Foreign Agricultural Service, 1955); Enrique Ochoa, *Feeding Mexico: The Political Uses of Food since 1910* (Wilmington, Del.: Scholarly Resources, 2000).

78. Mariano Chavez et al., "Cientos Se Plantan Frente a Oficinas de la Se en Durango, Hidalgo, Jalisco y Querétaro," *La Jornada,* May 20, 2006; W. D. Dobson, "How Mexico's Dairy Industry Has Evolved under the NAFTA—Implications for U.S. Dairy Exporters and Investors in Mexico's Dairy-Food Businesses," Discussion Paper 2002–1, Babcock Institute, University of Wisconsin, 2002, p. 22.

79. Some Mexican consumers also prefer raw milk and will go to considerable lengths to get it. Dairy farms in the eastern suburbs of Mexico City specialize in on-farm sales. In a 1998 survey, 8 percent of the capital's milk drinkers reported buying raw milk; 61 percent drank a subsidized brand made from imported powder. Hermenegildo Losada et al., "The Production of Milk from Dairy Herds in the Suburban Conditions of Mexico City: The Case of Iztapalapa," *Livestock Research for Rural Development* 8 (1996); Hermenegildo Losada et al., "The Mexico City Milk Supply System: Structure, Function, and Sustainability," *Agriculture and Human Values* 18 (2001): 305–317. On NAFTA's effects on dairying in both Mexico and the United States, see James McDonald, "NAFTA and Basic Food Production: Dependency and Marginalization on Both Sides of the U.S./ Mexico Border," in *Food in the USA: A Reader,* ed. C. Counihan (New York: Routledge, 2002), pp. 359–372; James McDonald, "Recon-

figuring the Countryside: Power, Control, and the (Re) Organization of Farmers in West Mexico," *Human Organization* 60 (2001): 247–258.

80. Powder made from whole milk has a shorter shelf life because of its fat content.

81. Richard Orr, "Vast Foreign Market Opens to U.S. Dairies," *Chicago Daily Tribune,* January 5, 1947; Hugh L. Cook and George H. Day, *The Dry Milk Industry: An Aid in the Utilization of Milk* (Chicago: American Dry Milk Institute, 1947).

82. Luis Arturo García Hernández, "Skim Milk Powder Imports and the Role of Conasupo," *Role of the State in Agricultural Trade Workshop* (Stanford, Calif.: Stanford University Press, 1998), pp. 16–17. On M. E. Franks, see also W. D. Dobson, "Competitive Strategies of Leading World Dairy Exporters," Babcock Institute Discussion Paper 95–1, University of Wisconsin, 1995. On DEIP, see Bailey, *Marketing and Pricing of Milk and Dairy Products in the United States.*

83. In France, UHT milk accounts for 95.5 percent of consumption, and in both Switzerland and Germany more than 60 percent. But it accounts for less than 10 percent of milk sales in the British Isles and Scandinavia. Valerie Elliot, "The UHT Route to Long-Life Planet," *TimesOnline,* October 15, 2007.

84. Andrea Wiley, "Transforming Milk in a Global Economy," *American Anthropologist* 109 (2007): 666–678.

85. Benjamin Jastrzembski, "The New Braceros: Mexican Dairy Workers in Vermont," B.A. thesis, Dartmouth College, 2008.

86. Ibid.

87. Sam Hemingway, "Farm Run by Governor's In-laws Employs Undocumented Workers," *Burlington Free Press,* February 18, 2007.

7. Fish

1. Harold McGee, *On Food and Cooking: The Science and Lore of the Kitchen* (New York: Scribner, 2004), pp. 189, 205; H. H. Huss, "Quality and Changes in Fresh Fish," Food and Agriculture Organization of the United Nations, 1995.

2. Elizabeth David and Jill Norman, *Harvest of the Cold Months: The Social History of Ice and Ices* (New York: Penguin, 1994); Robert David, "The Demise of the Anglo-Norwegian Ice Trade," *Business History* 37 (1995): 52–69.

3. James Davidson, "Fish, Sex and Revolution in Athens," *Classical Quarterly* 43 (1993): 53–66.

4. Felipe Fernández-Armesto, *Near a Thousand Tables: A History of Food* (New York: Free Press, 2002), pp. 210–211. Noel de la Moriniere and Simon Bartholemy Joseph, *Histoire générale des pêches anciennes et modernes, dans les mers et les fleuves des deux continents* (Paris: Impr. royale, 1815), p. 169.

5. Mark Kurlansky, *Salt: A World History* (New York: Walker and Company, 2002), p. 132.

6. Jules Haime, *The History of Fish-Culture in Europe from Its Earlier Records to 1854* (Washington, D.C., 1874), p. 467: David R. Montgomery, *King of Fish: The Thousand-Year Run of Salmon* (Boulder, Colo.: Westview Press, 2003), p. 64; C. Anne Wilson, "Preserving Food to Preserve Life: The Response to Glut and Famine from Early Times to the End of the Middle Ages," in *Waste Not, Want Not: Food Preservation from Early Times to the Present Day*, ed. C. A. Wilson (Edinburgh: Edinburgh University Press, 1991), p. 24.

7. Eugene N. Anderson, *The Food of China* (New Haven: Yale University Press, 1988), p. 55; N. Ishige, *The History and Culture of Japanese Food* (London: Kegan Paul, 2001).

8. Before the arrival of the railroad, Richard Pillsbury estimates, fish transported by horse and wagon, at around two or three miles an hour, could not travel much more than twenty miles from the water and stay fresh. Richard Pillsbury, *No Foreign Food: The American Diet in Time and Place* (Boulder, Colo.: Westview Press, 1998).

9. Sue Shepard, *Pickled, Potted, and Canned: How the Art and Science of Food Preserving Changed the World* (New York: Simon and Schuster, 2000); Kurlansky, *Salt*. See also Mark Kurlansky, *Cod: A Biography of the Fish That Changed the World* (New York: Penguin Books, 1998).

10. James Glass Bertram, *The Harvest of the Sea: A Contribution to the Natural and Economic History of the British Food Fishes* (New York: Appleton, 1866), p. 36.

11. Montgomery, *King of Fish*, p. 72.

12. Adam Smith, *An Inquiry into the Nature and Causes of the Wealth of Nations* (London: Methuen, 1776), vol. 1, ch. 11.

13. W. F. B. Massey Mainwaring, *The Preservation of Fish Life in Rivers by the Exclusion of Town Sewage* (London: W. Clowes and Sons, 1883).

14. Bertram, *The Harvest of the Sea*, ch. 7, pp. 205, 277.

15. Daniel L. Boxberger, "Ethnicity and Labor in the Puget Sound Fishing Industry, 1880–1935," *Ethnology* 33 (1994); Natalie Fobes, *Reaching Home: Pacific Salmon, Pacific People* (Anchorage: Alaska Northwest Books, 1994); Dan Landeen and Allen Pinkham, *Salmon and His People: Fish and Fishing in Nez Perce Culture* (Lewiston, Me.: Confluence Press, 1999); Courtland L. Smith, *Salmon Fishers of the Columbia* (Corvallis: Oregon State University Press, 1979), p. 13.

16. Montgomery, *King of Fish,* p. 47.

17. Jim Lichatowich, *Salmon without Rivers: A History of the Pacific Salmon Crisis* (Washington, D.C.: Island Press, 2001), pp .82–83; Kurlansky, *Salt,* p. 114.

18. Smith, *Salmon Fishers of the Columbia,* p. 13.

19. Martin Bruegel, "How the French Learned to Eat Canned Food, 1809–1930s," in *Food Nations: Selling Taste in Consumer Societies,* ed. Warren Belasco and Philip Scranton (New York: Routledge, 2002), pp. 113–130.

20. Otis W. Freeman, "Salmon Industry of the Pacific Coast," *Economic Geography* 11 (1935): 125; Lichatowich, *Salmon without Rivers,* p. 87.

21. Smith, *Salmon Fishers of the Columbia,* p. 63.

22. Lichatowich, *Salmon without Rivers,* p. 101; Boxberger, "Ethnicity and Labor in the Puget Sound Fishing Industry." Only decades later did the area's tribes win back their fishing rights.

23. Lichatowich, *Salmon without Rivers,* p. 94.

24. Ibid., p. 90; Daniel Jack Chasan, *The Water Link: A History of Puget Sound as a Resource* (Seattle: Washington Sea Grant Program, University of Washington, 1981), p. 47.

25. Lichatowich, *Salmon without Rivers,* p. 204.

26. Homer Ewart Gregory and Kathleen Barnes, *North Pacific Fisheries, with Special Reference to Alaska Salmon* (San Francisco: American Council, Institute of Pacific Relations, 1939), p. 308; "Pacific Coast Fish," *Fortune* 11 (1935): 110.

27. "Salmon Pack Is Short," *New York Times,* December 27, 1903; "Pacific Coast Fish." While the United Kingdom remained by far the largest export market for Alaskan canned salmon, the salmon was also shipped to other countries in Western Europe as well as to South America (Chile, Columbia, Venezuela, Argentina) and Asia (China, Sri Lanka, the Philippines). John N. Cobb, *Pacific Salmon*

Fisheries (Washington, D.C.: U.S. Government Printing Office, 1917), p. 66.

28. Alaska Packers' Association, *Interesting Facts about Canned Salmon* (San Francisco: Alaska Packers' Association, 1908).

29. Ward Taft Bower, *Alaska Fishery and Fur Seal Industries in 1939* (Washington, D.C.: U.S. Government Printing Office, 1940), p. 122.

30. J. K. Kilbourn, *Fish Preservation and Refrigeration* (London: W. Clowes and Sons, 1883).

31. Harden Franklin Taylor, *Refrigeration of Fish* (Washington, D.C.: U.S. Government Printing Office, 1927), p. 506; Donald Kiteley Tressler and Clifford F. Evers, *The Freezing Preservation of Foods* (New York: Avi Publishing, 1947), ch. 17.

32. Cobb, *Pacific Salmon Fisheries,* p. 140.

33. James Critchell, "Imports of Refrigerated Food Products of the United Kingdom, 1880–1907: Progress and Statistics," in Premier congrès international du froid Paris, 5 au 12 octobre 1908, J.D. Loverdo, ed. (Paris: Secrétariat général de l'Association internationale du froid, 1908), pp. 322–323. Mrs. Edmund Burke, "Domestic Science," *Chicago Daily Tribune,* October 1, 1903.

34. A. W. Ponsford, "Refrigeration in the Tuna Fishing Industry," *Ice and Refrigeration* 82 (1932): 23–25; Oscar E. Anderson, *Refrigeration in America: A History of a New Technology and Its Impact* (Princeton, N.J.: Princeton University Press, 1953), p. 271.

35. Taylor, *Refrigeration of Fish.*

36. R. Uglow, Personal communication, May 1, 2007.

37. Rudi Volti, "How We Got Frozen Food," *American Heritage of Invention and Technology* 9 (1994): 46–56.

38. *Birdseye Handbook for Salesmen* (New York: Frozen Foods Sales Corporation, 1947).

39. Robert Martin, "Fish Kept Fresh 1500 Miles from the Sea by Scientific Refrigeration," *Popular Science Monthly* 115 (1929): 57–58.

40. Clarence Birdseye, "Preservation of Foods by New Quick Freezing Methods," *Refrigerating Engineering* 25 (1933): 188.

41. Volti, "How We Got Frozen Food."

42. Tressler and Evers, *The Freezing Preservation of Foods,* p. 562.

43. A 1951 Fish and Wildlife Service survey found that while consumers appreciated the convenience and price of frozen fish, they preferred fresh for its taste: "If it cost the same amount to serve fresh fish or fro-

zen fish, most households would prefer to serve fresh fish." *Fish and Shellfish Preferences of Household Consumers-1951,* Fishery Leaflet 408 (Washington, D.C.: United States Department of the Interior, Fish and Wildlife Service, 1953).

44. Paul Greenberg, "Green to the Gills," *New York Times Magazine,* June 18, 2006.

45. Gunnar Knapp et al., *The Great Salmon Run: Competition between Wild and Farmed Salmon* (Washington, D.C.: TRAFFIC North America, 2007), p. 85.

46. Ibid., p. 82.

47. Ibid., p. 64.

48. Rosamond L. Naylor et al., "Effect of Aquaculture on World Fish Supplies," *Nature* 405 (2000): 1017–1024; Rosamond Naylor et al., "Fugitive Salmon: Assessing the Risks of Escaped Fish from Net-Pen Aquaculture," *BioScience* 55 (2005): 427–437; Josh Eagle, et al., "Why Farm Salmon Outcompete Fishery Salmon," *Marine Policy* 28, 3 (2004); Martin Krkošek, et al., "Epizootics of Wild Fish Induced by Farm Fish," *Proceedings of the National Academy of Sciences* 103, 42 (2006): 15506–15510.

49. Knapp et al., *The Great Salmon Run,* pp. 29, 105, 222.

50. Ibid., pp. 231–232.

51. Chris Carrel, "Killer Salmon," *Seattle Weekly,* 1998.

52. Knapp et al., *The Great Salmon Run,* pp. 84–85, 126, 131; Roseanne Harper, "Salmon Soars Whether Copper River Wild or Tasmanian Tame, Salmon Leaps Past Shrimp as the Seafood of Choice for American Consumers Fishing for Good Health and Great Taste," *Supermarket News,* (2001), p. 29.

53. John Gillie, "Seafood Market Takes Off; Niche Fish Business Pays Off for Alaska Airlines. Restaurants and Businesses across the Country Clamor for Fresh, Wild-Caught Fish. Alaska Airlines Does Its Part to Get Alaskan Seafood to the Masses," *The News Tribune,* October 2, 2005; interview, Shannon Stevens, May 31, 2007.

54. Hot Dish, "Safe Passage for Salmon," *Seattle Weekly,* December 23, 2006. www.seattleweekly.com/food/0520/050518_food_hotdish.php.

55. Charlie Ess, "North Pacific," *National Fisherman* 85 (2005): 18.

56. Renee M. Covino, "A Favorable Fish Forecast: The Time Is Right for Consumers to Fall for Frozen Seafood, Hook, Line and Sinker," *Frozen Food Age* 53: 29; Jill Shepherd, "Wild Salmon, Wily Women," *Alaska* 67 (2001): 38–43.

57. Sasha Issenberg, *The Sushi Economy: Globalization and the Making of a Modern Delicacy* (New York: Gotham, 2007), pp. 12–13, 214–215; Julia Moskin, "Sushi Fresh from the Deep . . . The Deep Freeze," *New York Times,* April 8, 2004. See also Theodore Bestor, *Tsujiki: Fish Market at the Center of the World* (Berkeley: University of California Press, 2004).

58. Eugene N. Anderson and Marja Anderson, "Modern China: South," in *Food in Chinese Culture: Anthropological and Historical Perspectives,* ed. B. Chang (New Haven: Yale University Press, 1977), p. 360.

59. Frederick J. Simoons, *Food in China: A Cultural and Historical Inquiry* (Boca Raton, Fl.: CRC Press, 1991), p. 339; Anderson, *Food of China,* p. 208; I-Ling Kog-Hwang, *Symbolism in Chinese Food* (Singapore: Graham Brash, 1991), p. 56; Lawrence W. C. Lai et al., "Marine Fish Production and Marketing for a Chinese Food Market: A Transaction Cost Perspective," *Aquaculture Economics & Management* 9 (2005): 292.

60. Charles Hugh Stevenson, *The Preservation of Fishery Products for Food* (Washington, D.C.: United States Fish Commission, 1899), pp. 339–342; Stanley Todd, "Bring 'Em to Market Alive!" *Fishing Gazette* 48 (1931): 45–50.

61. Eugene N. Anderson, *The Floating World of Castle Peak Bay* (Washington, D.C.: American Anthropological Association, 1970), p. 64.

62. W. W. L. Cheung and Y. Sadovy, "Retrospective Evaluation of Data-Limited Fisheries: A Case from Hong Kong," *Reviews in Fish Biology and Fisheries* 14 (2004): 181–206.

63. Y. J. Sadovy et al., *While Stocks Last: The Live Reef Food Fish Trade,* Report (Asian Development Bank, 2003), p. 32.

64. Robert Johannes and Michael Riepen, *Environmental, Economic and Social Implications of the Live Reef Fish Trade in Asia and the Western Pacific* (Nature Conservancy, 1995), pp. 11–13.

65. Annie Chan, "Live-in Foreign Domestic Workers and Their Impact on Hong Kong's Middle-Class Families," *Journal of Family and Economic Issues* 26 (2005): 509–528.

66. Martin Wong and Lilian Goh, "Turbot Banned from Fish Markets," *South China Morning Post,* November 23, 2006; P. Redmayne, "China's Giant Aquaculture Industry Experiences Growing Pains," *Aquaculture Magazine* (September–October 2005): 27–30.

67. Otherwise known as *umami,* or the "fifth taste." Julia Moskin, "Yes, MSG, the Secret Behind the Savor," *New York Times,* March 5, 2008.

68. Noel Chan, "An Integrated Attitude Survey on Live Reef Food Fish Consumption in Hong Kong," *SPC Live Reef Fish Information Bulletin* 8 (2000).

69. Thierry Chan, *Proposed Legislative Amendment in Hong Kong: Should Live Fish Be Regarded as Food?* (Hong Kong: Civic Exchange, 2006).

70. Sadovy et al., *While Stocks Last,* p. 69; Don E. McAllister et al., "Cyanide Fisheries: Where Did They Start?" *SPC Live Reef Fish Information Bulletin* 5 (1999); Celia Lowe, "Who Is to Blame? Logics of Culpability in the Live Reef Food Fish Trade in Sulawesi, Indonesia," *SPC Live Reef Fish Information Bulletin* (2002): 7–16.

71. Johannes and Riepen, *Environmental, Economic and Social Implications of the Live Reef Fish Trade;* Mark V. Erdmann and Lida Pet-Soede, "How Fresh Is Too Fresh? The Live Reef Food Fish Trade in Eastern Indonesia," *Naga, The ICLARM Quarterly* (January 1996); Sadovy et al., *While Stocks Last,* pp. 65–68.

72. Sadovy et al., *While Stocks Last,* pp. 65–68.

73. Ibid., ch. 8. WWF-Hong Kong is among the NGOs most active in promoting consumer awareness. A downloadable seafood guide is available on its website: www.wwf.org.hk/eng/index.php

74. Florence Chong, "MCT Fishes Hungry Waters from Top Floor," *Weekend Australian,* March 31, 2007.

Epilogue

1. John Cloud, "Eating Better Than Organic," *Time,* March 2, 2007; Michael Pollan, *The Omnivore's Dilemma: A Natural History of Four Meals* (New York: Penguin Press, 2006), p. 409.

2. Wal-Mart, "Locally Grown Products," walmartstores.com/Sustainability/7985.aspx (accessed June 9, 2008).

3. Donald E. Pitzer, *America's Communal Utopias* (Chapel Hill: University of North Carolina Press, 1997); S. Tarlow, "Excavating Utopia: Why Archaeologists Should Study 'Ideal' Communities of the Nineteenth Century," *International Journal of Historical Archaeology* 6 (2002): 299–323; Krishan Kumar, *Utopianism* (Minneapolis: University of Minnesota Press, 1991).

4. This description of utopian food owes much to Warren James Belasco, *Meals to Come: A History of the Future of Food* (Berkeley: University

of California Press, 2006), pp. 95–118. As Belasco notes, women utopian writers were especially interested in getting rid of domestic drudgery, usually through some combination of technology and socialism. See, for example Charlotte Perkins Gilman, *Herland* (New York: Pantheon Books, 1979); Mary E. Bradley Lane and Jean Pfaelzer, *Mizora: A Prophecy* (Syracuse, N.Y.: Syracuse University Press, [1880] 2000). See also Carol Farley Kessler, *Daring to Dream: Utopian Stories by United States Women, 1836–1919* (Boston: Pandora Press, 1984).

5. Members of the mid-century francophone Icarian community in Nauvoo, Illinois, for example, were dissatisfied with their daily diet, which was abundant (café au lait and eggs for breakfast, meat and vegetables at dinner, soup for supper) but perhaps monotonous. Their leader banned complaints about the food, but some members boycotted it anyway. Robert Sutton, "An American Elysium: The Icarian Communities," in *America's Communal Utopias*, ed. D. E. Pitzer (Chapel Hill: University of North Carolina Press, 1997), pp. 231–244.

6. On the problems of utopias that tried to freeze time in small spaces, see David Harvey, *Spaces of Hope* (Berkeley: University of California Press, 2000).

7. Kim Severson, "Some Good News on Food Prices," *New York Times,* April 2, 2008.

8. On the gentrification of the New England countryside, see Bob Rakoff, "The Changing Meanings of New England Farmland," *New England Watershed* (August–September, 2006): 14–17. For a discussion of inequities within local food movements and how they might be addressed, see C. Clare Hinrichs, "The Practice and Politics of Food System Localization," *Journal of Rural Studies* 19 (2003): 33–45; Patricia Allen, *Together at the Table: Sustainability and Sustenance in the American Agrifood System* (State College: Pennsylvania State University Press, 2004); E. M. DuPuis and D. Goodman, "Should We Go 'Home' to Eat?: Toward a Reflexive Politics of Localism," *Journal of Rural Studies* 21 (2005): 359–371.

9. For an ambitious attempt to link these various crises, see Rajeev Patel, *Stuffed and Starved: The Hidden Battle for the World Food System* (Brooklyn, N.Y.: Melville House, 2008).

Bibliography

Adams, W. Bridges. "Letter on the Preservation of Food," *Journal of the Society of Arts* 13 (1865): 339.

"Advertising Cold Storage," *Ice and Refrigeration* 73 (1927): 347–349.

Alaska Packers' Association. *Interesting Facts about Canned Salmon.* San Francisco: Alaska Packers' Association, 1908.

Albers, Jan. *Hands on the Land: A History of the Vermont Landscape.* Cambridge, Mass.: MIT Press, 2000.

Allen, Patricia. *Together at the Table: Sustainability and Sustenance in the American Agrifood System.* State College: Pennsylvania State University Press, 2004.

Alvord, Henry E., and Raymond A. Pearson. *The Milk Supply of Two Hundred Cities and Towns.* Washington, D.C.: U.S. Department of Agriculture, Bureau of Animal Industry, 1903.

American Warehousemen's Association. "Proceedings of the 25th Annual Meeting of the American Warehousemen's Association," New York, 1915, 233–239.

Anderson, Burton. *America's Salad Bowl: An Agricultural History of the Salinas Valley.* Salinas, Calif.: Monterey County Historical Society, 2000.

Anderson, Eugene N. *The Floating World of Castle Peak Bay.* Washington, D.C.: American Anthropological Association, 1970.

————. *The Food of China*. New Haven: Yale University Press, 1988.

Anderson, Eugene N., and Marja Anderson. "Modern China: South." In *Food in Chinese Culture: Anthropological and Historical Perspectives,* ed. B. Chang. New Haven: Yale University Press, 1977.

Anderson, Oscar E. *Refrigeration in America: A History of a New Technology and Its Impact*. Princeton, N.J.: Princeton University Press, 1953.

Anderson, Virginia DeJohn. *Creatures of Empire: How Domestic Animals Transformed Early America*. New York: Oxford University Press, 2004.

Apple, Rima D. "Science Gendered: Nutrition in the United States, 1840–1940." In *The Science and Culture of Nutrition, 1840–1940,* ed. Harmke Kamminga and Andrew Cunningham. Atlanta: Rodopi, 1995, pp. 129–154.

————. *Vitamania: Vitamins in American Culture*. New Brunswick, N.J.: Rutgers University Press, 1996.

Armour, J. Ogden. *The Packers, the Private Car Lines and the People*. Philadelphia: Henry Altemus, 1906.

Armstrong, Paul. "Sunkist Advertising—How It Sells California Oranges and Lemons," *California Citrograph* (1923): 3–14.

Aspinwall, Grace. "The Joys of Raw Food," *Good Housekeeping* 50 (January 1910): 110–112.

"Assurez la conservation parfaite des oeufs," *Vie à la Campagne* (1911): 283.

"Attractive and Effective Newspaper Advertisements of Ice," *Ice and Refrigeration* 71 (1926): 159–160.

Atwater, W. O. "Pecuniary Economy of Food," *The Century* 25 (1888): 437–446.

Austen, Ralph. *A Treatise of Fruit Trees*. Oxford: Printed by William Hall for Amos Curteyne, 1665. Available at EEBO, Early English Books Online.

Bailey, Kenneth W. *Marketing and Pricing of Milk and Dairy Products in the United States*. Ames: Iowa State University Press, 1997.

Baker, John Randal. "Increasing Winter Egg-Production in Spain More Than a Hundred Years Ago," *Nature* 143 (1936): 477.

Baltet, Charles. *Les fruits populaires*. Paris: Bibliotheque du Jardin, 1889.

Baretta, Silvio R., and John Markoff. "Civilization and Barbarism: Cattle Frontiers in Latin America," *Comparative Studies in Society and History* 20 (1978): 587–620.

Barnett, L. Margaret. "Every Man is His Own Physician: Dietetic Fads, 1890–1914." In *The Science and Culture of Nutrition, 1840–1940,* ed. Harmke Kamminga and Andrew Cunningham, 155–178. Atlanta: Rodopi, 1995.

Barr, Ann, and Paul Levy. *The Official Foodie Handbook: Be Modern— Worship Food.* New York: Timbre Books, 1984.

Barrett, James R. *Work and Community in the Jungle: Chicago's Packing-house Workers, 1894–1922.* Urbana: University of Illinois Press, 1987.

Bassett, T. D. Seymour. "500 Miles of Trouble and Excitement: Vermont Railroads, 1848–1861," *Vermont History* 49 (1981): 133–153.

Baxter, Leonora. "The New Ice Age," *Golden Book Magazine* (January 1931): 83–86.

Beal, George Max, and Henry Harrison Bakken. *Fluid Milk Marketing.* Madison, Wisc.: Mimir Publishers, 1956.

Belasco, Warren James. *Meals to Come: A History of the Future of Food.* Berkeley: University of California Press, 2006.

Bell, Donald. "Forces That Have Helped Shape the U.S. Egg Industry," *Poultry Tribune* (1995): 30–43.

Bengy-Puyvallée, C. A. de. *Mémoire sur la culture du pêcher.* Paris: Librairie Agricole de la Maison Rustique, 1860.

Bentley, Amy Lynn. "Eating for Victory: United States Food Rationing and the Politics of Domesticity during World War Two," Ph.D. diss., University of Pennsylvania, 1992.

Bernardi, John L. "A Century and a Quarter of Milk Contracting in New England: The Hood Company, 1846–1970," Ph.D. diss., University of Pennsylvania, 1971.

Bernstein, Alexis. "The Land of Milk and Money: Philanthropy, Milk and Infant Mortality, 1893–1914," B.A. Thesis, Harvard University, 2005.

Bertram, James Glass. *The Harvest of the Sea: A Contribution to the Natural and Economic History of the British Food Fishes.* New York: Appleton, 1866.

Bestor, Theodore. *Tsujiki: Fish Market at the Center of the World.* Berkeley: University of California Press, 2004.

Bilger, Burkhard. "Salad Days: How a Lowly Leaf Became a High-End Delicacy," *New Yorker,* September 6, 2004, 136+.

Bird, Thomas A. "The Ice Man as an Advertiser," *Ice and Refrigeration* 38 (1910): 144–145.

Birdseye, Clarence. "Preservation of Foods by New Quick Freezing Methods," *Refrigerating Engineering* 25 (1933): 185+.

Birdseye Handbook for Salesmen. New York: Frozen Foods Sales Corporation, 1947.

Blanchard, I. "The Continental European Cattle Trades, 1400–1600," *Economic History Review* 39 (1986): 427–460.

Block, Daniel. "Protecting and Connecting: Separation, Connection and the U.S. Dairy Economy, 1840–2002," *Journal for the Study of Food and Society* 6 (2002): 22–30.

Block, D., and E. M. DuPuis. "Making the Country Work for the City: Von Thünen's Ideas in Geography, Agricultural Economics and the Sociology of Agriculture," *American Journal of Economics and Sociology* 60 (2001): 79–98.

Bonnechaux, Emile. "L'industrie du froid en Asie, Afrique, Australie, et aux Etats-Unis." In *Premier congrès international du froid Paris, 5 au 12 octobre 1908,* ed. International Congress of Refrigeration, under the direction of J. de Loverdo, 482+. Paris: Secrétariat général de l'Association internationale du froid, 1908.

Bonnefons, Nicolas de. *Les délices de la campagne. Suite du jardinier François, ou est enseigné à preparer pour l'usage de la vie tout ce qui croist sur la terre, & dans les eaux.* Paris: Compagnie des marchands libraires du Palais, 1665.

Booher, J. M., ed. *Scientific Weight Control: An Improved System for Reducing or Increasing Weight, Together with an Explanation of the Benefits to Be Gained from Weighing Daily.* Chicago: Chicago Continental Scale Works, 1925.

Boston Chamber of Commerce. *Investigation and Analysis of the Production, Transportation, Inspection and Distribution of Milk and Cream in New England.* Boston: Chamber of Commerce, 1915.

Bournon, Fernand. *Etat des communes à la fin du Xix siècle: Montreuil.* Paris: Conséil General, 1906.

Bové, José, and François Dufour, trans. Anna de Casparis. *The World Is Not for Sale: Farmers against Junk Food.* New York: Verso, 2001.

Bower, Ward Taft. *Alaska Fishery and Fur Seal Industries in 1939.* Washington, D.C.: U.S. Government Printing Office, 1940.

Boxberger, Daniel L. "Ethnicity and Labor in the Puget Sound Fishing Industry, 1880–1935," *Ethnology* 33 (1994): 179–191.

Bradley, Alice. *Electric Refrigerator Menus and Recipes. Recipes Prepared*

Especially for the General Electric Refrigerator. Cleveland, Ohio: General Electric Co., 1927.

Bremer, D. B. "Fruit in Cold Storage," *Ice and Refrigeration* 3 (1892): 33–34.

Briley, George. "A History of Refrigeration," *ASHRAE Journal,* (2004): S31–34.

Broadway, M. J. "Following the Leader: IBP and the Restructuring of Canada's Meatpacking Industry," *Culture & Agriculture* 18 (1996): 3–8.

Brown, David A. "Advertising Ice," *Ice and Refrigeration* 38 (1910): 212–213.

Brown, Dona. *Inventing New England: Regional Tourism in the Nineteenth Century.* Washington, D.C.: Smithsonian Institution Press, 1995.

Browne, Daniel J. *The American Poultry Yard.* New York: C. M. Saxton, 1850.

Browning, Frank. *Apples.* London: Allan Lane, Penguin Press, 1998.

Bruegel, Martin. "How the French Learned to Eat Canned Food, 1809–1930s." In *Food Nations: Selling Taste in Consumer Societies,* ed. Warren Belasco and Philip Scranton, 113–130. New York: Routledge, 2002.

Brun, H. "Les entrepôts frigorifiques," *L'Industrie Frigorifique* 1 (1903): 17–26.

Brunet, Jacques, and Nicole Savard. *Les Savards: histoires des vies.* Paris: Valette Editions, 2006.

Burks, J. D. "Clean Milk and Public Health," *Annals of the American Academy of Political and Social Science* 37 (1911): 192–206.

"California Fruits." *The Horticulturist* 14 (1859): 23–25.

Campbell, Gwyn. *An Economic History of Imperial Madagascar, 1750–1895.* New York: Cambridge University Press, 2005.

Carpenter, Stephanie A. *On the Farm Front: The Women's Land Army in World War II.* DeKalb: Northern Illinois University Press, 2003.

Carrière, E-A. *Montreuil-aux-pêches.* Paris, 1890.

Cato, Marcus, trans. William Davis Hooper. *On Agriculture.* Loeb Classical Library, volume 283. Cambridge, Mass.: Harvard University Press, 1935.

Chan, Annie. "Live-in Foreign Domestic Workers and Their Impact on Hong Kong's Middle-Class Families," *Journal of Family and Economic Issues* 26 (2005): 509–528.

Chan, Noel. "An Integrated Attitude Survey on Live Reef Food Fish Consumption in Hong Kong," *SPC Live Reef Fish Information Bulletin* 8 (2000).

Chan, Thierry. *Proposed Legislative Amendment in Hong Kong: Should Live Fish Be Regarded as Food?* Hong Kong: Civic Exchange, 2006.

Chandler, Alfred Dupont. *The Visible Hand: The Managerial Revolution in American Business.* Cambridge, Mass.: Belknap Press of Harvard University Press, 1977.

Chasan, Daniel Jack. *The Water Link: A History of Puget Sound as a Resource.* Seattle: Washington Sea Grant Program, University of Washington, 1981.

Chase, Alvin Wood. *Dr. Chase's Recipe Book.* Detroit: F. B. Dickerson Company, 1891.

Cheke, Valerie Essex. *The Story of Cheese-Making in Britain.* London: Routledge and Kegan Paul, 1959.

Chemla, Guy. *Les ventres de Paris: les Halles, la Villette, Rungis: l'histoire du plus grand marché du monde.* Grenoble: Glénat, 1994.

Cheung, W. W. L., and Y. Sadovy. "Retrospective Evaluation of Data-Limited Fisheries: A Case from Hong Kong," *Reviews in Fish Biology and Fisheries* 14 (2004): 181–206.

Chipman, N. P. *Report upon the Fruit Industry of California.* California State Board of Trade, 1889.

Christian, Eugene, and Mollie Griswold Christian. *Uncooked Foods & How to Use Them.* New York: The Health-Culture Company, 1904.

Claflin, Kyri. "Culture, Politics and Modernization in Paris Provisioning, 1880–1920," Ph.D. diss., Boston University, 2006.

Clark, Walter E. *The Cost of Living.* Chicago: A. C. McClurg, 1915.

Clemen, Rudolf A. *The American Livestock and Meat Industry.* New York: Ronald Press, 1923.

Cloud, John. "Eating Better Than Organic," *Time,* March 2, 2007.

Cobb, Ernest. *The Hen at Work: A Brief Manual of Home Poultry Culture.* New York: G. P. Putnam's Sons, 1919.

Cobb, John N. *Pacific Salmon Fisheries.* Washington, D.C.: U.S. Government Printing Office, 1917.

"Cold Storage Eggs Good," *Ice and Refrigeration* 40 (1911): 35–36.

"Cold Storage Ordinance," *Ice and Refrigeration* 31 (1906).

"Cold Storage Prejudice Declining," *Ice and Refrigeration* 43 (1912): 56–57.

Collins, J. H. "Vegetables Will Be Dressed Up," *Western Grower and Shipper* 17 (December 1945): 31+.

Combles, Charles-Jean de, and Antoine Boudet. *Traité de la culture des pêchers*. Paris: Chez Antoine Boudet, 1745.

Connolly, Joel. "Difficulties Encountered in the Control of Mechanical Refrigeration," *American Journal of Public Health* 20 (1930): 252–256.

Cook, Hugh L., and George H. Day. *The Dry Milk Industry: An Aid in the Utilization of Milk*. Chicago: American Dry Milk Institute, 1947.

Cooke, Kathy. "From Science to Practice, or Practice to Science? Chickens and Eggs in Raymond Pearl's Agricultural Breeding Research," *Isis* 88 (1997): 62–86.

Cotterill, Ronald W., Adam N. Rabinowitz, Michael A. Cohen, Melanie R. Murphy, and Charles R. Rhodes. *Toward Reform of Fluid Milk Pricing in Southern New England: Farm Level, Wholesale and Retail Prices in the Fluid Milk Marketing Channel: 2003–2006*. Storrs: Food Marketing Policy Center, University of Connecticut, 2007.

Covino, Renee M. "A Favorable Fish Forecast: The Time Is Right for Consumers to Fall for Frozen Seafood, Hook, Line and Sinker," *Frozen Food Age*, 53 (2004): 29.

Cowan, Ruth Schwartz. *More Work for Mother*. New York: Basic Books, 1983.

———. "How the Refrigerator Got Its Hum." In *The Social Shaping of Technology*, ed. Donald Mackenzie and Judy Wajcman, 203–218. Philadelphia: Open University Press, 1985.

Critchell, James, and Joseph Raymond. *A History of the Frozen Meat Trade*. London: Constable and Company, 1912.

Cronon, William. *Nature's Metropolis: Chicago and the Great West*. New York: W. W. Norton, 1991.

Crosby, M. J. "Refrigeration Cookery by Electricity," *Ladies' Home Journal* (1927): 129.

Cruess, William V. *Home and Farm Food Preservation*. New York: Macmillan, 1918.

Cullen, James. "To Ice!" *Ice and Refrigeration* 73 (1927): 162.

Cummings, Richard O. *The American and His Food: A History of Food Habits in the United States*. Chicago: University of Chicago Press, 1941.

Dairy Task Force Report. *Vermont State Agricultural Overview 2005*.

Montpelier: National Agricultural Statistics Service, U.S. Department of Agriculture; Vermont Department of Tourism and Marketing, The Travel and Tourism Industry in Vermont, 2006.

Dale, Edward Everett. *Cow Country.* Norman: University of Oklahoma Press, 1965.

Dalrymple, Dana G. "The Development of an Agricultural Technology: Controlled-Atmosphere Storage of Fruit," *Technology and Culture* 10 (1969): 35–48.

David, Elizabeth, and Jill Norman. *Harvest of the Cold Months: The Social History of Ice and Ices.* New York: Penguin, 1994.

David, Robert. "The Demise of the Anglo-Norwegian Ice Trade," *Business History* 37 (1995): 52–69.

Davidson, James. "Fish, Sex and Revolution in Athens," *Classical Quarterly* 43 (1993): 53–66.

De Voe, Thomas F. *The Market Assistant: Containing a Brief Description of Every Article of Human Food Sold in the Public Markets of the Cities of New York, Boston, Philadelphia, and Brooklyn.* New York: Orange Judd, 1866.

DeWitt, Howard A. "The Filipino Labor Union: The Salinas Lettuce Strike of 1934," *Amerasia Journal* 5 (1978): 1–21.

Dexter, Miriam. "The Housekeeping Club," *Good Housekeeping* 50 (1910): 263–267.

Dobson, W. D. "Competitive Strategies of Leading World Dairy Exporters." Discussion Paper 95–1. Babcock Institute, University of Wisconsin, 1995.

———. "How Mexico's Dairy Industry Has Evolved under the NAFTA— Implications for U.S. Dairy Exporters and Investors in Mexico's Dairy-Food Businesses." Discussion Paper 2002–1. Babcock Institute, University of Wisconsin, 2002.

Donaldson, Barry, and Bernard Nagengast. *Heat and Cold: Mastering the Great Indoors.* Atlanta: American Society of Heating, Refrigerating and Air-Conditioning Engineers, 1994.

Dowsett, C. F. *A Start in Life: A Journey across America: Fruit Farming in California.* London: Dowsett, 1891.

"Dressed Beef," *Ice and Refrigeration* 5 (1894): 397–398.

Drews, George J. *Unfired Food and Hygienic Dietetics: For Prophylactic (Preventive) Feeding and Therapeutic (Remedial) Feeding: (Treats on Food in the Cause, Prevention and Cure of Disease).* Chicago: G. J. Drews, 1909.

Drummond, Jack Cecil, and Anne Wilbraham. *The Englishman's Food: A History of Five Centuries of English Diet*. London: J. Cape, 1958.

Dunthorne, Gordon. *Flower and Fruit Prints of the 18th and Early 19th Centuries, Their History, Makers and Uses, with a Catalogue Raisonné of the Works in Which They Are Found*. Washington, D.C.: The author, 1938.

DuPuis, E. M., and D. Goodman. "Should We Go 'Home' to Eat?: Toward a Reflexive Politics of Localism," *Journal of Rural Studies* 21 (2005): 359–371.

DuPuis, E. Melanie. *Nature's Perfect Food: How Milk Became America's Drink*. New York: New York University Press, 2002.

DuPuis, E. Melanie, and Daniel Block. "Sustainability and Scale: U.S. Milk-Market Orders as Relocalization Policy," *Environment and Planning* 40 (2008): 1987–2005.

Durand, Loyal, Jr. "The Migration of Cheese Manufacture in the United States," *Annals of the Association of American Geographers* 42 (1952): 263–282.

———. "The Historical and Economic Geography of Dairying in the North Country of New York State," *Geographical Review* 57 (1967): 24–47.

Eagle, Josh, Rosamond Naylor, and Whitney Smith. "Why Farm Salmon Outcompete Fishery Salmon," *Marine Policy* 28, 3 (2004): 259–270.

Economics Research Service. "Milk Costs and Returns." Washington, D.C.: U.S. Department of Agriculture, 2005.

"Editorial Notes," *Ice and Refrigeration* 11 (1896): 172.

"Eggs from China: Millions Frozen for Shipment," *Literary Digest* 122 (1936): 38.

Elliot, Charles. "The Preservation of Food," *Journal of the Society of Arts* 9 (1861): 95–97.

Engel, Charles. "Cold in Modern Life: The Importance of Refrigeration," *Scientific American Supplement* 68 (1909): 168.

Engels, Friedrich. *The Condition of the Working Class in England*. New York: Penguin, 1967 [1887].

"Entrepôts frigorifiques sous les tropiques, Les." *La Glace et les industries du froid* 4 (1907).

Erba, E., and A. M Novakovic. *The Evolution of Milk Pricing and Government Intervention in Dairy Markets*. Ithaca, N.Y.: Department of Agricultural, Resource, and Managerial Economics, Cornell University, 1995.

Erdmann, Mark V., and Lida Pet-Soede. "How Fresh Is Too Fresh? The Live Reef Food Fish Trade in Eastern Indonesia," *Naga, The ICLARM Quarterly* (January 1996).

Eskin, N. A. M. "Biochemical Changes in Raw Foods: Fruits and Vegetables." In *Biochemistry of Foods,* ed. N. A. M. Eskin, 69–165. San Diego: Academic Press, 1990.

Ess, Charlie. "North Pacific," *National Fisherman* 85 (2005): 18.

"Étiquetage des fruits conservés par le froid et le phobie du froid, L'." *La Revue Générale du Froid* 3 (1911): 536–538.

Fearne, Andrew, and David Hughes. "Success Factors in the Fresh Produce Supply Chain: Insights from the U.K." *British Food Journal* 102 (2000): 760–772.

Ferber, Edna. "Maymeys from Cuba." In *Buttered Side Down: Stories,* ed. E. Ferber. New York: Grosset and Dunlap, 1911.

Fernández-Armesto, Felipe. *Near a Thousand Tables: A History of Food.* New York: Free Press, 2002.

Ferns, Henry S. *Britain and Argentina in the Nineteenth Century.* Oxford: Clarendon Press, 1960.

Ferriss, Susan, and Ricardo Sandoval. *The Fight in the Fields: Cesar Chavez and the Farmworkers Movement.* New York: Harcourt Brace, 1997.

Fielding, Gordon J. "Dairying in Cities Designed to Keep People Out," *Professional Geographer* 14 (1962): 12–17.

Finlay, Mark. "Early Marketing of the Theory of Nutrition: The Science and Culture of Liebig's Extract of Meat." In *The Science and Culture of Nutrition, 1840–1940,* ed. Harmke Kamminga and Andrew Cunningham, 48–76. Atlanta: Rodopi, 1995.

Fischer, Edward F., and Peter Benson. *Broccoli and Desire: Global Connections and Maya Struggles in Postwar Guatemala.* Stanford, Calif.: Stanford University Press, 2006.

Fitzgerald, D. K. *Every Farm a Factory: The Industrial Ideal in American Agriculture.* New Haven: Yale University Press, 2003.

Fitzsimmons, Margaret. "The New Industrial Agriculture: The Regional Integration of Specialty Crop Production," *Economic Geography* 62 (1986): 334–353.

Flandrin, Jean Louis. "The Early Modern Period." In *Food: A Culinary History from Antiquity to the Present,* ed. J. L. Flandrin, M. Montanari, and A. Sonnenfeld, 349–373. New York: Penguin, 1999.

———. "Seasoning, Cooking and Dietetics in the Late Middle Ages." In *Food: A Culinary History from Antiquity to the Present,* ed. J. L. Flandrin, M. Montanari, and A. Sonnenfeld, 2313–2327. New York: Penguin, 1999.

Fletcher, Horace. *The New Glutton or Epicure.* New York: Frederick A. Stokes, 1906.

Flexner, James. "The Battle for Pure Milk in New York City." In *Is Loose Milk a Health Hazard? The Report of the Commission Appointed by Dr. Shirley W. Wynne, Commissioner of Health of the City of New York, to Study the Public Health Aspects of the Sale of Loose Milk in New York City and to Make Recommendations,* ed. E. F. Brown, 161–196. New York: The Commission, 1931.

Florea, James Howard. "Education." In *American Poultry History 1823–1873,* ed. O. A. Hanke, J. L. Skinner, and J. H. Florea, 50–102. Madison, Wisc.: American Poultry Historical Society, 1974.

Fobes, Natalie. *Reaching Home: Pacific Salmon, Pacific People.* Anchorage: Alaska Northwest Books, 1994.

Foucart, Georges. *Le commerce et la colonisation à Madagascar.* Paris: Augustin Challamel, 1894.

Fox, B. J. "Selling the Mechanized Household: 70 Years of Ads in *Ladies Home Journal,*" *Gender and Society* 4, 1 (1990): 25–40.

Fox, Nicols. *Spoiled: The Dangerous Truth about a Food Chain Gone Haywire.* New York: Basic Books, 1997.

Francisco, Don. *Cooperative Advertising: A Social Service as Well as a Powerful Sales Force.* Los Angeles: California Fruit Growers Exchange, 1920.

———. "How the 'Sunkist' Campaign Was Sold to the Association," *Western Advertising* 1 (1920): 28–32.

———. "The Plans Behind Sunkist Advertising," *California Citrograph* (1920): 2–14.

———. "The Advertising of Agricultural Specialties," *American Cooperation* 2 (1928): 134–148.

Franklin, I. C. "The Service of Cold Storage in the Conservation of Foodstuffs," *Yearbook of Agriculture,* U.S. Department of Agriculture (1917): 363–369.

Frazier, R. T. "The Household Ice Refrigerator," *Refrigerating Engineering* 18 (1929).

Frederick, C. "Your Health Depends on Your Eating—So Why Not Get

Better Acquainted with Fruit?" *Ladies' Home Journal* 36 (February 1919): 53.

Frederick, Christine. "Improving Our Restaurants," *The American Restaurant* (1928): 68.

"Frederic Tudor: Ice King," *Journal of the Business Historical Society* 6, 4 (1932): 1–8.

Freeman, Otis W. "Salmon Industry of the Pacific Coast," *Economic Geography* 11 (1935): 109–129.

Freidberg, Susanne. "Cleaning up Down South: Supermarkets, Ethical Trade and African Horticulture," *Social and Cultural Geography* 4 (2003): 353–368.

———. *French Beans and Food Scares: Culture and Commerce in an Anxious Age*. New York: Oxford University Press, 2004.

French, Michael, and Jim Phillips. *Cheated Not Poisoned? Food Regulation in the United Kingdom, 1875–1938*. Manchester: Manchester University Press, 2000.

Frey, Sherry. "New Rules for Perishables," *FMI Show*, 2007.

Friedland, William H., Amy E. Barton, and Robert J. Thomas. *Manufacturing Green Gold: Capital, Labor, and Technology in the Lettuce Industry*. New York: Cambridge University Press, 1981.

Friedman, B. A. "Vacuum Cooling Upheld in Tests," *Western Grower and Shipper* 23 (July 1952): 21+.

Friedmann, Karen. "Victualling Colonial Boston," *Agricultural History*, 47 (1973): 189–205.

Frigidaire. *Frigidaire Recipes: Prepared Especially for Frigidaire Automatic Refrigerators Equipped with the Frigidaire "Cold Control."* Dayton, Ohio: Frigidaire Corporation, 1929.

"'Frigorifique' á Buenos-Ayres, Le," *L'Illustration*, March 10, 1877, 151.

"Frigoriphobie, Le," *L'Industrie Frigorifique* 11 (1913): 126.

Furlough, Ellen, and Carl Strikwerda. *Consumers against Capitalism?: Consumer Cooperation in Europe, North America, and Japan, 1840–1990*. Lanham, Md.: Rowman and Littlefield, 1999.

Fussell, G. E. *The English Dairy Farmer, 1500–1900*. New York: A. M. Kelley, 1966.

Galarza, Ernesto. *Merchants of Labor: The Mexican Bracero Story; an Account of the Managed Migration of Mexican Farm Workers in California, 1942–1960*. Charlotte, Calif.: McNally and Loftin, 1964.

Gale, Sarah F. "Radlo Foods Hatches a High-Tech Egg Safety Plan," *Food Safety Magazine* 11 (2005/6): 70–74.

Ganzel, Bill. "IBP, Boxed Beef and a New 'Big Four,'" Wessels Living History Farm, www.livinghistoryfarm.org/farmingithe50s/money_17.html.

García Hernández, Luis Arturo. "Skim Milk Powder Imports and the Role of Conasupo," *Role of the State in Agricultural Trade Workshop,* Stanford University, 1998.

Garnier, Edmond. *L'agriculture dans le département de la Seine et le marché Parisien du point de vue ravitaillement alimentaire.* Poitiers: Imprimerie l'Union, 1939.

Gast, Ross H. "My Own Page," *Western Grower and Shipper* (1931).

———. "Competition Comes in Cans, Too," *Western Grower and Shipper* (1937): 8.

———. "Salads as a First Course an Old California Custom," *Western Grower and Shipper* (1938): 11.

———. "History of California Vegetables," *Western Grower and Shipper* 17 (1945): 27+.

Gener, Antoine. "De l'elevage du gros bétail à Madagascar," Faculté de Droit, Université d'Alger, 1927.

Geoffroy, M. "Le froid à Madagascar," *Revue Générale du Froid* 12 (1931): 261–263.

Gilbert, Arthur W. *The Food Supply of New England.* New York: Macmillan, 1924.

Gilman, Charlotte Perkins. *Herland.* New York: Pantheon Books, 1979.

Glamann, Kristoff. "The Cattle Trades." In *The Cambridge Economic History of Europe,* ed. M. M. Postan and H. J. Habakkuk, 232–240. Cambridge: Cambridge University Press, 1966.

Glass, Judith C. "Conditions Which Facilitate Unionization of Agricultural Workers: A Case Study of the Salinas Valley Lettuce Industry," Ph.D. diss., University of California, Los Angeles, 1966.

Gollner, Adam. *The Fruit Hunters: A Story of Nature, Adventure, Commerce, and Obsession.* New York: Scribner, 2008.

Godfrey, Hollis. *The Health of the City.* Boston: Houghton Mifflin, 1910.

Gontard, Maurice. *Madagascar pendant la Première Guerre Mondiale* (Tananarive: Imprint Société malgache d'édition, 1969).

Goodrich, Helen E. "Processed Foods Gain Vital Issue," *Western Grower and Shipper* 22 (1951): 25+.

Goudiss, C. Houston. *Eating Vitamines: How to Know and Prepare the Foods that Supply These Invisible Life-Guards, with Two Hundred*

H. J. Heinz Company. *Heinz Book of Salads.* Pittsburgh: 1925.

Holmes, George. "Consumers' Fancies." In *1904 Yearbook of Agriculture,* U.S. Department of Agriculture (1905): 417–434.

Horne, Frank A. "Legislation Affecting Cold Storage and Cold Stored Products," *Ice and Refrigeration* 41 (1911): 180–183.

Horowitz, Roger. *Putting Meat on the American Table: Taste, Technology, Transformation.* Baltimore: Johns Hopkins University Press, 2006.

Houllevigue, L. "Causerie scientifique: le congrès du froid," *Journal du Syndicat de la Boucherie de Paris* (1912): 2.

Hughes, Dora Morrell. *Thrift in the Household.* Boston: Lothrop, Lee and Shepard, 1918.

Humphrys, John. *The Great Food Gamble.* London: Hodder and Stoughton, 2002.

Huss, H. H. "Quality and Quality Changes in Fresh Fish," *Food and Agriculture Organization of the United Nations,* Technical Paper No. 348 (1995).

Hutchinson, Woods. "The Physical Basis of Brain-Work," *North American Review* 146 (1888): 522–531.

"Ice: How Much of It Is Used, and Where It Comes From," *De Bow's Review* 19 (1955): 709–712.

"Ice and the Ice Trade," *Hunt's Merchants' Magazine* (1855): 175.

"Iceberg Lettuce on the Air," *Western Grower and Shipper* (1937): 9.

"Ice in Europe," *Ice and Refrigeration* 3 (1892): 359–362.

Isenstadt, Sandy. "Visions of Plenty: Refrigerators in America around 1950," *Journal of Design History* 11 (1998): 311–321.

Ishige, N. *The History and Culture of Japanese Food.* London: Kegan Paul, 2001.

Issenberg, Sasha. *The Sushi Economy: Globalization and the Making of a Modern Delicacy.* New York: Gotham, 2007.

"It's a Process—Not a Fad," *Western Grower and Shipper* (1946): 8–9.

J.C.C. "Ice Made by Mechanical Power," *Scientific American,* September 22, 1849, 3.

J. Walter Thompson Company. "Presentation on California Fresh Bartlett Pears." Giannini Foundation for Agricultural Economics, University of California, Berkeley, 1936.

Jackson, W. Turrentine. "British Interests in the Range Cattle Industry." In *When Grass Was King,* ed. Maurice Frink, 135–334. Boulder: University of Colorado Press, 1956.

Janson, H. Frederic. *Pomona's Harvest: An Illustrated Chronicle of Antiquarian Fruit Literature.* Portland, Oreg.: Timber Press, 1996.

Jasper, William. *Poultry Farm Practices and Egg Quality.* U.S. Department of Agriculture Production and Marketing Administration. Marketing Research Report 22, 1952.

———. "Marketing." In *American Poultry History 1823–1873,* ed. O. A. Hanke, J. L. Skinner, and J. H. Florea. Madison, Wisc.: American Poultry Historical Society, 1974, pp. 306–369.

Jastrzembski, Benjamin. "The New Braceros: Mexican Dairy Workers in Vermont." B.A. thesis, Dartmouth College, 2008.

Jelinek, Lawrence J. *Harvest Empire: A History of California Agriculture.* San Francisco: Boyd and Fraser, 1982.

Jellison, Katherine. *Entitled to Power: Farm Women and Technology, 1913–1963.* Chapel Hill, N.C.: University of North Carolina Press, 1993.

Johannes, Robert, and Michael Riepen. *Environmental, Economic and Social Implications of the Live Reef Fish Trade in Asia and the Western Pacific.* Nature Conservancy, 1995.

Johns, Michael. "Industrial Capital and Economic Development in Turn of the Century Argentina," *Economic Geography* 68 (1992): 188–204.

———. "The Antinomies of Ruling Class Culture: The Buenos Aires Elite, 1880–1910," *Journal of Historical Sociology* 6 (1993): 74–101.

Johnson, Joshua. *The Art of Cheese Making Reduced to Rules, and Made Sure and Easy, from Accurate Observation & Experience, Published for the Help of Dairy Women.* Albany, N.Y.: Charles R. and George Webster, 1801.

Johnson, Nathanael. "The Revolution Will Not Be Pasteurized: Inside the Raw Milk Underground," *Harper's* (April 2008): 71–78.

Jones, H. A. *The Head-lettuce Industry of California.* Berkeley, Calif.: 1932.

Jones, William O. "The Salinas Valley: Its Agricultural Development," Ph.D. diss., Stanford University, 1947.

———. "A Case Study in Risk Distribution: The California Lettuce Industry," *Journal of Farm Economics* 33 (1951): 235–241.

Joret. "Le transport des produits coloniaux au moyens du froid," *L'Industrie Frigorifique* 1 (1903): 27–30.

Jull, M. A., Lawrence W. C. Lai, K. W. Chau, S. K. Wong, N. Matsuda, and F. T. Lorne. "The Poultry Industry." In *Yearbook of the Department of Agriculture,* U.S. Department of Agriculture (1925): 377–456.

Kamminga, Harmke, and Andrew Cunningham, eds. *The Science and Culture of Nutrition, 1840–1940.* Atlanta: Rodopi, 1995.

Kann, Kenneth L. *Comrades and Chicken Ranchers: The Story of a California Jewish Community.* Ithaca, N.Y.: Cornell University Press, 1993.

Kelvinator. *New Delights from the Kitchen.* Detroit, Mich.: 1930.

Kennard, D. C., and V. D. Chamberlin. *All-Night Light for Layers.* Bulletin 476, Ohio Agricultural Experiment Station, 1931.

Kessler, Carol Farley. *Daring to Dream: Utopian Stories by United States Women, 1836–1919.* Boston: Pandora Press, 1984.

Kilbourn, J. K. *Fish Preservation and Refrigeration.* London: W. Clowes and Sons, 1883.

Kimball, E. Smith. "Characteristics of U.S. Poultry Statistics," *Journal of Farm Economics* 22 (1940): 359–366.

Kindstedt, Paul. *American Farmstead Cheese.* White River Junction, Vt.: Chelsea Green, 2005.

Kingsley, M. "Household Refrigeration Bureau," *Ice and Refrigeration* 70 (1926): 376–377.

Kingsolver, Barbara, with Stephen L. Hopp and Camille Kingsolver. *Animal, Vegetable, Miracle: A Year of Food Life.* New York: HarperCollins, 2007.

Knapp, Gunnar, et al. *The Great Salmon Run: Competition between Wild and Farmed Salmon.* Washington, D.C.: TRAFFIC North America, 2007.

Knightley, Phillip. *The Rise and Fall of the House of Vestey: The True Story of How Britain's Richest Family Beat the Taxman and Came to Grief.* London: Warner Books, 1993.

———. "Curse of the Spam Clan," *Daily Mail,* August 14, 1999.

Kog-Hwang, I-Ling. *Symbolism in Chinese Food.* Singapore: Graham Brash, 1991.

Kriger, T. J. "Syndicalism and Spilled Milk: The Origins of Dairy Farmer Activism in New York State, 1936–1941," *Labor History* 38 (1997): 266–286.

Krkošek, Martin, Mark A. Lewis, Alexandra Morton, L. Neil Fraser, and John P. Volpe. "Epizootics of Wild Fish Induced by Farm Fish,"

Proceedings of the National Academy of Sciences, 103, 42 (2006) 15506–15510.

Kujovich, Mary Yeager. "The Refrigerator Car and the Growth of the American Dressed Beef Industry," *Business History Review* 44 (1970): 460–482.

Kumar, Krishan. *Utopianism.* Minneapolis: University of Minnesota Press, 1991.

Kupferman, Eugene. "The Early Beginnings of Controlled Atmosphere Storage," *Post Harvest Pomology Newsletter* 7 (1989): 3–4.

Kurlansky, Mark. *Cod: A Biography of the Fish That Changed the World.* New York: Penguin Books, 1998.

———. *Salt: A World History.* New York: Walker and Company, 2002.

Lafond, Georges. *L'industrie frigorifique Argentine et la crise de "la vie chere."* Paris: Société d'Editions Internationales, 1912.

Lai, Lawrence W. C., et al. "Marine Fish Production and Marketing for a Chinese Food Market: A Transaction Cost Perspective," *Aquaculture Economics & Management* 9 (2005): 289–316.

Lamb, Helen B. "Industrial Relations in the Western Lettuce Industry," Ph.D., Harvard University, 1942.

Landeen, Dan, and Allen Pinkham. *Salmon and His People: Fish and Fishing in Nez Perce Culture.* Lewiston, Me.: Confluence Press, 1999.

Lane, Mary E. Bradley, and Jean Pfaelzer. *Mizora: A Prophecy.* Syracuse, N.Y.: Syracuse University Press, (1880) 2000.

Langlois, Hippolyte. *Le livre de Montreuil-aux-pêches.* Paris: Librairie Firmin-Didot, 1875.

Langworthy, C. F. "Food and Diet in the United States," *Yearbook of Agriculture,* U.S. Department of Agriculture (Washington, D.C.: Government Printing Office, 1907): 361–378.

———. "Green Vegetables and Their Uses in the Diet," *Yearbook of Agriculture,* U.S. Department of Agriculture (Washington, D.C.: Government Printing Office, 1911): 439–452.

Latour, Bruno. *The Pasteurization of France.* Cambridge, Mass.: Harvard University Press, 1988.

Leahy, S. C., D. G. Higgins, G. F. Fitzgerald, and D. van Sinderen. "Getting Better with Bifidobacteria," *Journal of Applied Microbiology* 98 (2005) 1303–1315.

Lears, T. J. Jackson. "American Advertising and the Reconstruction of the Body, 1880–1930." In *Fitness in American Culture: Images of*

Health, Sport, and the Body, 1830–1940, ed. Kathryn Grover, 47–66. Amherst: University of Massachusetts Press, 1989.

———. *Fables of Abundance: A Cultural History of Advertising in America.* New York: Basic Books, 1994.

———. *No Place of Grace: Antimodernism and the Transformation of American Culture, 1880–1920.* Chicago: University of Chicago Press, 1994.

———. "From Salvation to Self-Realization: Advertising and the Therapeutic Roots of the Consumer Culture, 1880–1930," *Advertising & Society Review* 1 (2000).

Legrand d'Aussy, Pierre Jean-Baptiste. *Histoire de la vie privée des Français: depuis l'origine de la nation jusqu'à nos jours.* Paris: P. D. Pierres, 1782.

Le Roy, G. *La mort de Charles Tellier: ses obseques.* Paris: Association Francaise du Froid, 1913.

Lesage, Robert. *Charles Tellier, le père du froid.* Paris: A. Giraudon, 1928.

Levenstein, Harvey A. "'Best for Babies' or 'Preventable Infanticide'? The Controversy over Artificial Feeding of Infants in America, 1880–1920," *Journal of American History* 70 (1983): 75–94.

———. *Paradox of Plenty: A Social History of Eating in Modern America.* New York: Oxford, 1993.

———. *Revolution at the Table: The Transformation of the American Diet.* Berkeley: University of California Press, 2003.

Lewis, Harry R. *Productive Poultry Husbandry.* Philadelphia: J. B. Lipincott, 1919.

Lichatowich, Jim. *Salmon without Rivers: A History of the Pacific Salmon Crisis.* Washington, D.C.: Island Press, 2001.

Liebman, Ellen. *California Farmland: A History of Large Agricultural Landholdings.* Totowa, N.J.: Rowman and Allanheld, 1983.

"Locally Grown Products," Wal-Mart, http://walmartstores.com/ Sustainability/7985.aspx. (June 9, 2008).

Locamus, P. *Madagascar et l'alimentation Européenne: céréales et viandes.* Paris: Augustin Challamel, 1896.

———. *Madagascar et ses richesses.* Paris: Augustin Challamel, 1896.

Logan, Adrian E. "Dairying in Vermont: Farming and the Changing Face of Agriculture, 1945–1992." Ph.D. diss., University of Vermont, 1998.

Loiseau, Léon. *De l'ensachage des fruits.* Paris: Librairie et imprimerie horticoles, 1903.

———. "Du role des appareils frigorifiques dans la conservation des fruits." *Industrie Frigorifique* 4 (1906): 334–338.

Losada, H., R. Bennett, J. Cortés, J. Vieyra, and R. Soriano. "The Mexico City Milk Supply System: Structure, Function, and Sustainability," *Agriculture and Human Values* 18 (2001): 305–317.

Losada, H., J. Cortés, D. Grande, J. Rivera, R. Soriano, J. Vieyra, A. Fierro, and L. Arias. "The Production of Milk from Dairy Herds in the Suburban Conditions of Mexico City: The Case of Iztapalapa," *Livestock Research for Rural Development* 8 (1996).

Loverdo, J. de. *Le froid artificiel et ses applications industrielles, commerciales et agricoles.* Paris: Dunod, 1903.

———. *Monographie sur l'etat actuel de l'industrie du froid en France.* Paris: Association Francaise du Froid, 1910.

Lowe, Celia. "Who Is to Blame? Logics of Culpability in the Live Reef Food Fish Trade in Sulawesi, Indonesia," *SPC Live Reef Fish Information Bulletin* (2002): 7–16.

Luce, Robert, *Report on the Commission of the Cost of Living.* Boston: Wright and Potter Printing Company, 1910.

Lucier, Gary, Susan Pollack, Mir Ali, and Agnes Perez. *Fruit and Vegetable Backgrounder.* VGS-313-012222. Economic Research Service, U.S. Department of Agriculture, 2006.

Lysaght, Patricia. *Milk and Milk Products from Medieval to Modern Times: Proceedings of the Ninth International Conference on Ethnological Food Research, Ireland, 1992.* Edinburgh: Canongate Academic Press, 1994.

Macdonald, James. *Food from the Far West.* London: W. P. Nimmo, 1878.

MacNutt, Joseph Scott. *The Modern Milk Problem in Sanitation, Economics, and Agriculture.* New York: Macmillan, 1917.

Mainwaring, W. F. B. Massey. *The Preservation of Fish Life in Rivers by the Exclusion of Town Sewage.* London: W. Clowes and Sons, 1883.

Manchester, Alden Coe. *The Public Role in the Dairy Economy: Why and How Governments Intervene in the Milk Business.* Boulder, Colo.: Westview Press, 1983.

Manchester, Alden C., and Don P. Blayney. *Milk Pricing in the United States.* Agriculture Information Bulletin 761. Market and Trade Economics Division, Economic Research Service, U.S. Department of Agriculture, 2001.

"Manifestion du 15 fevrier en l'honneur du Charles Tellier," *Industrie Frigorifique* 11 (1913): 67–73.

Marchand, Roland. *Advertising the American Dream: Making Way for Modernity, 1920–1940*. Berkeley: University of California Press, 1985.

Martin, A. L. "Who'll Do the Pre-Packaging?" *Western Grower and Shipper* 17 (1946): 88+.

Martin, Robert. "Fish Kept Fresh 1500 Miles from the Sea by Scientific Refrigeration," *Popular Science Monthly* 115 (1929): 57–58.

Marx, Leo. *The Machine in the Garden: Technology and the Pastoral Ideal in America*. New York: Oxford University Press, 1964.

Massachusetts Commission on Cold Storage of Food. *Report of the Commission to Investigate the Subject of the Cold Storage of Food and of Food Products Kept in Cold Storage*. Boston: 1912.

Massé, Alfred. *Le troupeau français et la guerre: viande indigène, viande importée*. Paris: Librairie Agricole de la Maison Rustique, 1915.

Maumené, Albert. "Ni fardage, ni maquillage: la beauté vraie," *Vie à la Campagne*, June 15, 1934.

McAllister, Don E., Ning L. Caho, and C.-T. Shih. "Cyanide Fisheries: Where Did They Start?" *SPC Live Reef Fish Information Bulletin 5* (1999).

McCann, Alfred W. *Thirty Cent Bread: How to Escape a Higher Cost of Living*. New York: George H. Doran, 1917.

McClelland, Gordon T., and Jay T. Last. *California Orange Box Labels: An Illustrated History*. Beverly Hills, Calif.: Hillcrest Press, 1985.

McCollum, Elmer Verner. "What to Teach the Public Regarding Food Values?" *Journal of Home Economics* 10 (1918): 195–206.

———. *A History of Nutrition; the Sequence of Ideas in Nutrition Investigations*. Boston: Houghton Mifflin, 1957.

McCollum, Elmer Verner, Elsa Orent-Keiles, and Harry G. Day. *The Newer Knowledge of Nutrition*. 5th ed. New York: Macmillan, 1939.

McCollum, Elmer Verner, and Nina Simmonds. *The American Home Diet: An Answer to the Ever Present Question, What Shall We Have for Dinner*. Detroit: Frederick C. Mathews, 1920.

McComas, Paul S. "The New England Dairy Industry, with Special Reference to Inter-Regional Competition," Ph.D. diss., Harvard University, 1947.

McCormick, Finbar. "The Distribution of Meat in a Hierarchical Society: The Irish Evidence." In *Consuming Passions and Patterns of Con-*

sumption, ed. P. Miracle and N. Milner, 25–31. Cambridge, England: McDonald Institute for Archaeological Research, 2002.

McDonald, James. "Reconfiguring the Countryside: Power, Control, and the (Re) Organization of Farmers in West Mexico," *Human Organization* 60 (2001): 247–258.

———. "NAFTA and Basic Food Production: Dependency and Marginalization on Both Sides of the U.S./Mexico Border." In *Food in the USA: A Reader,* ed. C. Counihan, 359–372. New York: Routledge, 2002.

McGee, Harold. *On Food and Cooking: The Science and Lore of the Kitchen.* New York: Scribner, 2004.

McMahon, Sarah F. "A Comfortable Subsistence: The Changing Composition of Diet in Rural New England, 1620–1840," *The William and Mary Quarterly* 42 (1985): 26–65.

———. "'All Things in Their Proper Season': Seasonal Rhythms of Diet in Nineteenth-Century New England," *Agricultural History* 63 (1989): 130.

McWilliams, Carey. *Factories in the Field: The Story of Migratory Farm Labor in California.* Boston: Little, Brown and Company, 1939.

Menalque, E. "Les frigorifiques dans les abbatoirs," *Industrie Frigorifique* 2 (1904): 273–278.

Meusy, J.-J., ed. *La Bellevilloise (1877–1939): un page de l'histoire de la coopération et du mouvement ouvrier français.* Paris: Créaphis, 2001.

Miller, Julie. "To Stop the Slaughter of the Babies: Nathan Straus and the Drive for Pasteurized Milk," *New York History* 73 (1993): 159–184.

Miller, O. D. "'All Work and No Play—,'" *Western Grower and Shipper* 1 (1930): 14.

Mintz, Sidney. *Sweetness and Power: The Place of Sugar in Modern History.* New York: Penguin, 1986.

Mishra, Vijay K., and T. V. Gamage. "Post-Harvest Physiology of Fruit and Vegetables." In *Handbook of Food Preservation,* 2nd ed., ed. M. Shafiur Rahman, 19–48. Taylor and Francis, 2007.

Mitchell, Don. *The Lie of the Land: Migrant Workers and the California Landscape.* Minneapolis: University of Minnesota Press, 1996.

"Model Abattoir, A," *Ice and Refrigeration Illustrated* 8 (1895).

Montgomery, David R. *King of Fish: The Thousand-Year Run of Salmon.* Boulder, Colo.: Westview Press, 2003.

Montgomery, E. G. *Points Brought out in the Canned Food Survey.* Washington, D.C.: U.S. Department of Commerce, 1926.

Montgomery, E. G., and C. H. Kardell. *Apparent Per Capita Consumption of Principal Foodstuffs in the United States.* Department of Commerce, Domestic Commerce Series 38. Washington, D.C., 1930.

Mooallem, Jon. "Twelve Easy Pieces," *New York Times Magazine,* February 12, 2006.

Moore, C. B. "Consumer Packaging and Research," *Western Grower and Shipper* (1947): 22+.

———. "Labor Trends," *Western Grower and Shipper* (1950): 25+.

———. "'Evolution' of Packing Industry." *Western Grower and Shipper* (July 1953).

Moore, Jason W. "The Modern World-System as Environmental History? Ecology and the Rise of Capitalism," *Theory and Society* 32 (2003): 307–377.

Moriniere, Noel de la, and Simon Bartholemy Joseph. *Histoire générale des pêches anciennes et modernes, dans les mers et les fleuves des deux continents.* Paris: Impr. royale, 1815.

Moses, H. Vincent. "G. Harold Powell and the Corporate Consolidation of the Modern Citrus Enterprise, 1904–1922," *Business History Review* 69 (1995): 119–155.

Moussu, M. "L'exploitation de la richesse en gros bétail dans nos colonies Africaines (Madagascar, Ouest Africaine) par l'industrie frigorifique," *Industrie Frigorifique* 12 (1914): 3–8.

Moyer, Judith. "From Dairy to Doorstep: The Processing and Sale of New Hampshire Dairy Products, 1860s to 1960s," *Historical New Hampshire* 58 (2003): 101–122.

Myers, R. Kelly. *Survey Report of Current Use Practices in New Hampshire.* Durham, N.H.: Institute for Policy and Social Science Research, 1993.

Nason, Arthur L. *Report of the Commission to Investigate the Subject of the Cold Storage of Food and of Food Products Kept in Cold Storage.* Boston: Wright and Potter Printing, 1912.

"National Educational and Goodwill Advertising Campaign for the Ice Industry, The," *Ice and Refrigeration* 73 (1927): 158–161.

"National Educational Publicity for Ice," *Ice and Refrigeration* 70 (1926): 546–547.

"Nation-wide Newspaper Campaign Now Underway," *Western Grower and Shipper* 1 (February 1930): 12–13.

Naylor, Rosamond L., Rebecca J. Goldburg, Jurgenne H. Primavera, Nils Kautsky, Malcolm C.M. Beveridge, Jason Clay, Carl Folke, Jane

Lubchenco, Harold Mooney, and Max Troell. "Effect of Aquaculture on World Fish Supplies," *Nature* 405 (2000): 1017–1024.

Naylor, Rosamond, Kjetil Hindar, Ian A. Fleming, Rebecca Goldburg, John Volpe, Fred Whoriskey, Josh Eagle, Dennis Kelso, Marc Mangel, and Susan Williams. "Fugitive Salmon: Assessing the Risks of Escaped Fish from Net-Pen Aquaculture," *BioScience* 55 (2005): 427–437.

Neff, Peter. "Domestic Refrigerating Machine," *Ice and Refrigeration* 49 (1915): 143–144.

Neil, M. "The Vegetables You Have Grown," *Ladies' Home Journal* (October 1917): 30.

Nerdeux, Leon. "Les nouvelles installations dans l'alimentations et l'industrie fruitiere," *Industrie Frigorifique* 4 (1906): 244–245.

Nestle, Marion. *What to Eat.* New York: North Point Press, 2006.

"New Look for Lettuce, A," *Western Grower and Shipper* 31 (1962): 11+.

Newman, L. F. "Some Notes on Foods and Dietetics in the Sixteenth and Seventeenth Centuries," *The Journal of the Royal Anthropological Institute of Great Britain and Ireland* 76 (1946): 39–49.

Nierenberg, Danielle. *Happier Meals: Rethinking the Global Meat Industry.* WorldWatch paper. Washington, D.C.: WorldWatch, 2005.

Novakovic, Andrew. "Pricing Milk in the U.S.—Spatial Values of Plant Pay Prices and Price Received by Dairy Farmers," National Workshop for Dairy Economists and Policy Analysts, Charleston, S.C., 2007.

"Novels, Plays, Books, Radio Spotlight California's Agricultural Labor Problems," *Western Grower and Shipper* (1939): 14+.

Nutting, William R. *California Views in Natural Colors.* San Francisco: California View Publishing, 1889.

Nye, David E. *Electrifying America: Social Meanings of a New Technology, 1880–1940.* Cambridge, Mass.: MIT Press, 1990.

Oakes, Elinor. "A Ticklish Business: Dairying in New England and Pennsylvania, 1750–1812," *Pennsylvania History* 47 (1980): 195–212.

Ochoa, Enrique. *Feeding Mexico: The Political Uses of Food since 1910.* Wilmington, Del.: Scholarly Resources, 2000.

O'Connell, Jock, Bert Mason, and John Hagen. *The Role of Air Cargo in California's Agricultural Export Trade.* Fresno: Center for Agricultural Business, California State University, 2005.

O'Keefe, W. M. "Cold Storage Division A. W. A," *Ice and Refrigeration* 78 (1930): 513–515.

Ortiz, Paul. "From Slavery to Cesar Chavez and Beyond: Farmworker Organizing in the United States." In *The Human Cost of Food: Farmworkers' Lives, Labor, and Advocacy*, ed. C. D. Thompson and M. Wiggins, 249–276. Austin: University of Texas Press, 2002.

"Our Culinary Vegetables," *Hovey's American Gardener* 31, 8 (1865): 225–233.

"Pacific Coast Fish," *Fortune* 11 (1935): 106+.

"Packaging: Who Wants What?" *Western Grower and Shipper* (1955): 14+.

Padfield, Harland, and William Edwin Martin. *Farmers, Workers and Machines: Technological and Social Change in Farm Industries of Arizona*. Tucson: University of Arizona Press, 1965.

Park, W. H., and L. E. Holt. "Report upon the Results with Different Kinds of Pure and Impure Milk in Infant Feeding in Tenement Houses and Institutions of New York City: A Clinical and Bacteriological Study," *Archives of Pediatrics* 20 (1903): 881–910.

Parker, Horatio Newton. *City Milk Supply*. New York: McGraw-Hill, 1917.

Parrish, F. C. Aging of Beef. Chicago: National Cattlemen's Beef Association. www.askthemeatman.com/dry_aged_beef.htm.

Patel, Rajeev. *Stuffed and Starved: The Hidden Battle for the World Food System*. Brooklyn, N.Y.: Melville House, 2008.

Pearson, Andrea G. "'Frank Leslie's Illustrated Newspaper' and 'Harper's Weekly': Innovation and Imitation in Nineteenth-Century American Pictorial Reporting," *Journal of Popular Culture* 23 (1990): 81.

Pearson, H. "Frederic Tudor, Ice King," *Proceedings of the Massachusetts Historical Society* 65 (1933): 183.

Peneau, S., E. Hoehn, H.-R. Roth, F. Escher, and J. Nuessli. "Importance and Consumer Perception of Freshness of Apples," *Food Quality and Preference* 17 (2006): 9–19.

Pennington, Mary E. "Relation of Cold Storage to the Food Supply and the Consumer," *Annals of the American Academy of Political and Social Science* 48 (1913): 154–163.

———. "Address to National Convention of the United Master Butchers of America," *Ice and Refrigeration* 59 (1920): 98.

———. "Better Food for the Masses," *Ice and Refrigeration* 75 (1928): 33–35.

———. *Journeys with Refrigerated Foods: Eggs.* Chicago: National Association of Ice Industries, 1928.

———. "What Is an Egg?" In *Eggs,* ed. P. Mandeville, 37–48. Chicago: Progress Publications, 1933.

———. "Fifty Years of Refrigeration in the Egg and Poultry Industry," *Ice and Refrigeration* 101 (1941): 43–48.

Pennington, M. E., H. C. Pierce, and H. L. Shrader. "The Egg and Poultry Demonstration Car Work in Reducing Our 50,000,000 Waste in Eggs." In *Yearbook of Agriculture,* U.S. Department of Agriculture (1914): 363–379.

Pentzer, W. T. "The Giant Job of Refrigeration," *Yearbook of Agriculture: Protecting Our Food,* ed. J. Hayes, U.S. Department of Agriculture (1966): 123–138.

"Perishable Produce," *Progressive Grocer Associates,* May 1, 1997.

Perkin, M. R. "Unpasteurized Milk: Health or Hazard?" *Clinical & Experimental Allergy* 37 (2007): 627–630.

Perren, Richard. *The Meat Trade in Britain, 1840–1914.* London: Routledge and Kegan Paul, 1978.

Petaluma Chamber of Commerce. *Petaluma, Sonoma County, California: The Largest Poultry Center in the World.* Petaluma: City of Petaluma, 1916.

Petrick, Gabriella. "'Like Ribbons of Green and Gold': Industrializing Lettuce and the Quest for Quality in the Salinas Valley, 1920–1965," *Agricultural History* 80 (2006): 269–295.

Pierce, Bessie Louise, and Joe Lester Norris. *As Others See Chicago: Impressions of Visitors, 1673–1933.* Chicago, Ill.: University of Chicago Press, 2004.

Piettre, Maurice. *Les bases d'un grand elevage colonial.* Paris: Association Colonies Sciences, 1929.

———. *L'avenir des industries animales aux colonies.* Centre de Perfectionnement Technique, 1944.

Pillsbury, Richard. *No Foreign Food: The American Diet in Time and Place.* Boulder, Colo.: Westview Press, 1998.

Pisani, Donald J. *From the Family Farm to Agribusiness: The Irrigation Crusade in California and the West, 1850–1931.* Berkeley: University of California Press, 1984.

Pitzer, Donald E. *America's Communal Utopias.* Chapel Hill: University of North Carolina Press, 1997.

"Plans for Advertising the Ice Industry in 1928," *Ice and Refrigeration* 73 (1927): 211–218.

Platt, Frank. "Poultry Keeping: An Art, a Science, an Industry." In *Eggs,* ed. P. Mandeville, 49–170. Chicago: Progress Publications, 1933.

Pollan, Michael. *The Omnivore's Dilemma: A Natural History of Four Meals.* New York: Penguin Press, 2006.

———. *In Defense of Food: An Eater's Manifesto.* New York: Penguin Press, 2008.

Ponsford, A. W. "Refrigeration in the Tuna Fishing Industry," *Ice and Refrigeration* 82 (1932): 23–25.

"Power Cuts Growing Costs," *Western Grower and Shipper* (1932): 9.

Prescott, Samuel. "The Milk Supply of Boston," *Science Conspectus* 2 (1911): 22–28.

Prilleray, A. "La refrigeration et l'approvisionnement en fruits du marché de Londres," *Industrie Frigorifique* 2 (1904).

———. "Transports frigorifiques des fruits coloniaux," *Industrie Frigorifique* 3 (1905): 123–127.

Prugh, A. E. "Why Western Vegetables Are Popular," *Western Grower and Shipper* 1 (August 1930): 8+.

Putnam, George. *Supplying Britain's Meat.* London: George Harrap, 1923.

Putnam, Judy, and Jane Allshouse. "Trends in U.S. Per Capita Consumption of Dairy Products, 1909–2001," *Amber Waves* 1 (2003) 12–13.

Quellier, Florent. *Des fruits et des hommes: l'arboriculture fruitière en Ile-de-France.* Rennes: Presses Universitaires de Rennes, 2003.

"Quick Chill for Salinas Lettuce." *P. G. & E. Progress* 32, 9 (1955).

Rakoff, Bob. "The Changing Meanings of New England Farmland," *New England Watershed* (August–September 2006): 14–17.

Randall, Richard. "Making Markets for 1925 Crops of Prunes and Apricots," *Western Advertising* 1 (1920): 7+.

"Rate of Production Mounting," *American Egg and Poultry Review* (1941): 106.

Read, F. W. "'Blue Anchor' Grade Specifications, Season 1927," *Blue Anchor* 4 (1927): 7+.

———. "The Objectives of Standardization," *Blue Anchor* (1928): 9+.

"Real Issue, The," *Western Grower and Shipper* (1936): 5.

Redmayne, P. "China's Giant Aquaculture Industry Experiences Growing Pains," *Aquaculture Magazine* (September–October 2005): 27–30.

Redouté, Pierre Joseph, and J. A. Guillemin. *Choix des plus belles fleurs: prises dans différentes familles du ráegne végétal et de quelques branches des plus beaux fruits.* Paris: Chez l'auteur, 1827.

Rees, Jonathan. "'I Did Not Know . . . Any Danger Was Attached': Safety Consciousness in the Early American Ice and Refrigeration Industries," *Technology and Culture* 46 (2005): 541–560.

"Refrigeration Abroad," *Ice and Refrigeration* 26 (1904).

"Réfrigération des fruits et les préjugés courants, La," *La Revue Générale du Froid* 2 (1910): 154–155.

Renner, H. D. *The Origin of Food Habits.* London: Faber and Faber, 1944.

"Reprint of 'Conseil General de la Seine Paris, seance du 21 Juin,'" *La Revue Générale du Froid* 3 (1911): 373–385.

Rhode, Paul W. "Learning, Capital Accumulation and the Transformation of California Agriculture," *Journal of Economic History* 55 (1995): 773–800.

Richelet, Juan E. *A Defence of Argentine Meat.* Buenos Aires, 1930.

Rifkin, Jeremy. *Beyond Beef: The Rise and Fall of the Cattle Culture.* New York: Penguin, 1992.

Rivière, Gustave. "Des moyens en usage pour accroître le volume des fruits et exalter leur coloris." In *Résumé des conférences agricoles,* ed. Gustave Rivière, 7–64. Versailles: Imprimerie Cerf et Cie, 1894.

Robinson, Lisa Mae. "Safeguarded by Your Refrigerator: Mary Engle Pennington's Struggle With the National Association of Ice Industries." In *Rethinking Home Economics: Women and the History of a Profession,* ed. Sarah Stage and Virginia B. Vincenti, 253–270. Ithaca, N.Y.: Cornell University Press, 1997.

Robson, John. "Fruit in the Human Diet: Fruit in the Diet of Prehistoric Man and of the Hunter-Gatherer," *Journal of Human Nutrition* 32 (1978): 19–26.

Rocques, Xavier. *Les industries de la conservation des aliments.* Paris: Gautier-Villars, 1906.

Rogers, Lore Alford. *Fermented Milks.* Circular. U.S. Department of Agriculture, Bureau of Animal Industry, 1911.

Rose, Mary Swartz. *Everyday Foods in Wartime.* New York: MacMillan, 1918.

Rosenau, M. J. *The Milk Question.* Boston: Houghton Mifflin, 1912.

Rubatzky, Vincent E., and Mas Yamaguchi. *World Vegetables: Principles,*

Production, and Nutritive Values. New York: Chapman and Hall, 1997.

Russell, Charles Edward. *The Greatest Trust in the World*. New York: Ridgway-Thayer, 1905.

Russell, Howard. *A Long Deep Furrow: Three Centuries of Farming in New England*. Hanover, N.H.: University Press of New England, 1976.

Ryder, E. J. "The New Salad Crop Revolution." In *Trends in New Crops and New Uses*, ed. J. Janick and A. Whipley, 408–412. Alexandria, Va.: ASHS Press, 2002.

Sackman, Douglas Cazaux. "'By Their Fruits Ye Shall Know Them': Nature Cross Culture Hybridization and the California Citrus Industry, 1893–1939," *California History* (1995): 83–99.

———. *Orange Empire: California and the Fruits of Eden*. Berkeley: University of California Press, 2005.

Sadovy, Y. J., T. J. Donaldson, T. R. Graham, F. McGilvray, G. J. Muldoon, M. J. Phillips, M. A. Rimmer, A. Smith, and B. Yeeting. *While Stocks Last: The Live Reef Food Fish Trade*. Report. Asian Development Bank, 2003.

Sanborn, A. F. "The Future of Rural New England," *Atlantic Monthly* (1897): 74–84.

Sanderson, S. "The Emergence of the 'World Steer': Internationalization and Foreign Domination in Latin American Cattle Production." In *Food, the State, and International Political Economy: Dilemmas of Developing Countries*, ed. F. L. Tullis and W. L. Hollist, 123–147. Lincoln: University of Nebraska Press, 1986.

Saunders, Caroline, Andrew Barber, and Greg Taylor. *Food Miles—Comparative Energy/Emissions Performance of New Zealand's Agriculture Industry*. Lincoln University Research Report 285 (July 2006).

Sawyer, Gordon. *The Agribusiness Poultry Industry*. New York: Exposition Press, 1971.

Sawyer, Richard C. *To Make a Spotless Orange: Biological Control in California*. Ames: Iowa State University Press, 1996.

Schabol, Jean-Roger, and Marie-Rose Simoni-Aurembou. *Parlers et jardins de la banlieue de Paris au Xviiie siècle*. Paris: Klincksieck, 1982.

Schudson, Michael. *Advertising, the Uneasy Persuasion: Its Dubious Impact on American Society*. New York: Basic Books, 1984.

Schuller, Philippe. "La Famille Vassout, le sens de la tradition," *Bulletin de la Société Régionale d'Horticulture de Montreuil* (2004): 9–15.

Schvarzer, Jorge. "The Argentine Riddle in Historical Perspective," *Latin American Research Review* 27 (1992): 169–181.

Schwartz, Hillel. *Never Satisfied: A Cultural History of Diets, Fantasies and Fat.* New York: Free Press, 1986.

Seaburg, Carl, and Stanley Paterson. *The Ice King: Frederic Tudor and His Circle.* Boston: Massachusetts Historical Society, 2003.

"Selling Western Sunshine," *Western Grower and Shipper* 1 (December 1929): 7+.

"Senate Committee Hearings," *Ice and Refrigeration* 38 (1910): 385–387.

Shapiro, Laura. *Perfection Salad: Women and Cooking at the Turn of the Century.* New York: Farrar, Straus, and Giroux, 1986.

Sharwell, S. "Lessons to Be Learned from an Inspection That Follows Milk from the Cow to the Consumer." In *Fourth Annual Report of the International Association of Dairy and Milk Inspectors,* ed. I. Weld, 194–197. Washington, D.C.: 1915.

Shepard, Sue. *Pickled, Potted, and Canned: How the Art and Science of Food Preserving Changed the World.* New York: Simon and Schuster, 2000.

Shepherd, Jill. "Wild Salmon, Wily Women," *Alaska* 67 (2001): 38–43.

Shircliffe, Arnold. *The Edgewater Beach Hotel Salad Book.* Chicago: Hotel Monthly Press, 1926.

Shumway, J. "The Woman and the War," *Ladies' Home Journal* (September 1917): 28+.

Sickel, William G. "Refrigeration on Ocean Steamships." In *Premier congrès international du froid Paris, 5 au 12 octobre 1908,* ed. International Congress of Refrigeration, under the direction of J. de Loverdo, 754–766. Paris: Secrétariat général de l'Association internationale du froid, 1908.

Simoons, Frederick J. *Food in China: A Cultural and Historical Inquiry.* Boca Raton, Fla.: CRC Press, 1991.

Sinclair, Robert O. "The Economic Effects of Bulk Milk Handling on the Dairy Industry of Vermont," M.S. thesis, University of Vermont, 1955.

Sinclair, Upton. *The Jungle.* New York: New American Library, 1906.

Skeel, Caroline. "The Cattle Trade between Wales and England from the Fifteenth to the Nineteenth Centuries," *Transactions of the Royal Historical Society, 4th Series,* 9 (1926): 135–158.

Smith, Adam. *An Inquiry into the Nature and Causes of the Wealth of Nations.* London: Methuen, 1776.

Smith, Alisa, and J. B. Mackinnon. *Plenty: One Man, One Woman, and a Raucous Year of Eating Locally.* New York: Harmony, 2006.

Smith, Andrew F. "The Origins of the New York Dairy Industry." In *Milk Beyond the Dairy: Proceedings of the Oxford Symposium on Food and Cookery 1999*, ed. H. Walker, 315–329. Totnes, Devon, England: Prospect Books, 2000.

Smith, Courtland L. *Salmon Fishers of the Columbia.* Corvallis: Oregon State University Press, 1979.

Smith, Page, and Charles Daniel. *The Chicken Book.* Boston: Little, Brown, 1975.

Smith, Peter H. *Politics and Beef in Argentina: Patterns of Conflict and Change.* New York: Columbia University Press, 1969.

Solano County Board of Supervisors. *Solano County: The Land of Fruit, Grain and Money.* Vallejo, Calif.: Press of the Vallejo Evening Chronicle, 1905.

"Spotlight: Product Shelf Life Extension," *Fresh Cut* (March 2005). www.freshcut.com/mar2005/productshelflife.htm.

Stadelman, William J., and Owen J. Cotterill. *Egg Science and Technology.* New York: Food Products Press/Haworth Press, 1995.

Steet, B. C. "On the Preservation of Food, Especially Fresh Meat and Fish, and the Best Form for Import and Provisioning Armies, Ships and Expeditions," *Journal of the Society of Arts* 13 (1865): 309–315.

Steinbeck, John. "Dubious Battle in California," *The Nation* 143 (1936): 302–304.

——. *The Grapes of Wrath.* New York: Viking Press, 1939.

Stevenson, Charles Hugh. *The Preservation of Fishery Products for Food.* Washington, D.C.: United States Fish Commission, 1899.

Stildolph, Dave. "Vacuum Cooling Continues to Grow," *Fruit and Vegetable Review* (1955): 30–32.

St. John de Crèvecoeur, J. Hector. *Letters from an American Farmer: Describing Certain Provincial Situations, Manners, and Customs, and Conveying Some Idea of the Late and Present Interior Circumstances of the British Colonies in North America.* London: T. Davies, 1782.

Stoll, Steven. *The Fruits of Natural Advantage: Making the Industrial Countryside in California.* Berkeley: University of California Press, 1998.

"Storing Your Vegetables: Your Government Tells You Exactly How to Do It," *Ladies' Home Journal* (October 1917).

Straus, Lina Gutherz. *Disease in Milk: The Remedy, Pasteurization; the Life Work of Nathan Straus*. New York: Dutton, 1917.

Street, R. S. "The 'Battle of Salinas': San Francisco Bay Area Press Photographers and the Salinas Valley Lettuce Strike of 1936," *Journal of the West* 26, 2 (1987): 41–51.

Strobel, D. R., and C. J. Babcock. *Recombined Milk: A Dependable Supply of Fluid Milk Far from the Cow*. U.S. Department of Agriculture Foreign Agricultural Service, 1955.

Stull, Donald D., and Michael J. Broadway. *Slaughterhouse Blues: The Meat and Poultry Industry in North America*. Belmont, Calif.: Thomson/Wadsworth, 2004.

Sustain. *Eating Oil: Food Supply in a Changing Climate*. London: Sustain/Elm Farm Research Centre Report, December 2001.

Sutton, Robert. "An American Elysium: The Icarian Communities." In *America's Communal Utopias,* ed. Donald E. Pitzer, 231–244. Chapel Hill: University of North Carolina Press, 1997.

Sweet, George A. "A Grower Looks at the Future," *Western Grower and Shipper* 3 (1932): 8.

Swett, Frank. "Collective Selling by Pear Growers," *Pacific Rural Press,* January 1, 1921.

Swift, Louis Franklin. *The Yankee of the Yards: The Biography of Gustavus Franklin Swift*. Chicago: A. W. Shaw, 1927.

Switzler, R. H. "Refrigerated Warehousing over the Years." In *Proceedings of the American Warehousemen's Association* (1941), pp. 62–69.

Tarlow, S. "Excavating Utopia: Why Archaeologists Should Study 'Ideal' Communities of the Nineteenth Century," *International Journal of Historical Archaeology* 6 (2002): 299–323.

Tavernetti, A. A. "The Salinas Deal in the Roaring Twenties," *Western Grower and Shipper* 21 (1950): 18+.

Taylor, Frederick Winslow. *The Principles of Scientific Management*. New York: Harper, 1911.

Taylor, Harden Franklin. *Refrigeration of Fish*. Washington, D.C.: U.S. Government Printing Office, 1927.

Taylor, William A. "The Influence of Refrigeration on the Fruit Industry." In *Yearbook of Agriculture,* U.S. Department of Agriculture (1900): 561–579.

Tellier, Charles. "Conservation de la viande et autres substances alimentaires," *Usine frigoforique d'Auteuil*, 1871.

———. "Communication aux actionnaires de la société fondatrice pour la conservation de la viande fraiche par le froid," 1877.

———. *Histoire d'une invention moderne: le frigorifique*. Paris: Delgrave, 1910.

Tellus, Barbara. "Salad Month Western Style," *Western Grower and Shipper* (1966): 13+.

Teutenberg, H. J. "History of Cooling and Freezing Techniques and Their Impact on Nutrition in Early Twentieth Century Germany." In *Food Technology, Science, and Marketing: European Diet in the Twentieth Century*, ed. Adel P. den Hartog, 51–65. East Linton: Tuckwell Press, 1995.

Theiss, Lewis Edwin. "What Shall We Eat to Be Well?" *Good Housekeeping* (July–August 1920): 153–157.

Thomas, Robert J. *Citizenship, Gender, and Work: The Social Organization of Industrial Agriculture*. Berkeley: University of California Press, 1985.

Thompson, Virginia, and Richard Adloff. *The Malagasy Republic*. Stanford, Calif.: Stanford University Press, 1965.

Thorne, J. Frederic. *Chinese Eggs: Conditions of Egg Production and Competition in Pacific Coast Markets, Prices, Imports Exports*. Eugene: University of Oregon Press, 1916.

Thünen, Johann Heinrich von, and Peter Geoffrey Hall. *Isolated State; an English Edition of Der Isolierte Staat*. Oxford: Pergamon Press, 1966.

Tobey, Ronald, and Charles Wetherell. "The Citrus Industry and the Revolution of Corporate Capitalism in Southern California, 1887–1944," *California History* (1995): 6–21.

Todd, Stanley. "Bring 'Em to Market Alive!" *Fishing Gazette* 48 (1931): 45–50.

Tressler, Donald Kiteley, and Clifford F. Evers. *The Freezing Preservation of Foods*. New York: Avi Publishing, 1947.

Trollope, Frances Milton. *Domestic Manners of the Americans*. New York: Penguin Books, 1997 [1832].

Tryon, T. *The good house-wife made a doctor*. London: 1692.

———. *A pocket-companion, containing things necessary to be known by all that values their health and happiness being a plain way of nature*. 1694.

Twamley, J. *Dairying Exemplified, or the Business of Cheese-Making.* Providence: Carter and Wilkinson, 1796.

United States National Resources Committee, Science Committee, headed by W. F. Ogburn. *Technological Trends and National Policy, Including the Social Implications of New Inventions, June 1937.* Washington, D.C.: U.S. Government Printing Office, 1937.

U.S. Department of Agriculture Economic Research Service. "U.S. Lettuce: Per Capita Use, 1960–2005." U.S. Department of Agriculture, 2006.

U.S. Department of the Interior, Fish and Wildlife Service. *Fish and Shellfish Preferences of Household Consumers—1951.* Fishery Leaflet 408. Washington, D.C., 1953.

U.S. Food and Drug Administration. Public Meeting on Use of the Term "Fresh" on Foods Processed with Alternative Technologies. Chicago, July 21, 2000. www.cfsan.fda.gov/~dms/flfresh.html.

———. "Subpart F—Specific Requirements for Descriptive Claims That Are Neither Nutrient Content Claims nor Health Claims, Sec. 101.95 'Fresh,' 'Freshly Frozen,' 'Fresh Frozen,' 'Frozen Fresh,'" *Code of Federal Regulations, Title 21,* 2006.

Vallejo, Guadeloupe. "Gold Hunters of California. Ranch and Mission Days in Alta California," *The Century* 41 (December 1890): 183–192.

Van deer Vaart, S. S. "Growth and Present Status of the Refrigerating Industry in the United States." In *Premier congrès international du froid Paris, 5 au 12 octobre 1908,* vol. 3, ed. International Congress of Refrigeration, under the direction of J. de Loverdo, 299–327. Paris: Secrétariat général de l'Association internationale de froid, 1908.

Vartan, Starre. "Happy Eggs: 'Free Range,' 'Cage Free,' 'Organic'— What's the Story?—Eating Right," *The Environmental Magazine* (May–June 2003).

Vaught, David. *Cultivating California: Growers, Specialty Crops and Labor, 1875–1920.* Baltimore: Johns Hopkins University Press, 1999.

"Vente de pêches aux halles, La," *Journal du syndicat des cultivateurs de département de la Seine,* September 1, 1900.

Vera, D. "Critical Problems Dictate Industry Harmony," *Western Grower and Shipper* (November 1960): 52–53.

Vermont Dairy Promotion Council. "Dairy Farm Numbers," www.vermontdairy.com/dairy_industry/farms/numbers. (July 19, 2007).

Vest, George. *Report of the Select Committee on the Transportation and Sale of Meat Products, Senate Report No. 829.* 51st Congress, 1st Session, United States Senate, 1890.

Volti, Rudi. "How We Got Frozen Food," *American Heritage of Invention and Technology* 9 (1994): 46–56.

W. Weddel and Company. *Review of the Chilled and Frozen Meat Trade.* 1922.

Waserman, Manfred J. "Henry L. Coit and the Certified Milk Movement in the Development of Modern Pediatrics," *Bulletin of the History of Medicine* 46 (1972): 359–390.

Weightman, Gavin. *The Frozen Water Trade.* New York: Hyperion, 2004.

Wellman, H. R. *Lettuce.* Berkeley, Calif.: 1926.

Wentworth, Edward. "Mystery of the Egg and Chick." In *Eggs,* ed. P. Mandeville, 171–186. Chicago: Progress Publications, 1933.

"What's Fresh in Fresh-cut Packaging?" *Brand Packaging* 9, 1 (2005).

Whitaker, James. *Feedlot Empire: Beef Cattle Feeding in Illinois and Iowa, 1840–1900.* Ames: Iowa State University Press, 1975.

White, John H. *The Great Yellow Fleet: A History of American Railroad Refrigerator Cars.* San Marino, Calif.: Golden West Books, 1986.

Whitehorne, E. "Household Refrigeration," *The House Beautiful* (September 1921).

Whorton, James C. *Before Silent Spring: Pesticides and Public Health in Pre-DDT America.* Princeton, N.J.: Princeton University Press, 1974.

"Why a Campaign for Real Milk?" Weston Price Foundation, www.realmilk.com. (June 8, 2008).

Wickson, Edward J. *The Vacaville Early Fruit District of California.* San Francisco: California View Publishing, 1888.

———. *The California Vegetables in Garden and Field; a Manual of Practice, with and without Irrigation, for Semitropical Countries.* San Francisco: Pacific Rural Press, 1910.

———. *The California Fruits and How to Grow Them.* San Francisco: Pacific Rural Press, 1926.

Wiebe, Robert H. *The Search for Order, 1877–1920.* New York: Hill and Wang, 1967.

Wiessner, Pauline Wilson, and Wulf Schiefenhövel. *Food and the Status Quest: An Interdisciplinary Perspective.* Providence, R.I.: Berghahn Books, 1996.

Wiley, Andrea. "Transforming Milk in a Global Economy," *American Anthropologist* 109 (2007): 666–678.

Willard, Frances. *Occupations for Women.* New York: Success Company, 1897.

Wilson, Bee. "In the Pink," *New Statesman,* February 12, 2001.

Wilson, C. Anne. "Preserving Food to Preserve Life: The Response to Glut and Famine from Early Times to the End of the Middle Ages." In *Waste Not, Want Not: Food Preservation from Early Times to the Present Day,* ed. C. A. Wilson, 5–31. Edinburgh: Edinburgh University Press, 1991.

———., ed. *Waste Not, Want Not: Food Preservation from Early Times to the Present Day.* Edinburgh: Edinburgh University Press, 1991.

———. "From Garden to Table: How Produce Was Prepared for Immediate Consumption." In *The Country House Kitchen Garden, 1600–1950: How Produce Was Grown and How It Was Used,* ed. C. A. Wilson, 144–161. London: Sutton Publishers in association with the National Trust, 1998.

Wilson, Harold F. *The Hill Country of Northern New England: Its Social and Economic History, 1790–1930.* New York: AMS Press, 1967.

Wilson, W. O. "Housing." In *American Poultry History 1823–1973,* ed. O. A. Hanke, J. L. Skinner, and J. H. Florea, 218–247. Madison, Wisc.: American Poultry Historical Society, 1974.

Winterrowd, W. H. "Design and Construction of Refrigeration Cars," *Ice and Refrigeration* 63 (1922): 144–148.

Woehlke, E. "In the Orange Country: Where the Orchard Is a Mine—The Human Factor among the Gold-bearing Trees of California," *Sunset* 26 (March 1911): 263.

Wolf, Jacqueline. *Don't Kill Your Baby: Public Health and the Decline of Breastfeeding in the 19th and 20th Centuries.* Columbus: Ohio State University Press, 2001.

Wolf, Jacqueline, and Leslie C. Frank. "A State-Wide Milk Sanitation Program: Commentary," *Public Health Reports* 121 (2006 [1924]): 174–189.

Woods, May. *Glass Houses: A History of Greenhouses, Orangeries and Conservatories.* London: Aurum Press, 1988.

Woolrich, Willis Raymond. *The Men Who Created Cold: A History of Refrigeration.* New York: Exposition Press, 1967.

Wright, L. K. *The Next Great Industry: Opportunities in Refrigeration and Air Conditioning.* New York: Funk and Wagnalls, 1939.

Zahara, M., S. Johnson, and R. Garrett. "Labor Requirements, Harvest Costs and the Potential for Mechanical Harvest of Lettuce," *Hortscience* 99, 6 (1974): 535–537.

Acknowledgments

The research for this book was funded by the American Council of Learned Societies (ACLS); Harvard University's Schlesinger Library; and Dartmouth College's Rockefeller Center, John D. Dickey Center for International Understanding, and Leslie Center for the Humanities. I am especially grateful to the ACLS for supporting a year's leave, and to the Dickey Center for arranging an invaluable review of the manuscript. A nonresident fellowship at the W. E. B. Du Bois Institute at Harvard University also proved immensely useful.

The Schlesinger's librarians were among those who helped me find and navigate the archives, periodical collections, and other materials consulted for this book. Deserving special mention are Philippe Schuller at the Musée des Murs à Pêches in Montrueil and Mona Gudgel at the Monterey County Historical Society. Thanks also to the staff at Harvard's Baker Library Historical Collections and at the University of California at Berkeley's Bancroft, NRLF, and Giannini libraries. For help with images, I am indebted to the people at Historic New England, the Petaluma Historical Society, the Owen D. Young Library at Saint Lawrence University, and especially to Gordon McClelland and his orange crate label collection.

At Harvard University Press, I owe thanks to Kathleen McDermott, for all her help in turning *Fresh* from an idea into a real book, to Jill Breitbarth, for work on the illustrations, and to Christine Thorsteinsson, for sharp eyes and a light touch.

Many people offered advice and ideas over the past few years. Those whose expertise rescued me from utter ignorance include Patrick Christie, Celia Lowe, Yvonne Sadovy, Roger Uglow, Cheung Siu Keung, Bonifacio Comandante, and Pete Knutson (on fish); Antoine Jacobsohn (on fruit and fruit trees); and James McDonald, Pat and Mary McNamara, Lourdes Smith, and Melanie Dupuis (on milk).

I am particularly grateful to a few people whose input shaped the course of the book: Sidney Mintz, who told me to go to Hong Kong; Sidney Cheung and Emily Ting, Hongkongers who helped me find my way around the city, once I got there; and Andrew Berry, who first convinced me "just" to write about freshness.

At Dartmouth, students Katie Greenwood, Adeline Yong, Jenna Smith, Chelsea Voake (and her sister Devon), Miguel Peralta, Laura MacGregor, and Lissa Goldstein all assisted with the research, contributing ideas and experiences and often long and late hours. Christine Wohlforth, Ned Lebow, and Colleen Boggs not only read the manuscript but devoted half a day to discussing it. And in the Geography Department, I am grateful to my colleagues Chris Sneddon and Richard Wright, for also reading the thing; Mona Domosh and Frank Magilligan, for letters and levity; and Kelly Woodward, for all-around superhuman competence.

In addition, friends and neighbors offered ideas, distraction, and companionship. In Paris, I am thankful to Martin Bruegel, Arouna Ouedraego, and Ed Cahill, for years of hospitality, and to Maurice Bloch, for introducing me to the *Frigos*. Stateside, Heather Paxson, Karen Stern, Paul O'Connell, and Erik Brockett lent me their ears and books (I will return them, Erik!), while Daniel Levitt, Michael Vassallo, and Peter Walsh always poured just enough wine. I owe special thanks to Kyri Claflin, Carolyn de la Pena, Julie Guthman, and Aaron Bobrow-Strain for their smart, spot-on comments on earlier drafts, especially those sent on short notice and from remote locations. Not least, I'm indebted to Sam Stoloff, agent and ally extraordinaire, for reading the first letter and the last paragraph, and for savvy and kindness all along the way.

Finally, great thanks and love to my parents and siblings, Stan, Colleen, Mark, and Jill Freidberg, for always making me glad to go home (wherever home may be), and to Tom Hegg, for an off-the-cuff remark about Foucault and freshness, and just about everything since.

Index